Cognitive Systems Monographs

Volume 12

Editors: Rüdiger Dillmann · Yoshihiko Nakamura · Stefan Schaal · David Vernon

Tijana T. Ivancevic, Bojan Jovanovic,
Sasa Jovanovic, Milka Djukic,
Natalia Djukic and Alexandar Lukman

Paradigm Shift for Future Tennis

The Art of Tennis Physiology, Biomechanics and Psychology

Rüdiger Dillmann, University of Karlsruhe, Faculty of Informatics, Institute of Anthropomatics, Humanoids and Intelligence Systems Laboratories, Kaiserstr. 12, 76131 Karlsruhe, Germany

Yoshihiko Nakamura, Tokyo University Fac. Engineering, Dept. Mechano-Informatics, 7-3-1 Hongo, Bukyo-ku Tokyo, 113-8656, Japan

Stefan Schaal, University of Southern California, Department Computer Science, Computational Learning & Motor Control Lab., Los Angeles, CA 90089-2905, USA

David Vernon, Department of Robotics, Brain and Cognitive Sciences, Italian Institute of Technology, Genoa, Italy

Authors

Dr. Tijana T. Ivancevic
Citech Research IP Pty Ltd.
QLIWW IP Pty Ltd. & Tesla Science Evolution Institute, Mawson Lakes
P.O. Box 342, Burnside, 5066
Adelaide, Australia
E-mail: Tijana.ivancevic@alumni.adelaide.edu.au

Dr. Bojan Jovanovic
Sports Academy
Save Kovacevica 3/5
21000 Novi Sad, Serbia
E-mail: jovanboj@yahoo.com

Sasa Jovanovic
University of Novi Sad
Art Academy QLIWW IP Pty Ltd
Fruskogorska 30/143
21000 Novi Sad, Serbia
E-mail: 2plus2studio@gmail.com

Dr. Milka Djukic
Alfa University
School of Management in Sport
Palmira Toljotija 3
Novi Beograd, Serbia
E-mail: milka_djukic@yahoo.com

Natalia Djukic
QLIWW IP Pty Ltd.
Tesla Science Evolution Institute
South Australia
P.O. Box 342, Burnside, 5066
Australia
E-mail: Natalia.Djukic@quantumleapintowealthyworld.com

Dr. Alexandar Lukman
Sports Academy
St. Deligradska 27
11000 Belgrade, Serbia
E-mail: spak@sbb.rs

ISBN 978-3-642-42329-1 ISBN 978-3-642-17095-9 (eBook)

DOI 10.1007/978-3-642-17095-9

Cognitive Systems Monographs ISSN 1867-4925

Softcover re-print of the Hardcover 1st edition 2011

Typeset & Cover Design: Scientific Publishing Services Pvt. Ltd., Chennai, India.

Printed on acid-free paper

5 4 3 2 1 0

springer.com

Preface

The book “Paradigm Shift for Future Tennis: The Art of Tennis Physiology, Biomechanics and Psychology” is a sequel to our previous book “Complex Sports Biodynamics with Practical Applications in Tennis”, Springer, Cognitive Systems Monographs, Vol. 2, 2009.

In a historic Wimbledon 2009 final, the celebrated Swiss player Roger Federer became the greatest men’s tennis player of all time, beating the popular American Andy Roddick in an epic five sets: 5-7, 7-6(8-6), 7-6(7-5), 3-6, 16-14. It was a battle of the two top serves in the game. Roddick won the first set. The second set went into a tiebreak; Roddick had 4 set points, yet he lost them all, and with them, the set. Federer proved successful in the next tiebreak, as well. However, Roddick came back very strong in the fourth set, broke Federer, and won it, six games to three. The American continued, starting the fifth set absolutely on fire, yet could not endure the four hour and sixteen minute marathon than followed. Federer won this battle of muscles and nerves to make tennis history.

This was Federer’s 15th Grand Slam title, the total list consisting of: Wimbledon 2003, Australian Open 2004, Wimbledon 2004, US Open 2004, Wimbledon 2005, US Open 2005, Australian Open 2006, Wimbledon 2006, US Open 2006, Australian Open 2007, Wimbledon 2007, US Open 2007, US Open 2008, French Open 2009, and Wimbledon 2009. It was also Federer’s sixth Wimbledon crown (2003, 2004, 2005, 2006, 2007, 2009).

Pete Sampras, holder of the previously unbeaten record of 14 Grand Slam titles and seven-time Wimbledon champion (1993, 1994, 1995, 1997, 1998, 1999, 2000), was watching the match from the Royal Box, along with fellow tennis legends; Sweden Bjorn Borg, a five-time Wimbledon champion (1976, 1977, 1978, 1979, 1980), and Australian Rod Laver, two-time Wimbledon champion of an Open Era (1968, 1969). In addition, the TV commentators were none other than American John McEnroe and German Boris Becker, both three-time Wimbledon champions (respectively, 1981, 1983, 1984 and 1985, 1986, 1989).

What's more, this was Federer's fourth Grand Slam finals victory over Andy Roddick. Previously, Roger beat Andy at the finals of Wimbledon 2004 (46, 75, 76(3), 64); Wimbledon 2005 (6-2, 7-6(2), 6-4); and US Open 2006 (62, 46, 75, 61). With his latest victory, the Swiss managed to extend his overall mastery over the American to 19-2, including 8-0 at Grand Slams. And all this against an ex-number-one player in the world. How can this astonishing resulting factor, 19-2, be possible? Is it due to Federer's forehand? His backhand? Or, perhaps his commonly underestimated serve? Could it simply be the entire *Federer package*? And if so, then what makes up this outstanding package? Notwithstanding the exceptional qualities of the ex-number-one, in this book we will reveal the secret weapon of Roger Federer. And much more: we will reveal *the blueprint of a future tennis champion.*

What is a *paradigm shift*? This beautiful term invented by Thomas Kuhn in 1962 in his book "The Structure of Scientific Revolution", implies on revolutionary changes in hard science theories. Today, popular and accepted in many areas *paradigm shift* depicts a dramatic change of view and perceptions. Is it time to radically change last hundred years of science development and come on track of unified field? Mechanics-oriented paradigm of today's science is coming to the "no through road" journey proving that we can't be treated as a computer. Human beings and human existence, as well as the whole creation have the matrix beyond the physical reality, closely connected with everything and everyone in the Universe.

A century ago, Nikola Tesla – the King of electricity – said: "When science begins the study of non-physical phenomena, it will make more progress in one decade then in all the previous centuries of its existence." Tesla is one example of scientist who had holistic and deep approach to our Universe and his achievements consequently dramatically improved the quality of our life.

This book starts with revelations that make obvious the limitations of today's tennis which does not use the Laws of Modern Biomechanics (see http://adsabs.harvard.edu/doi/10.2478/s11534-009-0148-z) and Neurophysiology. The second part of the book includes a new approach to the *quantum mind of a champion.* This paradigm shift started with the work of eminent physicists John A. Wheeler and David Bohm, and their ideas that we are active participants in the creation. Before them, Erwin Schrödinger, one of the fathers of quantum mechanics, mentioned the term "entanglement" as the very essential part of his wave mechanics. Today Guardian Science states: "Entanglements of two or more particles that have once interacted always remain bound in a very strange, hardly understandable way even when they are far apart and their connection being independent of distance." Very powerful quantum brain is the main characteristics of Roger Federer, as well as some other sport champions like Michael Schumacher.

July 2010

Adelaide and Novi Sad
Authors

Acknowledgments

We wish to express our sincere gratitude to *Springer* book series *Cognitive Systems Monographs* and especially to the Editors, Dr. Thomas Ditzinger and Dr. Dieter Merkle.

Contents

Part II: Modern Background for Tennis Science

Part I

The Laws of Core Future Tennis Science

1

The Federer Phenomenon and His Secret Weapon

The momentous five-setter in a historic Wimbledon 2009 final, eventually ended when Federer broke Roddick's serve for the first time all day (he himself was broken in the fourth set), with Roddick missing on a forehand. Federer had a total of 107 winners and 38 unforced errors (which is a W/E factor of 107/38=2.8). Roddick, on the other hand, had 74 winners and 33 unforced errors (a W/E factor of 2.2). Hence, the real difference in these statistics is the incredible number of Federer's winners. How does he make so many winners?

Firstly, the cheapest way to achieve a winner is by an ace. Federer served 50 aces, compared to Roddick's 27 aces (which is an A-factor of 50/27=1.85), to win this historic duel. On paper, his serve was more powerful than the most powerful serve in the game. But how can this be? Did Roddick serve poorly? No, Andy's serve was unbroken, repeatedly around 220 km/h and twice at 230 km/h (143 m/h), until the 77th – and last – game. Yet he only had 27 aces, while Federer, with a serve that was mostly under 200 km/h, hit an astounding 50 aces. So, does this mean that the speed of the ball is not important for aces? No, the ball-speed is actually the most important factor for an ace; but it is not the only important factor. Is it topspin then? Unlikely, as Federer and Roddick have about the same amount of topspin on both their first and second serves. What else is there? Well, this is the key question to ask, not just for this particular match, but to explain the whole Federer phenomenon. Before answering this question, let's biomechanically analyze the fastest serve in the game.

1.1 The Best Serve Ever: It's Biomechanics and Neuro-physiology

From Roddick's serve (see Figure 1.1), we can see that, as is properly advised by many coaches, at a certain point the racquet-head is parallel to Roddick's back. However, while most coaches assume (and hence teach their players) that the back should be upright at this instant, in Roddick's case, it is not. At this point, his torso is already leaning forward, as well as twisted around on the vertical

T.T. Ivancevic et al.: Paradigm Shift for Future Tennis, COSMOS 12, pp. 3–10.
springerlink.com

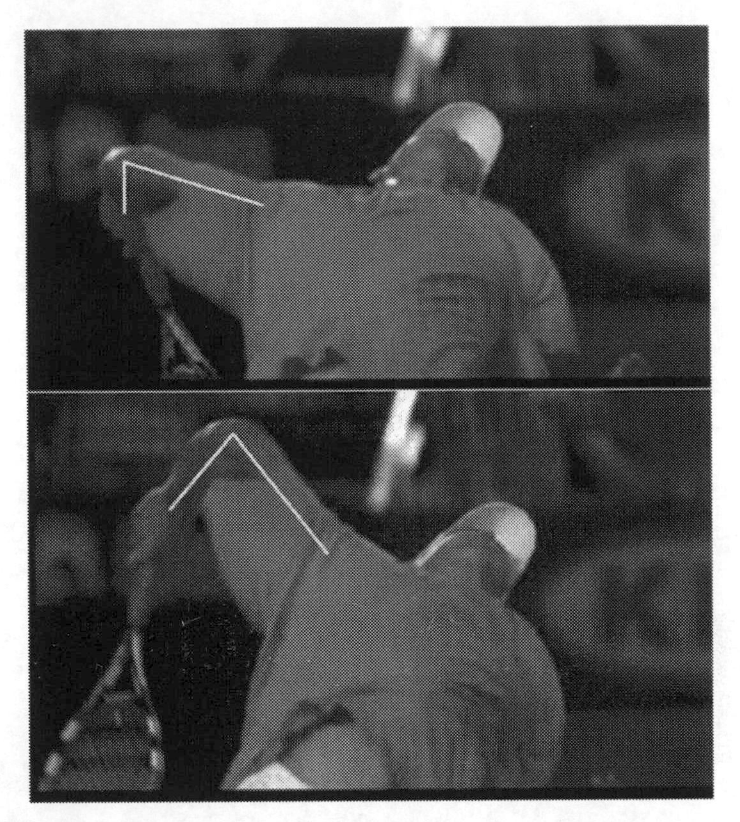

Fig. 1.1. The best serve ever of Andy Roddick: its biomechanical chain, whip-like movement and stretch-reflex in action.

body axis. This 'little' difference from the common (static) serve concept already shows the two most important physical (dynamical) factors of Roddick's serve:

1. A whip-like movement, the main biomechanical factor (proximal-to-distal kinetic chain, in which there is always a time-lag between two consecutive joints: first legs, then torso, then shoulder, then elbow, then wrist); and
2. The stretch-reflex in action, the main neuro-physiological factor (stretching the leg-extensors, torso, shoulder, elbow and wrist muscles prior to their ballistic contractions).

From both biomechanical and neuro-physiological point of view, Roddick's serve is by far the best ever in the tennis game. It is the only true whip-like serving action (the model of a proper serving movement) and the only true stretch-reflex-based serve in tennis.

In modern handball, volleyball, American football, rugby and water-polo, it is common knowledge that a high-performance throw (or smash) necessarily resembles the whip-like movement of an elite javelin throw. A biomechanical term for this whip-like movement is the kinetic chain: the sequential flow of energy and momentum from bigger segments to smaller ones. Tennis requires sequenced activation of muscles and movement of bones and joints to achieve the motions, positions, and velocities seen in a player. Never rotate the "hips and shoulder together", as you may often hear coaches advising while teaching the forehand. Instead, always first rotate the hips, then the shoulders, then the elbow, then the wrist. It is harder to hit the "sweet-spot" on a tennis racquet with this "loose" movement, but the result is unbelievable. All elite players do these shots – which are not in textbooks or coaching manuals – for their winners. Kinetic energy and momentum, as well as muscular power, are developed from the legs, hips and trunk muscles and transferred to the arm muscles. This allows the energy, momentum and power to be transferred efficiently to the hands, moving the racquet-head with maximum speed towards the ball. Furthermore, to achieve the highly-efficient whip-like technique of tennis serves, forehands and backhands, a proper sequencing of muscle stretch-reflex based actions must take place.

Andy Roddick holds the record for the world's fastest tennis serve: 246 km/h (153 m/h). His serve is a proper biomechanical chain, a full whip-like movement and the stretch-reflex in action. The stretch-reflex causes a stretched muscle to contract stronger and at the same time inhibits the antagonist muscle from contracting (that is, slowing the movement). Because this is an involuntary reflex response, the rate of contraction is significantly (several times) faster and more powerful than a completely voluntary muscular contraction. In fact, the faster the muscle is stretched eccentrically, the greater the force will be on the following concentric contraction.

When we train our skeletal muscles so that they efficiently utilize the stretch-reflex (and the related stretch-shortening cycle), then we effectively make powerful slingshots out of them. The prototype of such a muscular slingshot is a javelin throw; in particular, the world-record throw of Jan Zelezny (who was at that time only 80 kg!). We will talk more about Roddick's serve later.

1.2 The Secret behind Federer's Serve

So both biomechanically and neuro-physiologically, Andy Roddick has the best serve ever, clearly better than Federer's serve. However, it is better by only about 10% (or, even less), on average. And there is a hidden factor on Federer's side that is greater than 10%.

The explanation of the Ace-performance factor of 1.85 between Federer's and Roddick's serves is clearly non-physical: non-biomechanical and non-physiological. The explanation is cognitive. Roger Federer is simply the best blitz-chess player in the tennis game. This means three cognitive factors: He effectively plans his moves in advance; He effectively reads (anticipates) the opponent's moves and acts accordingly; and He effectively disguises his own moves. These three superior cognitive factors give Federer a huge mental advantage over his opponents, with the exception of his "nemesis", Rafael Nadal – the current world number one, who appears to have similar cognitive factors (together with an extraordinary physical fitness and the huge amount of topspin on all of his shots).

In particular, this tennis blitz-chess is the key to Federer's aces. He plans the ball placement, anticipates Roddick's return before even serving, and then disguises the motion of the serve. In the epical match at Wimbledon 2009, particularly in the tiebreaks and at the end of the fifth set, Andy could not read Roger's serve, while Roger was reading more and more of Andy's serves. And Federer is able to do the same in many other matches, against many other players.

Blitz-chess is the key to most of Federer's incredible winners. Again, he first plans, then anticipates the opponent's move, and finally disguises his own shot. And we can always expect another winner. In the next report, we will talk about Roger's forehand, backhand and volley winners.

1.3 Biomechanics of Federer's Shots

Federer's forehand, backhand and volleys (as well as his serve, which was addressed in the first part of the Wimbledon 2009 Finals report) are much more dynamic than the textbook ones. They are actually quite similar to the forehand, backhand and volleys (as well as serve) of Pete Sampras. In both the cases of Roger and Pete, you will never see the two compulsory phases of classical tennis ground-strokes: "loading" and "hitting". Instead, their most efficient winners are hit by apparently "loose" movements. They pretty much resemble the incredible 190 km/h (120 m/h) forehand of French Gael Monfils, which was hit not by strength, but by a combination of arm speed and joint flexibility.

Even more importantly, the most effective shots (of Roger and Pete, as well as Rafa Nadal) never use big loops or swings.

As this is a very common misconception, we need to elaborate on it a bit. The basic idea of textbook tennis (which means, used by a great majority of coaches and players), as well as the simplistic biomechanics of loops and swings in tennis shots (serve, forehand and backhand) are derived from the concept of a simple physical pendulum. An arm with a hand holding a racket has been seen as a more-or-less single rigid body, with no more than 3 degrees of freedom (DOF). And really, if you have a robotic arm with only 3 DOF, and you want to hit a ball with it, then you need to have a loop; and, even more, if you want to hit the ball hard, you need to have a big swing. That is absolutely true. In the language of modern biomechanics, "its phase space is a simple circle."

That is how we originally got our current loops – to play "nice tennis"; and swings – to be able to "hit hard". In particular, you need a big swing, as you want to use all of the potential energy of the racket's weight. Hence, as preparation for the shot (say, a forehand), you lift the racket-head as high as possible along the circle (that is, above your head). This is the common picture behind all loops and swings in tennis. Although simple to understand, it is not as easily implemented, which is where the many expensive lessons on the court come in. This system has produced thousands of young tennis players, all hitting the ball in virtually the same way, with the same distinguished tennis movements: loops and swings.

However, this picture is wrong! Our reality is much more complex than this simplistic model of a loop (circle) and a swing (potential energy along the circle), which produces only "tennis ballet".

Firstly, the human arm has 9 DOF (not including fingers). Just the shoulder, elbow and wrist together have 9 DOF. This is a redundant system, because the racket itself has only 6 DOF (three translations and three rotations). This means that there is an infinite number of possible ways to hit the ball, and we are able to choose the best way in each particular situation. Therefore, a real phase space for the human arm with a racket is much more complicated than the simplistic circle-loop model. Besides, every coach advises, quite correctly, not to use the arm only, but the whole body. So we actually have several hundred DOF at our disposal to perform each tennis stroke. Do you still believe that a simple circle and its associated loop is an appropriate model for the serve and ground-strokes of a tennis champion? Absolutely not, it is sufficient only for small children!

Secondly, there is the question of utilizing the racket's own potential energy – that is, raising the racket-head up high as preparation for the shot. This would make sense only for very weak players and very heavy rackets (in the same way that a weak player needs a long swing to accelerate a massive racket). However, with a true athlete (and every elite tennis player has to be one such athlete), the current 400 gram racket is just not heavy enough to make this potential-energy contribution to the shot significant. It would be significant with a 4 kg racket, but not with the current 400 gram one.

Put simply, loops are totally useless: it does not matter at all how you get into position for the shot. Swings are almost useless. They could be used, say, in the fifth hour of a five-set tennis match, to compensate for lack of strength, but this same fatigue would also make it hard to raise the racket-head high above your head in the first place.

In addition, we see that simplistic mechanics does not work with the human body. Robotics discovered this in the past three decades; tennis science has yet to learn it.

The only efficient type of movements in tennis shots (serve, forehand and backhand), as used by both Roger Federer and Pete Sampras, are whip-like movements, based on muscular slingshots. In other words, all the shots of Roger and Pete, just as Andy Roddick's serve, clearly show the following two components:

1. A *whip-like movement*, the main biomechanical factor (proximal-to-distal kinetic chain, in which there is always a time-lag between two consecutive joints: first legs, then torso, then shoulder, then elbow, then wrist); and
2. The *stretch-reflex in action*, the main neuro-physiological factor (stretching the leg-extensors, torso, shoulder, elbow and wrist muscles prior to their ballistic contractions).

1.4 The Mental Secret behind Federer's Winners

As already stated above (in the case of Federer's serve), Roger is simply the best *blitz-chess player* in the tennis game. Again, this means three cognitive factors:

1. He effectively *plans* his moves in advance;
2. He effectively *reads* (anticipates) the opponent's moves and acts accordingly; and
3. He effectively *disguises* his own moves.

These three superior cognitive factors give Federer a huge mental advantage over his opponents, with the exception of Rafa Nadal, who appears to have similar cognitive factors.

Similarly to Federer's aces, the *tennis blitz-chess* is the key to all of Federer's winners. He plans the ball placement, anticipates the opponent's return before executing a shot, and then disguises the motion of the stroke itself. He did it against Andy Roddick in the epical match at Wimbledon 2009, and he has done it in many other matches, against many other players.

The blitz-chess is the key to most of Federer's incredible winners. Again, he first plans, then anticipates the opponent's move, and finally disguises his own shot. And we can always expect another winner.

The so-called *complex reaction* has the highest importance in every sports duel, including tennis. In general, a complex reaction has two components: smart anticipation and lightning-fast reaction. You first anticipate the opponent's movement, and then you react by making your own movement.

The basic speed of a reaction is largely genetically predetermined and cannot be significantly improved. The best means for developing it is training based on the stretch-reflex.

On the other hand, good anticipation of the opponent's actions is the essential characteristic of a master in any sports competition. The best anticipation is actually called *mind reading*, and this is the difference between master and disciple. It is Federer's greatest strength. It can be learned, and it should be learned if you have high expectations in tennis.

Apart from getting experience from many tennis tournaments (at different competition levels), the best tool for developing anticipation in young players is the blitz-chess.

Actually, the best preparation for the *mental speed* in tennis is the so-called *lightning chess*, or *bullet chess*. It is the faster version of a blitz chess game, where each side has less than 3 minutes to complete all of their moves. Often, bullet chess is so fast that tactics and skill are secondary to quick moves.

The most common problems with a player's mental game are:

1. Confusion about strategy;
2. Trying to do too many things at once;
3. Being easily distracted;
4. Too much concern about winning or losing;
5. Perfectionism;

6. Complacency;
7. Having no clear plan or goal;
8. Too much spontaneity and creativity;
9. Lack of humility;
10. Inappropriate reactions to errors; Negative self-talk (self-criticism).

We could argue that Andy Roddick still feels some of these problems, and that by resolving them, he would easily be able to win Grand Slams. At the same time, both Roger and Rafa have already resolved them in a most effective way.

As a general advice, to improve your mental game you need to:

1. Set clear goals;
2. Create action steps that will take you closer to these goals (learn to visualize: try to "see" in your mind what you want to achieve);
3. Use positive self-talk (self-criticism);
4. Practice *yoga breathing* techniques;
5. Always put forth 100% effort;
6. Be process oriented;
7. Stay detached; Not dwell on the future or past – stay in the present moment;
8. Be non-reactive to the opinions of others;
9. Tolerate your inability to be perfect;
10. Do what it takes to have fun and smile;
11. Define winning in such a way that it includes more than just the final score.

Finally, you need to focus on the ball and yet still be aware of the opponent and the court. This is a big problem for most players, although Roger and Rafa seem to solve it consistently and routinely. Aside from a great tolerance for competitive pressure, focus is one of the key characteristics of a champion. You really need to "fix your eyes on the ball," as every coach advises, and yet have a full amount of the so-called *situational awareness* (SA), which is defined in modern psychology as "the perception of the elements in the environment within a volume of time and space, the comprehension of their meaning, and projection of their status in the near future". For example, SA is a key factor in the training of fighter pilots and Formula 1 car-racing drivers. The outcome of any complex situation critically depends on SA. Therefore, you need to develop:

1. A strong focus on the moving tennis ball;
2. The ability for strong SA, as defined above; and
3. The ability to focus on both the ball and SA at the same time.

This is similar to learning to play a hard piece on the piano: first you learn the right-hand part, then you learn the left-hand part, and then you learn how to combine them together. Obviously, physical and mental fatigue (including other disturbances, such as rain, wind, a noisy audience, the big overhead screen, etc) will gradually reduce both focus and SA. Therefore, besides physical speed and strength, you obviously need great mental strength and endurance to be able to win.

Roger Federer has all of these mental qualities. That is why he is the greatest tennis player ever. That is the secret behind the Federer phenomenon. It is not his forehand, his backhand, or his serve. The real secret is Federer's *powerful mind*.

Roger Federer is a phenomenon, no doubt about that. However, for all those talented kids who want to be the active participators in the tennis game – not just the passive observers – the most important question is: How can we go beyond Federer? Can we formulate the blueprint for the future Federer? In other words, Is there really such a thing as Tennis Science?

In the following Chapters, we will formulate the *main laws of tennis science*.

2

The Law of Muscular Structure and Function

2.1 Busting the Myth of "Muscle Memory"

From the scientific perspective, the common term "muscle memory," so popular among coaches and players, is *shear nonsense.* If neural motor pathways are damaged there is no any "muscle memory" left. This means that all the motor memory (containing all acquired motor skills) is in the neural system, not in the muscles. Muscles have their structure and function (of generating muscular force), they can be exercised and trained in strength, speed, endurance and flexibility, but they are still only dumb effectors (just like excretory glands). They only respond to neural command impulses. To understand the process of training motor skills one needs to know the basics of neural motor control. Within the motor control muscles are force generators, working in antagonistic pairs. They are controlled by neural reflex feedbacks and/or voluntary motor inputs. If any of these neural pathways are damaged, muscles are dead flesh without any memory left (see The Law of Neuro-Motor Control).

2.2 Structure and Function of Skeletal Muscles

Skeletal muscles are composed of muscle fibers wrapped by fascicles that also enveloped blood vessels. Muscles are connected to adjacent bones via elastic tendons, which are made up of epimysium, perimysium and endomysium (see Figure 2.1). Each muscle fiber (or muscle cell) is made up of many myofibrils. Each myofibril contains a series of contractile units called sarcomeres, made up from two types of protein filaments: thin actin filaments and thick myosin filaments (see Figure 2.2). See Appendix 1, for technical details.

The muscles are "slaves" that contract (shorten) only when their "masters", called *motor units*, fire (see Figure 2.3), otherwise they relax. Therefore, muscles can only pull (contract), never push. They usually work in mutually antagonistic pairs (see The Law of Basic Biomechanics).

T.T. Ivancevic et al.: Paradigm Shift for Future Tennis, COSMOS 12, pp. 11–15.
springerlink.com

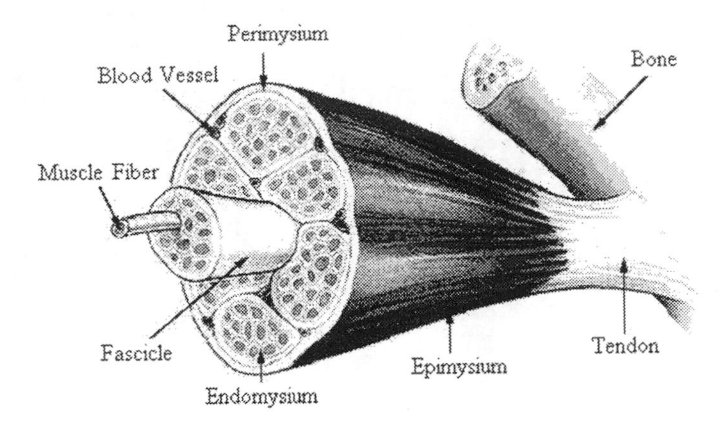

Fig. 2.1. Structure of human skeletal muscles. Epimysium consists of collagen fibers that surround muscle tissue. Perimysium is a connective tissue that separates adjacent fasciculi (small bundles of muscle fibers) in a skeletal muscle. Endomysium surrounds individual fibers. Epimysium and perimysium contain blood vessels and nerves.

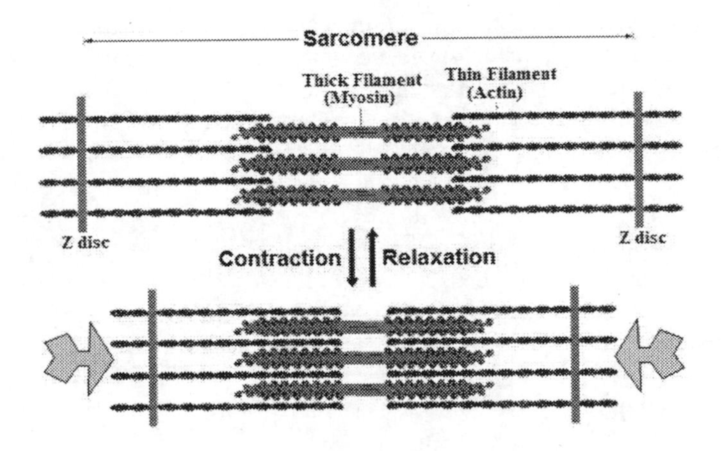

Fig. 2.2. Basic diagram of sarcomere shortening.

All muscles have the following three basic properties:

1. Contractility (shortening while generating a force);
2. Excitability (responding to neural stimuli);
3. Elasticity (recoiling to original resting length after being stretched).

There are three basic types of muscular contractions, closely related to the fundamental *force–velocity curve* (see Figure 2.4):[1]

[1] The force–velocity curve (Figure 2.4) shows the relationship between muscle tension (or, generated force) and the velocity of its shortening or lengthening. This fundamental muscular curve is used to analyze the effects of the speed–strength training, as well as to identify muscle fibre types used in different physical activities.

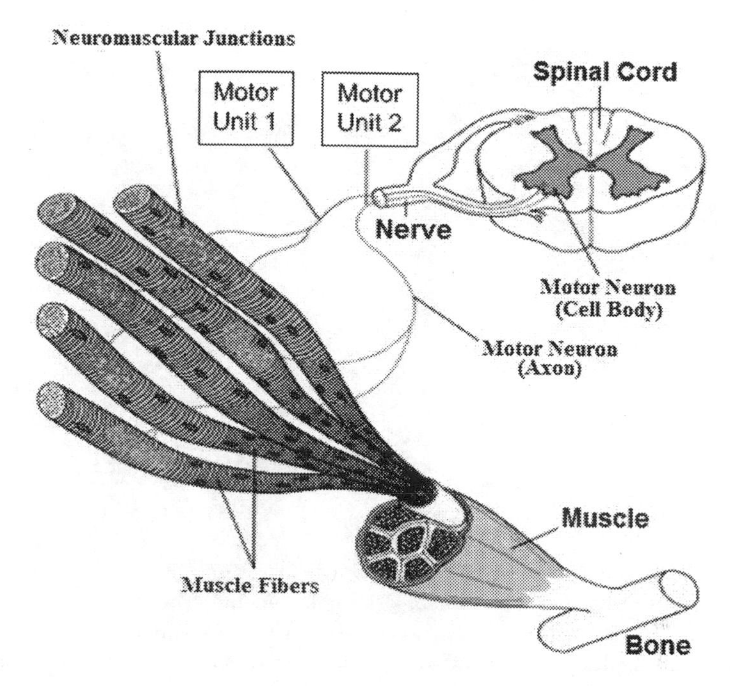

Fig. 2.3. One motor nerve/neuron and the muscle fibers that it controls via two motor units. Motor units fire fully or not at all (all or none Law).

1. Concentric contraction, in which muscle shortens and makes positive movement – its generated force is greater than the loading force;
2. Isometric contraction, in which muscle generates force without shortening (there is no visible movement) – its generated force is equal to the loading force;
3. Eccentric contraction, in which muscle generates force, yet the negative movement occurs because the loading force is greater than the muscle–generated force.

1. When a muscle is activated and required to lift a load which is less than the maximum tetanic tension it can generate, the muscle begins to shorten. Contractions that permit the muscle to shorten are referred to as concentric contractions. An example of a concentric contraction in the raising of a weight during a bicep curl. In concentric contractions, the force generated by the muscle is always less than the muscle's maximum. As the load that the muscle is required to lift decreases, contraction velocity increases. This occurs until the muscle finally reaches its maximum contraction velocity. By performing a series of constant velocity shortening contractions, a force-velocity relationship can be determined.

2. In isometric contraction the muscle is activated, but instead of being allowed to lengthen or shorten, it is held at a constant length. An example of an isometric

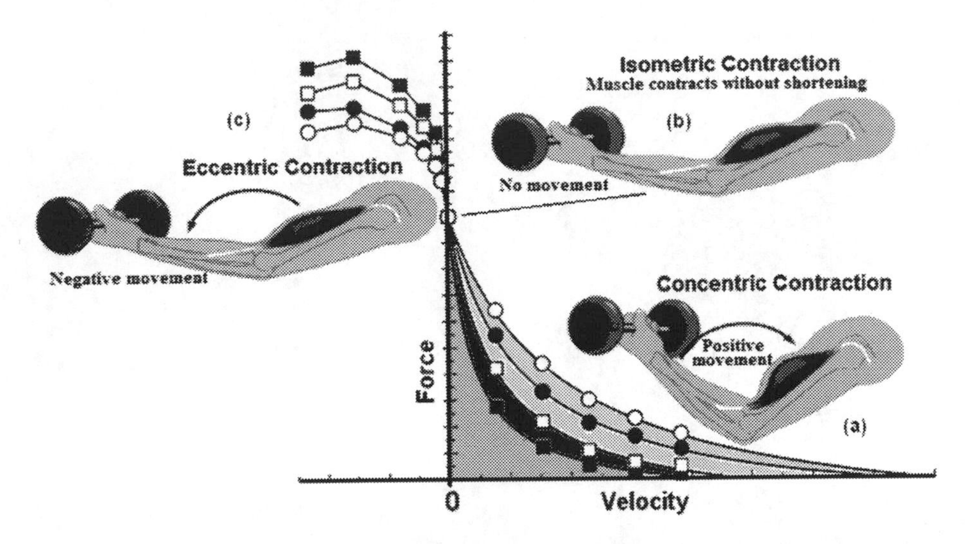

Fig. 2.4. Muscular force–velocity curve related to three types of muscular contractions: (a) concentric contraction (gives the positive movement), (b) isometric contraction (no movement), and (c) eccentric contraction (negative movement). The graph shows movements performed with four different speeds. The shaded area under the curve represents *muscular power*.

contraction would be carrying an object in front of you. The weight of the object would be pulling downward, but your hands and arms would be opposing the motion with equal force going upwards. Since your arms are neither raising or lowering, your biceps will be isometrically contracting. The force generated during an isometric contraction is wholly dependant on the length of the muscle while contracting.

3. During normal activity, muscles are often active while they are lengthening. Classic examples of this are walking, when the quadriceps (knee extensors) are active just after heel strike while the knee flexes, or setting an object down gently (the arm flexors must be active to control the fall of the object). As the load on the muscle increases, it finally reaches a point where the external force on the muscle is greater than the force that the muscle can generate. Thus even though the muscle may be fully activated, it is forced to lengthen due to the high external load. This is referred to as an eccentric contraction (please remember that contraction in this context does not necessarily imply shortening). There are two main features to note regarding eccentric contractions. First, the absolute tensions achieved are very high relative to the muscle's maximum tetanic tension generating capacity (you can set down a much heavier object than you can lift). Second, the absolute tension is relatively independent of lengthening velocity. This suggests that skeletal muscles are very resistant to lengthening.

2.3 Slow and Fast Muscles

Yes, some muscles are more suited for explosive speed–strength type activities, while others are adapted for endurance exercises. More precisely, there are three basic types of human skeletal muscles:

- Type I - slow twitch muscles, which are dark–red in color (under electronic microscope; they are predominately used in the aerobic energy system; they have high concentration of myoglobin, lots of capillaries and mitochondria.)
- Type IIA - intermediate fast twitch, which are reddish–white in color; their contraction is moderately fast; they have moderate myoglobin concentration; they are predominately used in anaerobic glycolysis energy system.
- Type IIB - ultra fast twitch, which are white in color; their contraction speed is the fastest; they have low myoglobin concentration; they are predominately used in anaerobic alactatic energy system.

There is *no* such thing as specialized "tennis muscles". What is crucial for the success in tennis is neuro–motor control, of which the most important is the stretch–reflex (we will talk about it later).

3

The Law of Functional Anatomy

3.1 Anatomical Description of an Ordinary Tennis Serve

Classical tennis serve has three stages:[1] (i) the ball toss, (ii) the jump, and (iii) the finishing smash.

(i) In the case of a right–handed player (like Federer), the ball toss is thrown with the left arm. The feet are apart, and the ball toss is performed with the contractions of the left deltoideus, the biceps and the palmar flexors muscles. This movement is done simultaneously with two other preparatory actions.

The first one of these preparatory actions is raising the right arm, "loading". The muscles used to carry this out are the right deltoideus, supraspinatus (a muscle going over the shoulder blade) and the biceps brachii. The second action is bending the knees, and thus preparing for the second stage of the serve (the jump). There are no flexor muscles used to bend the knees, for the bending of knees is accomplished by gravity alone (actually, leg–extensors are used in eccentric fashion).

(ii) The first–serve jump is performed high and forward. It is achieved by instantaneous actions of all the leg extensor muscles; left and right soleus, quadriceps femoris and gluteus maximus muscles. Jumping is the second part of "loading" in the serve. At the same time as the player lifts off, the racquet is placed behind the body, in a "back–scratching" position, and the right shoulder's rotation towards the ball begins. This movement involves the right biceps brachii and wrist extensor muscles. While in the air, the feet naturally join together (with Federer, the feet join in the air, not on the ground).

(iii) The finishing smash takes place in the air, before the player returns to the ground. To end the serve, the shoulders are rotated and the ball is hit simultaneously. By then, the shoulders should have been fully rotated and the feet prepared for landing. The internal and external obliques abdominal muscles complete the shoulder rotation. Hitting the ball is performed by the latissimus

[1] We will skip the notorious preparation here.

T.T. Ivancevic et al.: Paradigm Shift for Future Tennis, COSMOS 12, pp. 17–19.
springerlink.com

dorsi, then pectoralis major and finally triceps brachii muscles. To add a bit of spin or slice to the serve, the wrist is flicked slightly at the end, using the palmar flexors.

3.2 Anatomical Description of an Ordinary Tennis Forehand

Standard tennis forehand, in the case of a right–handed player (like Federer), basically has two phases: (i) preparation, or "loading", and (ii) hitting the ball.

(i) Preparation for the forehand includes two simultaneous actions. One is stepping into the right position, with the left leg forward. The other is the first half of the loop movement, lifting the racquet above the shoulders in a curved c-shaped movement. (This does not need to be too far back, like in Hewitt's forehand, but can be more to the side.) This is accomplished by the right deltoideus and biceps brachii muscles.

(ii) Hitting the ball includes four main movements. The first of these four is a right hip rotation towards the ball, while the feet are still on the ground. The right gluteus maximus and medius muscles carry out this action.

Secondly, a leap into the air is necessary to be able to hit the ball from a higher body position, so as not to hit the net. This is performed by all the leg extensor muscles: left and right soleus, quadriceps femoris and gluteus maximus – working together.

Thirdly, an arm swing of the racquet, the second part of the loop action. The right pectoralis major, deltoideus and biceps brachii muscles complete this.

Lastly, to create topspin; a slight twist of the wrist to just brush over the ball. This is done by the right palmar flexors.

3.3 Anatomical Description of an Ordinary Backhand

A common right–handed player's two-handed backhand (e.g., Andy Roddick's) basically has two stages: (i) preparation, or "loading", and (ii) hitting the ball.

(i) Preparation for the double-handed backhand includes two phases. One is stepping into the right position, with the right leg forward. The other is lifting the racquet in a loop movement similar to the forehand loop, only this time the racquet tends not to go further than about the shoulder–level. This is accomplished by both the left and right deltoideus and biceps brachii muscles.

(ii) Hitting the ball includes four main movements. The first of these four is a left hip rotation towards the ball. The left gluteus maximus and medius muscles carry out this action, helped by the left knee extension, which is performed by the quadriceps femoris. Secondly, a rotation of the left shoulder towards the ball, achieved by both the internal and external obliques abdominal muscles.

Thirdly, the arm swing of the racquet is performed by pulling of the right arm and pushing of the left arm. The pulling action of the right arm is accomplished by the right latissimus dorsi, pectoralis major and triceps brachii muscles. The pushing action of the left arm is accomplished by the left deltoideus, pectoralis major and biceps brachii muscles.

Lastly, to create topspin; a slight twist of both wrists. This is done by the left palmar flexors and right dorsal flexors.

4

The Law of Energy Flows

4.1 The Sources of Energy for Muscular Work

There are three different energetic resources for any kind of muscular work, including tennis (see Appendix 1):

1. ATP–CP (or, anaerobic[1] alactatic) system, which lasts 10–15 seconds, uses stored ATP[2] and creatine phosphate (CP), with no by–products. This energy source is related to speed and strength. It is essential for the serve and winning shots in tennis.
2. Glycolysis (or, anaerobic lactic) system, which lasts 15 seconds – 3 minutes, uses blood glucose and muscular glycogen to make ATP; its by–product is lactic acid. This energy source represents anaerobic endurance. It is essential in long, exhausting tennis relays.
3. Aerobic system,[3] which lasts from 2–3 minutes to several hours, uses glucose, glycogen, fats, and proteins to make ATP within the aerobic energy pathway; its by–products are carbon dioxide and water. This energy source represents aerobic endurance. It is essential in the fourth and fifth set of any serious tennis match.

4.2 Maximal Oxygen Consumption

In general, the maximal oxygen consumption, or VO_2-max, is the maximum volume of oxygen (O_2) that the human body can consume by breathing air at sea

[1] "Anaerobic" means without the use of oxygen, that is, none of its metabolic activity will involve 0_2.

[2] Adenosine–Triphosphate (ATP) is the fundamental energy source for muscular work, made in the mitochondria of muscular cells. It gives energy for muscular contraction according to the equation:

$$ATP \rightarrow ADP + P_i + Energy$$

where ADP is Adenosine–Diphosphate and P_i is inorganic phosphate. ATP is resynthesized both aerobically and anaerobically.

[3] "Aerobic" means in the presence of oxygen (0_2), that is, all of its metabolic activity will involve 0_2.

T.T. Ivancevic et al.: Paradigm Shift for Future Tennis, COSMOS 12, pp. 21–26.
springerlink.com

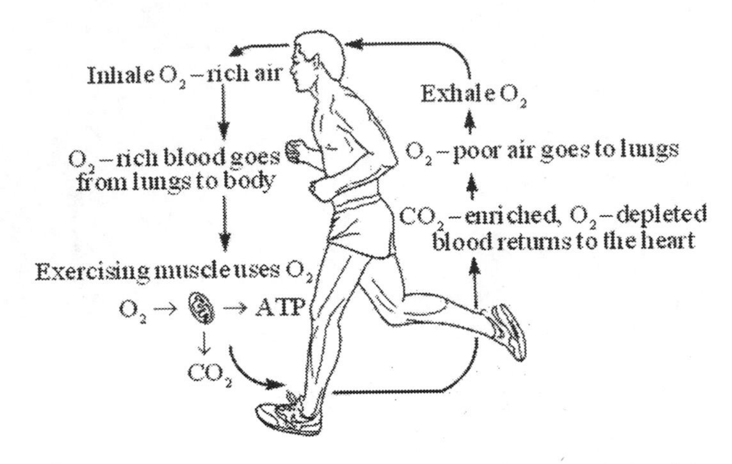

Fig. 4.1. O_2–consumption by skeletal muscles in mitochondrias.

level during an intense whole–body exercise. This volume is expressed as a rate in milliliters per kg bodyweight per minute (ml/kg/min). As O_2–consumption is linearly related to energy expenditure, when one measures O_2–consumption, they are indirectly measuring an individual's maximal capacity for aerobic work.

Every cell consumes O_2 in order to convert food energy to usable ATP for cellular work (see Figure 4.1). However, it is muscle that has the greatest range in oxygen O_2–consumption (at rest, muscle uses little energy; however, contracting muscle cells have high demands for ATP, proportionally to the work intensity). Endurance athletes have developed a strong cardiovascular system, as well as a strong oxidative capacity in their skeletal muscles. To receive this O_2 and use it to make ATP for muscular contraction, our muscle fibers depend on 2 things: (i) an external delivery system to bring O_2 from the atmosphere to the working muscle cells, and (ii) mitochondria in the muscle cells to carry out the process of aerobic energy transfer. In other words, one needs a big and efficient heart–pump to deliver O_2–rich blood to the muscles, and one also needs mitochondria–rich muscles to use the O_2 and support high rates of exercise.

4.3 The Problem of Lactic Acid

As the fourth or (even worse, fifth hour) of a five-set tennis match approaches its end, the main problem for both players becomes the accumulation of lactic acid. It feels like 'muscle soreness'. Every tennis player has experienced this painful fatigue. This is how it happens.

The carbohydrates you consume consist of several different sugar molecules; sucrose, fructose, glucose, etc. However, by the time the liver does it's job, all of these sugars are converted to glucose which can be taken up by all cells. Muscle fibers take up glucose and either use it immediately, or store it in the form of long glucose chains (polymers) called glycogen. During exercise, glycogen is broken back down to glucose which then goes through a sequence of enzymatic reactions that do *not* require oxygen to proceed. All of these reactions occur out in the cell fluid, or cytosol. These reactions proceed very rapidly and yield some energy for muscle work in the process. This glycogen/glucose breakdown pathway is called the *anaerobic glycolysis (or, glucose breakdown without oxygen) pathway*. Every single glucose molecule must go through this sequence of reactions for useful energy to be withdrawn and converted to ATP, the energy molecule that fuels muscle contraction, and all other cellular energy dependant functions.

In a single contracting muscle fiber the frequency and duration of contractions will determine ATP demand. ATP demand will be met by breaking down a combination of two energy sources: fatty acids and glucose molecules (ignoring the small contribution of protein for now). As ATP demand increases, the rate of glucose flux through glycolytic pathway increases. Therefore at high workloads within the single fiber, the rate of pyruvic acid production will be very high. If the muscle fiber is packed with lots of mitochondria, pyruvate will tend to be converted to Acetyl CoA and move into the mitochondria, with relatively little lactate production. Additionally, fatty acid metabolism will account for a higher percentage of the ATP need. Fat metabolism does not produce lactate, ever! If lactate is produced from glucose breakdown, it will tend to be transported from the area of high concentration inside the muscle cell to lower concentration out of the muscle fiber and into extra-cellular fluid, then into the capillaries.

Now let's look at an entire muscle, say the quadriceps muscle group during cycling. At a low workload, glycolytic flux is low (fatty acid breakdown is relatively high at low intensities) and the pyruvate produced is primarily shuttled into the mitochondria for oxidative breakdown. Since the intensity is low, primarily slow twitch muscle fibers are active. These fibers have high mitochondrial volume. As workload increases, more fibers are recruited and already recruited fibers have higher duty cycles (more work and less rest). Now ATP demand has increased in the previously active fibers, resulting in higher rates of pyruvic acid production. A greater proportion of this production is converted to lactic acid rather than entering the mitochondria, due to competition between the two enzymes LDH and PDH. Meanwhile, some fast twitch motor units are starting to be recruited. This will add to the lactate produced in and transported out from the working muscle due to the lower mitochondrial volume of these fibers. The rate of lactate appearance in the blood stream increases.

The quadriceps is just one of several muscles that are very active in cycling. With increasing intensity, increased muscle mass is called on to meet the force production requirements. All of these muscles are contributing more or less lactic acid to the extra-cellular space and blood volume, depending on their fiber type composition, training status and activity level. However, the body is not just producing lactate, but also consuming it. The heart, liver, kidneys, and inactive muscles are all locations where lactic acid can be taken up from the blood and either converted back to pyruvic acid and metabolized in the mitochondria or used as a building block to re-synthesize glucose (in the liver). These sites have low intracellular lactate concentration, so lactic acid is transported into these cells from the circulatory system. If the rate of uptake, or disappearance, of lactate equals the rate of production, or appearance, in the blood, then blood lactate concentration stays nearly constant. But, when the rate of lactate production exceeds the rate of uptake, lactic acid accumulates in the blood volume, then one can see the onset of blood lactate accumulation. This is the traditional "lactate threshold" (LT).

The following five factors influence the rate of lactate accumulation:

1. Exercise intensity;
2. Training status of active muscles (higher mitochondrial volume improves capacity for oxidative metabolism at high glycolytic flux rates);
3. Fiber type composition (slow twitch fibers produce less lactate at a given workload than fast twitch fibers, independent of training status);
4. Workload distribution (a large muscle mass working at a moderate intensity will develop less lactate than a small muscle mass working at a high intensity); and
5. Rate of blood lactate clearance (with training, blood flow to organs such as the liver and kidneys decreases less at any given exercise workload, due to decreased sympathetic stimulation; this results in increased lactate removal from the circulatory system by these organs).

4.4 Physiological Efficiency

Recall from above that high level endurance performance depends on two factors: (i) a high VO_2-max, and (ii) a high lactate threshold. Your VO_2-max sets the upper limit for your sustainable work potential. On the other hand, the lactate threshold tells us something about how much of the cardiovascular capacity you can take advantage of in a sustained effort; it is determined by skeletal muscle characteristics and training adaptations. Multiplying VO_2-max with LT gives us a measure of the effective size of your endurance engine. Now one comes to efficiency. What does efficiency have to do with endurance performance?

Physiological efficiency is defined as the percentage of energy expended by the body that is converted to mechanical work:

$$\textbf{Physiol. efficiency} = \textbf{Mechan. work/Chem. energy expended}$$

One can measure the mechanical work performed using an ergometer, like a bicycle ergometer, or rowing machine. One can measure the energy expended by the body indirectly via its oxygen consumption at sub maximal workloads. With some basic biochemistry, one can convert the oxygen consumption one measures during exercise to a standard measure of energy like kJoules, or Calories. And, one can do the same for the work one measures on the ergometer. Work divided by Time equals Power. Power is measured in watts and is a measure of the intensity of work. Intensity (watts) times exercise duration (minutes) gives us total work, again measured in kJoules or Calories.

4.5 A Proper Warm-Up for Tennis

Whilst the warm up for participation in any sporting or exercise activity is accepted as being essential for minimizing injuries and improving performance, the methods by which many sports attempt to achieve this are less than ideal. The warm up method used by many tennis players usually includes an initial jog around the field or court, followed by 10–15 minutes of static stretching. This is then followed by a few drills, and the players then begin their training session or game. Whilst the basis behind these methods may appear to be sound applications of current training principles, a closer analysis reveals major limitations with this method of preparing a player for tennis training.

The main physiological reason for a warm up include: to increase core temperature (an increase in rectal temperature of a least one to two degree Celsius appears to be sufficient); to increase heart rate and blood flow to skeletal tissues, which improves the efficiency of oxygen uptake and transport, carbon dioxide removal, and removal and breakdown of anaerobic byproducts (lactate); to increase the activation of the Central Nervous System (CNS; therefore increasing co-ordination, skill accuracy and reaction time); to increase the rate and force of muscle contraction and contractile mechanical efficiency (through increased muscle temperature); and to increase the suppleness of connective tissue (resulting in less incidence of musculo–tendinous injuries).

The major criticism against the "typical warm-up" is that it does not adequately prepare the athletes for the demands placed upon them in the ensuring session. Generally the initial jog is at a pace that has a minimal effect upon body temperature, and usually consists of jogging forwards, and in a straight line.

Similarly, the stretching performed is usually that of static stretching, with most stretches performed slowly and with the athletes either standing still or sitting on the ground. This method of stretching has been shown to be beneficial for the increase in limb range of motion, and aims to relax the muscles so that they are less resistant to passive stress for stretching. But this type of stretching does not prepare the muscle and connective tissue for the active contraction - relaxation process that will occur with any running, jumping or hitting movements as required in a tennis game situation.

During this stretching period (typically from 5-20 minutes), the body is very efficient in removing excess body heat, so the small increase in body temperature from the initial jog is quickly lost if the athlete does nothing but statically stretch for this time. This is even more prevalent in cold climates or cold seasons. In general, many injuries occur at the beginning of a competition due largely to an inadequate preparation for the activity. Also, inadequate warming-up can lead to less than optimal speed and skill levels that could result in quick scoring by the opponent or individual early in the game leading then to opponents having to catch-up placing more pressure on the player(s) involved.

The proper warm-up should be the complete physical and mental preparation for the dynamic actions to follow. The players should be able to begin the game or training session totally ready to perform at maximal intensity if required.

The initial jog is now replaced with a more dynamic series of running exercises that include regular alternation of running forwards, backwards, sideways, high knee drills, butt flicks, crossovers, bounding, jumps and progressive sprints. This component will only take 2-4 minutes depending on the climate. It is expected that the athletes are breathing quite heavily at the end of this short series of exercises.

With the stretching component, static stretching can still be included in the program, as many athletes still feel they need some static stretching to really prepare themselves. However, the dynamic stretching component is very important for the specific preparation of the musculature to dynamic movements. Dynamic stretching is defined as repetitive contractions of an agonist (prime mover) muscle to produce quick stretches of the antagonistic muscle, so any active callisthenic movement can be classified as dynamic stretching (jumping, body rotations, bending, etc).

4.6 Nutrition and Hydration

Tennis matches can last much longer than bouts in many other sports, so good nutrition and hydration are very important. Because of the amount of muscle exertion and stamina needed for tennis, the most important part of a player's diet is complex carbohydrates. Professional players eat up to eight servings of pasta, rice and bread every day. This is because the energy provided by carbohydrates is released slowly, compared with simple carbohydrates like sugars which provide a quick burst in energy.

Tennis players can lose up to 2.5 liters of fluid an hour through perspiration, so it is important to replace these fluids by drinking when you play. This means hydrating (drinking) before, during and after playing. Some drinks are better at hydrating you than others, and some like caffeinated drinks actually dehydrate you. The best things to drink when you're playing tennis, or any other sport, are energy drinks, water, and fruit juice. Because of their sugar content, soft drinks give a short-term energy boost for up to half an hour. Sports energy drinks have two big advantages over water. Firstly, they contain carbohydrates, which are very important for maintaining energy levels. Secondly, sports drinks replace electrolytes, such as sodium, which are lost through perspiration.

5

The Law of Basic Biomechanics

5.1 Mechanics of Topspin and Backspin Shots

A topspin shot is hit by sliding the racquet up and over the ball as it is struck. By dragging the racquet over the ball, the friction between the racquet's strings and the ball is used to make the ball spin forward, towards the opponent. The shot dips down after impact and also bounces at an angle lower to the ground than a shot hit with no topspin. As a ball travels towards a player after bouncing, it has natural topspin that is caused by the friction of the tennis court. When hitting a topspin shot, the player is reversing the spin of the ball, which requires more energy.

A backspin shot is hit in the opposite manner, by sliding the racquet underneath the ball as it is struck. This causes the ball to spin towards the player who just hit it as it travels away. Generating slice, or backspin, requires only about half the racket head speed compared to hitting topspin, because the player is not required to change the direction in which the ball is spinning. The oncoming ball bounces off the court with topspin, spinning from top to bottom as it comes toward the player. When a player returns the ball with a slice shot the direction in which the ball spins around the axis of rotation is maintained. The direction of the shot changes, but the ball continues to spin from top to bottom, from the player's perspective as it moves away from the player.

5.2 The Need for Biomechanics

In tennis, one transfers the energy from their body to the ball via a tennis racket to generate speed and spin of the ball. Energy can be either potential (stored energy) or kinetic energy (energy of movement). A specific type of potential energy is elastic energy (that is, the energy which causes, or is released by, the elastic distortion of a solid or a fluid). An example of elastic energy is the energy stored in a spring under tension. The human equivalent would be energy stored in muscles and their tendons under tension. On the other hand, kinetic energy

T.T. Ivancevic et al.: Paradigm Shift for Future Tennis, COSMOS 12, pp. 27–33.
springerlink.com

specifically refers to the work required to accelerate the ball from a resting position to a desired velocity.

Let's examine how the body transfers the necessary energy to the ball in a tennis stroke. Here, one thinks of the body as a series, or a chain, of linkages connected to one another and affecting each other in a specific sequence. For example, the foot is a link, which is connected to the leg by the ankle joint, which is in turn connected to the thigh by the knee joint and so on. During the initiation of a forehand ground stroke the feet are oriented for either an open stance or close stance position. The shoulder and torso are turned approximately 45 degrees, which in turn causes a "coiling" of the abdomen and pelvis, which in turn produce a slight knee bend. With the current forehand the racket is held fairly high at about head level. In this position there exists a great deal of potential energy, both in the form of gravity with the racket head up high and the form of elastic stored energy in the tensed muscles that are stretched in the coiled position (both internal and external abdominal obliques muscles, pectorals major, forearm muscles, hip girdle musculature, quadriceps femoris). This energy is released in sequence and there is an overlap in the sequence of linkages. As the racket starts to drop and begin an oval path (loop) the hips start to uncoil. The hips and knees begin to straighten. In sequence with the uncoiling of the hips the next event is the uncoiling of the torso and then the shoulders as the racket is brought forward to contact the ball. At the same time, the back leg is fully extended to powerfully drive the body up and forward. In fact, many professional players actually leave the ground during this point. At ball contact only medium grip pressure is required to guide and stabilize the racket. This is because the forward momentum will carry the racquet through the ball without much effort. After contact the shoulder and torso and hips naturally rotate towards the non-dominant side following the path of the racquet resulting in a stretch of the opposite side musculature which decelerates the racquet.

Naturally, all of this occurs in one fluid motion with precise timing so that maximum energy (and momentum) transfer occurs from loading to releasing. And, for maximum racket-head speed, some body segments may be slowed down to increase the speed of the racket, as in cracking a whip. Thus, we one see that not only one does need some basic biomechanics at all levels of his/her tennis maturity, but as one advances in tennis, one even needs a special biomechanics of whip–like movements, which is crucial to make every serve, forehand and backhand – efficient weapons (we will talk about this in detail later).

5.3 The Basic Biomechanical Unit

The basic biomechanical unit consists of a pair of mutually antagonistic muscles producing a common muscular torque, T_{Mus}, in the same joint, around the same axes. The most obvious example is the biceps–triceps pair (see Figure 5.1). Note that in the normal vertical position, the triceps downward action is supported by gravity, that is the torque due to the weight of the forearm and the hand (with the possible load in it).

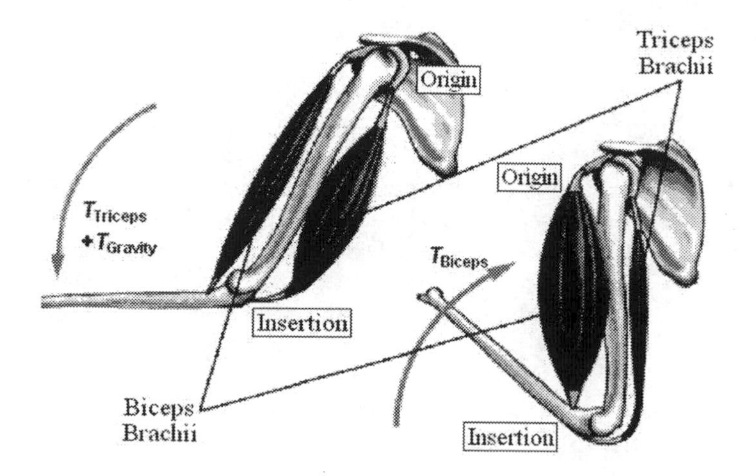

Fig. 5.1. Basic biomechanical unit: left – triceps torque T_{Triceps}; right – biceps torque T_{Biceps}.

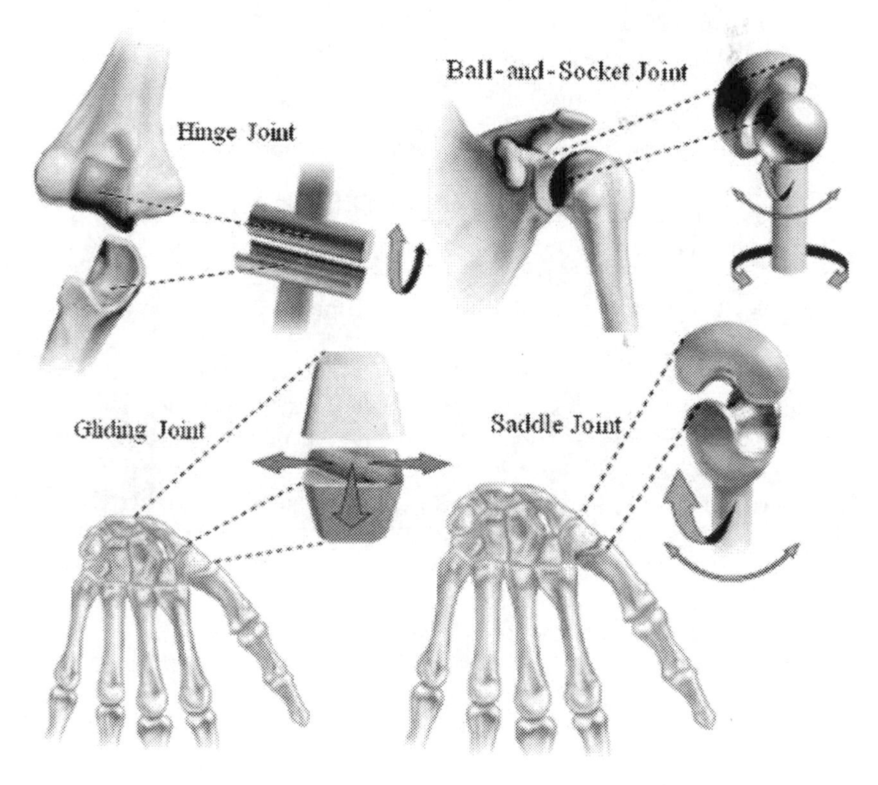

Fig. 5.2. Degrees-of-freedom in main human joints.

5.4 Degrees–of–Freedom in Human Joints

Some joints are only slightly movable, formed by two bones held together by cartilage, without joint cavity (e.g., an intervertebral joint in the spine consists of two vertebras and an intervertebral disc between them). On the other hand, major joints involved in human movement, like shoulder, hip, elbow and knee, are composed of several bones separated by a joint cavity, lubricated by synovial fluid and enclosed in a fibrous joint capsule. Different joints have different degrees–of–freedom (DOF) of movement: hinge joints have 1 DOF, gliding and saddle joints have 2 DOF, while ball-and-socket joints have 3 DOF (see Figure 5.2).

5.5 Minimum Dynamics Necessary for Tennis Biomechanics

Briefly, dynamics of human motion is governed by the Newton-Euler equations of a rigid body motion in our 3D space (see Figure 5.3 depicting a tennis ball modelled as a free rigid body[1]). A rigid body freely moving in space has 6 degrees-of-freedom: 3 translations (along the X,Y,Z-axes) and 3 rotations (around the X,Y,Z-axes). Translational motion is defined by 3 Newton's equations of motion of the type:

$$\text{force } (\mathbf{F}) = \text{mass } (\mathbf{m}) \times \text{acceleration } (\mathbf{a})$$

Similarly, rotational motion (labelled by superscript R) is governed by 3 Euler's equations of motion:

$$\text{torque } (\mathbf{T}) = \text{inertia-moment } (\mathbf{I}) \times \text{rotational-acceleration } (\mathbf{a}^R).$$

In Figure 5.3, with each axis (denoted by index $\mathbf{x}, \mathbf{y}, \mathbf{z}$) there is associated a translational **F**-equation and a rotational **T**-equation; **v** denotes velocity, while dot over a quantity denotes it's rate-of-change (with respect to time). Also,

$$\text{inertia-moment } (\mathbf{I}) = \text{mass } (\mathbf{m}) \times \text{moment-arm squared } (\mathbf{l}^2).$$

Finally,

$$\text{linear momentum } (\mathbf{p}) = \text{mass } (\mathbf{m}) \times \text{velocity } (\mathbf{v}),$$
$$\text{force } (\mathbf{F}) = \text{rate-of-change of linear momentum } (\dot{\mathbf{p}});$$

and similarly,

[1] It is obvious that this is only an approximation, as a tennis ball is not rigid, but rather a soft body. However, soft–body dynamics is much more complicated, so let's stick to our rigid–body approximation. Even true rigid–body dynamics is more complicated.

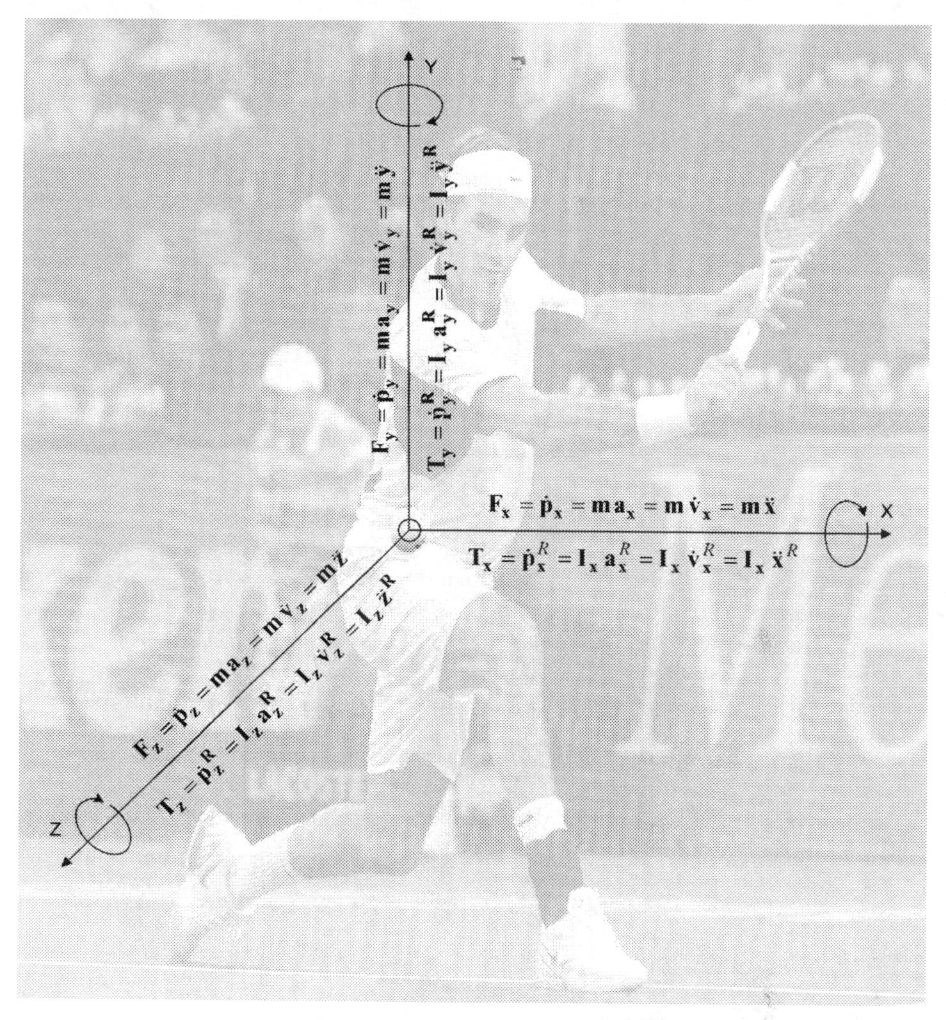

Fig. 5.3. Newton–Euler dynamics of a tennis ball modelled as a free rigid body.

$$\text{angular momentum } (\mathbf{p}^R) = \text{inertia-moment } (\mathbf{I}) \times \text{angular velocity } (\mathbf{v}^R),$$
$$\text{torque } (\mathbf{T}) = \text{rate-of-change of angular momentum } (\dot{\mathbf{p}}^R).$$

Newton's Causality Principle states: a force is a cause of acceleration, which is a cause of velocity, which is a cause of linear motion. Similarly, a torque is a cause of angular acceleration, which is a cause of angular velocity, which is a cause of angular motion.

In biomechanics, the only active force is muscular force. In every major human joint an antagonistic pair of muscles generates a driving torque. Any kind of human motion is a result of driving torques in major human joints.

From a muscular training perspective, the most important is *physical power*, which incorporates both strength and speed, mechanically defined as:

$$\text{power } (\mathbf{P}) = \text{force } (\mathbf{F}) \times \text{velocity } (\mathbf{v}).$$

Physiologically, it corresponds to the area under the force–velocity curve (see Figure 2.4). Muscular power is the key element of all *power sports*, including tennis.

5.6 The Difference between *Moment* and *Momentum*

To clarify any language ambiguities, we give here the main definitions:

- I: *Moment of Inertia:*

$$I \sim mr^2 : \text{mass} \times \text{moment–arm}^2 \qquad (\sim \text{ means “proportional to”})$$

- p^R: *Angular Momentum (or, Moment of Momentum):*

$$p^R = Iv^R : \text{moment of inertia} \times \text{angular velocity, analogous to: } p = mv$$
$$p^R = p \times r : \text{angular momentum} = \text{linear momentum} \times \text{moment–arm}$$

- T: *Torque (or, Moment of Force)*

$$T = \dot{p}^R = I\dot{v}^R : \text{torque} = \text{derivative of angular momentum; this is analogous to: } F = \dot{p} = m\dot{v} : \text{force} = \text{derivative of linear momentum}$$
$$T = F \times r : \text{torque} = \text{force} \times \text{moment–arm}$$

Note that in anatomical literature our *moment–arm* is called *lever–arm.*

5.7 Dynamical Balance of Tennis Players

The above Newton-Euler dynamics (depicted in Figure 5.3) has its neural sensors. The *vestibular organ* in the inner ear helps to maintain equilibrium by sending the brain information about the motion (both linear and angular) and position of the head. The vestibular organs consist of three membranous semicircular canals (SCC), and two large sacs, the utricle and saccule. All the vestibular organs share a common type of receptor cell, the hair cell.

The SCC within the vestibular organ of each ear contain fluid and hair receptor cells encased inside a fragile membrane called the cupula. The cupula is located in a widened area of each canal called the ampulla. When you move your head, the fluid in the ampulla lags behind, pushing the cupula a very tiny bit which causes the hairs to also bend a very tiny bit. The bending hairs stimulate the hair cells, which in turn trigger sensory impulses in the vestibular nerve going to the brain to “report” the movement. Hair cells are amazingly sensitive.

For example, a cupula movement of even a thousandth of an inch is detected by the brain as a big stimulus.

The SCC are positioned roughly at right angles to one another in the three planes of space. Thus, the canals react separately and in combination to detect different types of angular head movement. They detect when one nods in an up and down motion (pitch), when one tilts their head to the side towards their shoulder (roll), and when one shakes their head "no" in a side-to-side motion (yaw). The SCC are responsible for detecting any kind of rotational motion in the head, thus effectively sensing the Euler's dynamics (Figure 5.3).

Two other vestibular organs are located in membranous sacs called the utricle and the saccule. On the inside walls of both the utricle and the saccule is a bed (a macula) of several thousand hair cells covered by small flat piles of calcium carbonate crystals which look like sand, imbedded in a gel-like substance. The crystals are called otoliths, a word which literally means "ear stones." In fact, the utricle and the saccule are often called the *otolith organs*.

When a person's head is in the normal erect position, the hair cells in the utricle lie approximately in a horizontal plane. When the head is tilted to one side, the stones want to slide "downhill." This moves the gel just enough to bend the sensory hairs. The bending hairs stimulate the hair cells, which in turn send a signal to the brain about the amount of head tilt. The stones also move if the person is accelerated forward and back, or side to side. Similarly, the hair cells in the saccule are oriented in somewhat of a vertical position when the head is erect. When a person tilts their head, or is accelerated up and down (as in an elevator), or moved forward and back, the otoliths move and a signal is sent to the brain. The signals from the otoliths in the saccule and the utricle complement each other and give us an integrated signal about our movement. The otolith organs are primarily responsible for detecting any degree of linear motion of the head, thus effectively sensing the Newton's dynamics (Figure 5.3).

6

The Law of Neuro-motor Control

6.1 Control of Tennis Movements by the Brain

All of the body's voluntary movements are controlled by the brain. One of the brain areas most involved in controlling these voluntary movements is the motor cortex (see Figure 6.1).

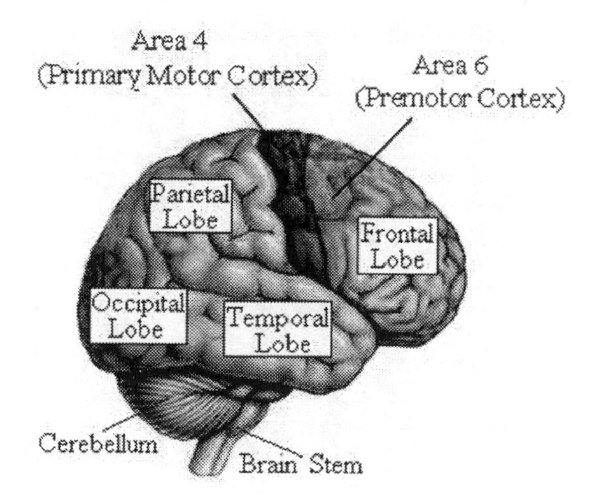

Fig. 6.1. Main lobes (parts) of the brain, including motor cortex.

In particular, to carry out goal–directed tennis movements, your motor cortex must first receive various kinds of information from the various lobes of the brain: information about the body's position in space, from the the parietal lobe; about the goal to be attained and an appropriate strategy for attaining it, from the anterior portion of the frontal lobe; about memories of past strategies, from the temporal lobe; and so on.

T.T. Ivancevic et al.: Paradigm Shift for Future Tennis, COSMOS 12, pp. 35–39.
springerlink.com

There is a popular "motor map", a pictorial representation of the primary motor cortex (or, Brodman's area 4), called *Penfield's motor homunculus*[1] (see Figure 6.2). The most striking aspect of this motor map is that the areas assigned to various body parts on the cortex are proportional not to their size, but rather to the complexity of the movements that they can perform. Hence, the areas for the hand and face are especially large compared with those for the rest of the body. This is no surprise, because the speed and dexterity of human hand and mouth movements are precisely what give us two of our most distinctly human faculties: the ability to use tools and the ability to speak. Also, stimulations applied to the precentral gyrus trigger highly localized muscle contractions on the contralateral side of the body. Because of this *crossed neural control*, this motor center normally controls the voluntary movements on the opposite side of the body.

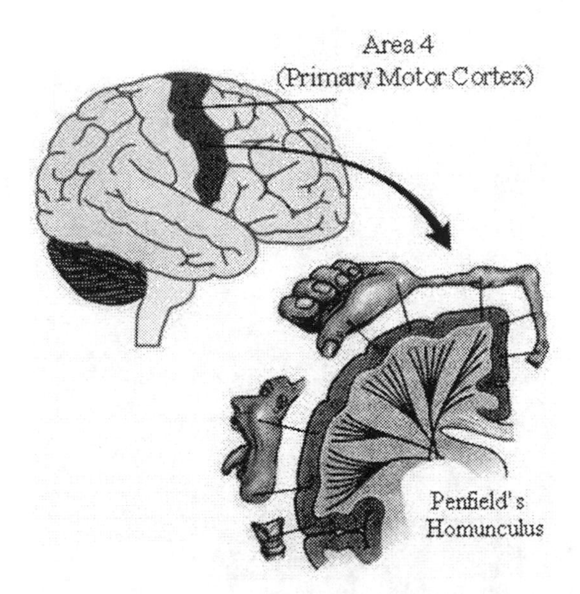

Fig. 6.2. Primary motor cortex and its motor map, the Penfield's homunculus.

Planning for any tennis movement is done mainly in the forward portion of the frontal lobe of the brain (see Figure 6.3). This part of the cortex receives information about the player's current position from several other parts. Then, like the ship's captain, it issues its commands, to Brodman's area 6, the premotor cortex. Area 6 acts like the ship's lieutenants. It decides which set of muscles to contract to achieve the required tennis movement, then issues the corresponding orders to the primary motor cortex (Area 4). This area in turn activates specific muscles or groups of muscles via the motor neurons in the spinal cord.

[1] Note that there is a similar "sensory map", with a similar complexity–related distribution, called *Penfield's sensory homunculus*.

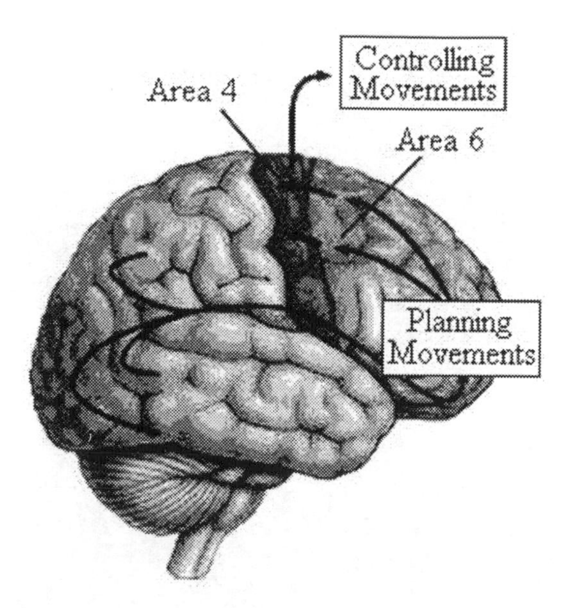

Fig. 6.3. Planning and control of movements by the brain.

6.2 The 'Missing Link' between Brain and Muscles

Between brain (that plans the movements) and muscles (that execute the movements), the most important link is the *cerebellum* (see Figure 6.1). For you to perform even so simple a gesture as touching the tip of your nose, it is not enough for your brain to simply command your hand and arm muscles to contract. To make the various segments of your hand and arm deploy smoothly, you need an internal "clock" that can precisely regulate the sequence and duration of the elementary movements of each of these segments. That clock is the cerebellum.

The cerebellum performs this fine coordination of movement in the following way. First it receives information about the intended movement from the sensory and motor cortices. Then it sends information back to the motor cortex about the required direction, force, and duration of this movement (see Figure 6.4). It acts like an air traffic controller who gathers an unbelievable amount of information at every moment, including (to return to our original example) the position of your hand, your arm, and your nose, the speed of their movements, and the effects of potential obstacles in their path, so that your finger can achieve a "soft landing" on the tip of your nose.

Therefore, to ensure efficiency of tennis movements, i.e., that all the movements are fast, precise, and well coordinated, the nervous system must constantly receive sensory information from the outside world and use this information to adjust and correct the hand's trajectory. The nervous system achieves these adjustments chiefly by means of the cerebellum, which receives information about the positions in space of the joints and the body from the proprioceptors. Even for a movement as simple as picking up a glass of water, one can scarcely

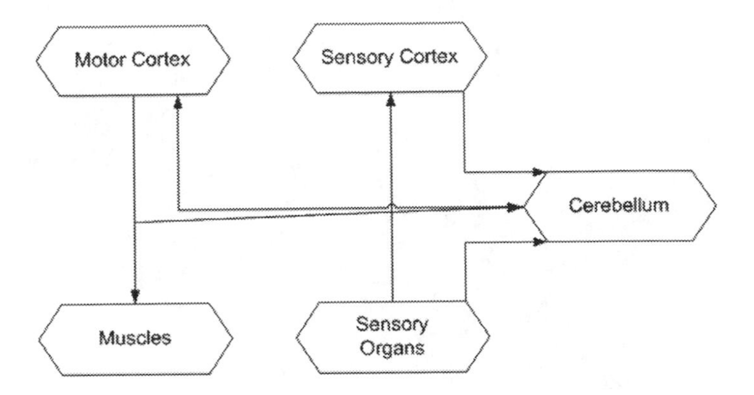

Fig. 6.4. The cerebellum loop for movement coordination.

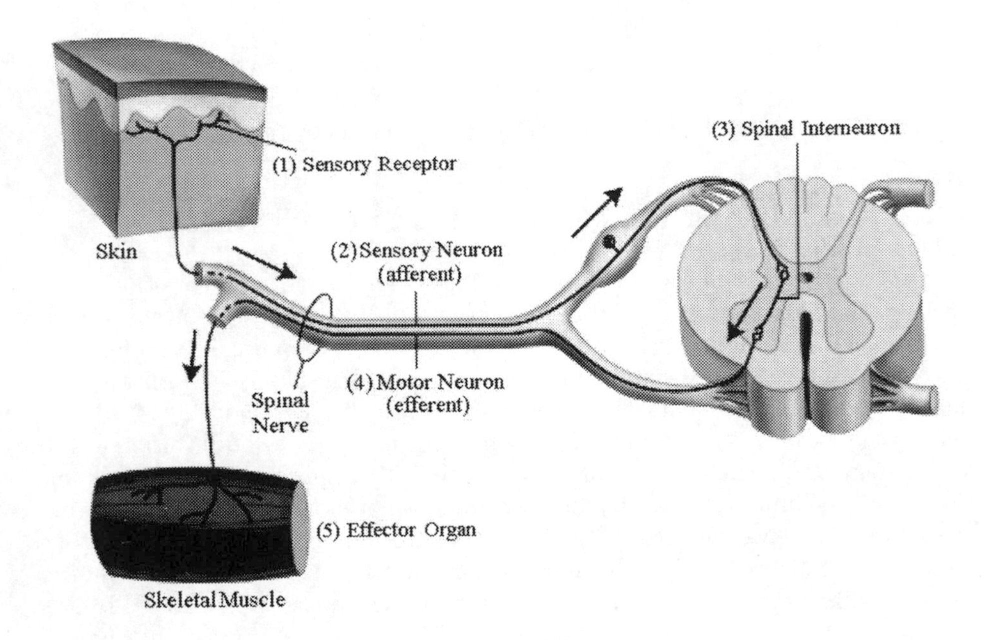

Fig. 6.5. Five main components of a reflex arc: (1) sensory receptor, (2) sensory (afferent) neuron, (3) spinal interneuron, (4) motor (efferent) neuron, and (5) effector organ – skeletal muscle.

imagine trying to consciously specify the sequence, force, amplitude, and speed of the contractions of every muscle concerned. Therefore, the cerebellum clearly does a very important job in tennis.

6.3 The Reflex Arc

A reflex is a spinal neural feedback, the simplest functional unit of a sensory–motor control. Its *reflex arc* involves five basic components (see Figure 6.5): (1) sensory receptor, (2) sensory (afferent) neuron, (3) spinal interneuron, (4) motor (efferent) neuron, and (5) effector organ – skeletal muscle. Its purpose is the quickest possible reaction to the potential threat.

6.4 The Role of Spinal Reflexes in Movement

While the motor cortex plans the movements (top control level), the cerebellum makes them efficient (middle control level), at the most basic level, movement is controlled by the *spinal cord* alone, with no help from the brain. The neurons of the spinal cord thus take charge of the *reflex movements* as well as the rhythmic movements involved in walking. We will talk about reflexes later. For the lightning–speed of the future tennis game, they are the most important neural part.

7 The Law of Cognition

7.1 The Role of Sports Psychology

The main topics of sports psychology are: (i) motor (and general) learning; (ii) behavioral patterns (e.g., fight-or-flight); (iii) visualization; (iv) concentration; (v) relaxation strategies; (vi) self talk (introspective thought) strategies; (vii) arousal strategies; and (viii) stress management. We will talk about each of them in some detail in this book.

7.2 Learning

Learning is a relatively permanent change in behavior that marks an increase in knowledge, skills, or understanding thanks to recorded memories. A memory[1] is the fruit of this learning process, the concrete trace of it that is left in your neural networks.

More precisely, learning is a process that lets us retain acquired information, affective states, and impressions that can influence our behavior. Learning is the main activity of the brain, in which this organ continuously modifies its own structure to better reflect the experiences that one has had. Learning can also be equated with encoding, the first step in the process of memorization. Its result, memory, is the persistence both of autobiographical data and of general knowledge.

But memory is not entirely faithful. When you perceive an object, groups of neurons in different parts of your brain process the information about its shape, color, smell, sound, and so on. Your brain then draws connections among these different groups of neurons, and these relationships constitute your perception of the object. Subsequently, whenever you want to remember the object, you must reconstruct these relationships. The parallel processing that your cortex does for this purpose, however, can alter your memory of the object.

[1] "The purpose of memory is not to let us recall the past, but to let us anticipate the future. Memory is a tool for prediction" – Alain Berthoz.

T.T. Ivancevic et al.: Paradigm Shift for Future Tennis, COSMOS 12, pp. 41–48.
springerlink.com

Also, in your brain's memory systems, isolated pieces of information are memorized less effectively than those associated with existing knowledge. The more associations between the new information and things that you already know, the better you will learn it.

If you show a chess grand master a chessboard on which a game is in progress, he can memorize the exact positions of all the pieces in just a few seconds. But if you take the same number of pieces, distribute them at random positions on the chessboard, then ask him to memorize them, he will do no better than you or I. Why? Because in the first case, he uses his excellent knowledge of the rules of the game to quickly eliminate any positions that are impossible, and his numerous memories of past games to draw analogies with the current situation on the board.

Psychologists have identified a number of factors that can influence how effectively memory functions, including:

1. Degree of vigilance, alertness, attentiveness,[2] and concentration.
2. Interest, strength of motivation,[3] and need or necessity.
3. Affective values associated with the material to be memorized, and the individual's mood and intensity of emotion.[4]
4. Location, light, sounds, smells..., in short, the entire context in which the memorizing takes place is recorded along with the information being memorizes.[5]

Forgetting is another important aspect of memorization phenomena. Forgetting lets you get rid of the tremendous amount of information that you process every day but that your brain decides it will not need in future.

7.3 Memory

Human memory is fundamentally associative. You can remember a new piece of information better if you can associate it with previously acquired knowledge that is already firmly anchored in your memory. And the more meaningful the association is to you personally, the more effectively it will help you to remember. Memory has three main types: sensory, short–term and long–term (see Figure 7.1).

[2] Attentiveness is often said to be the tool that engraves information into memory.

[3] It is easier to learn when the subject fascinates you. Thus, motivation is a factor that enhances memory.

[4] Your emotional state when an event occurs can greatly influence your memory of it. Thus, if an event is very upsetting, you will form an especially vivid memory of it. The processing of emotionally-charged events in memory involves norepinephrine, a neurotransmitter that is released in larger amounts when one is excited or tense. As Voltaire put it, "That which touches the heart is engraved in the memory."

[5] Our memory systems are thus contextual. Consequently, when you have trouble remembering a particular fact, you may be able to retrieve it by recollecting where you learnt it or the book from which you learnt it.

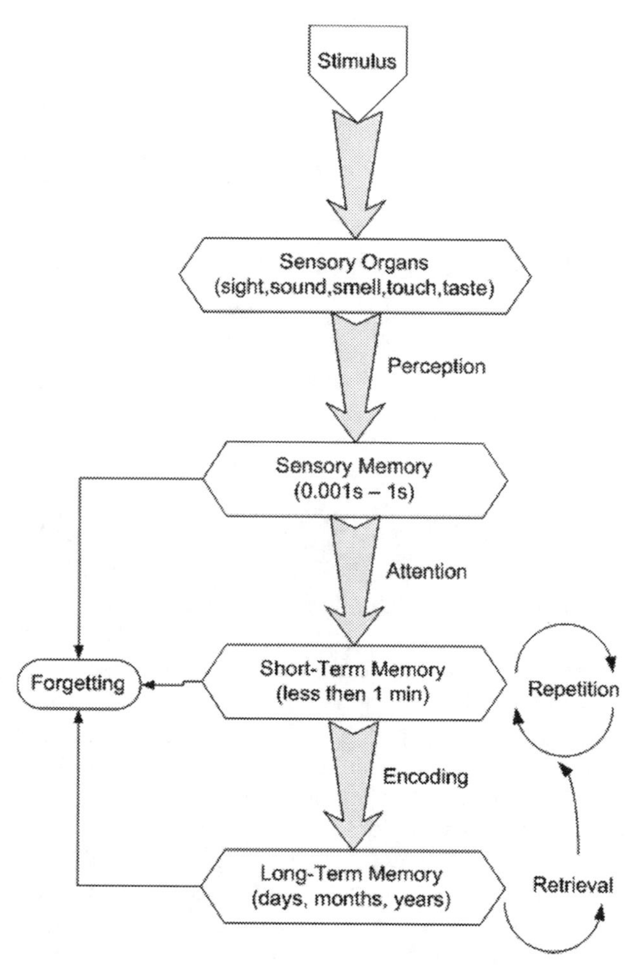

Fig. 7.1. Human cognitive memory.

The *sensory memory* is the memory that results from our perceptions automatically and generally disappears in less than a second.

The *short–term memory* depends on the attention paid to the elements of sensory memory. Short–term memory lets you retain a piece of information for less than a minute and retrieve it during this time.

The *working memory* is a novel extension of the concept of short–term memory; it is used to perform cognitive processes (like reasoning) on the items that are temporarily stored in it. It has several components: a control system, a central processor, and a certain number of auxiliary "slave" systems.

The *long–term memory* includes both our memory of recent facts, which is often quite fragile, as well as our memory of older facts, which has become more consolidated (see Figure 7.2). It consists of three main processes that take place consecutively: encoding, storage, and retrieval (recall) of information. The

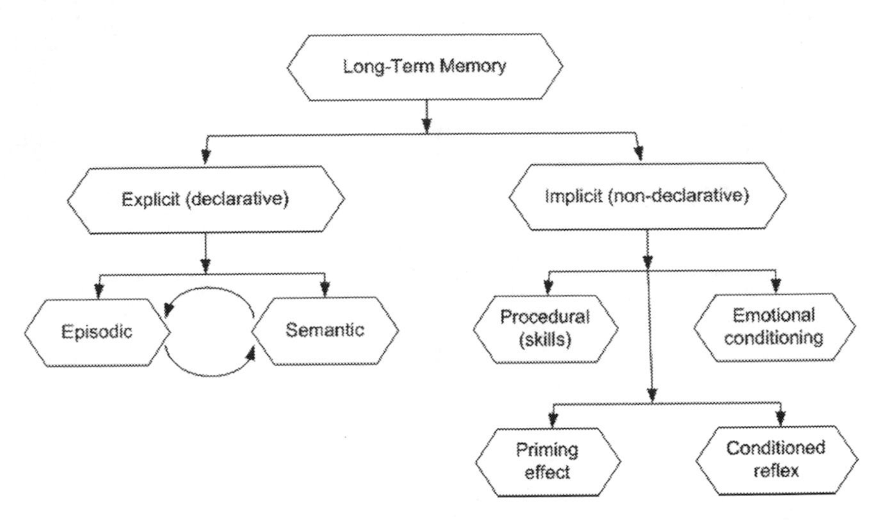

Fig. 7.2. Long–term memory.

purpose of *encoding* is to assign a meaning to the information to be memorized. *Storage* can be regarded as the active process of consolidation that makes memories less vulnerable to being forgotten. Lastly, retrieval (recall) of memories, whether voluntary or not, involves active mechanisms that make use of encoding indexes. In this process, information is temporarily copied from long-term memory into working memory, so that it can be used there.

Retrieval of information encoded in long-term memory is traditionally divided into two categories: recall and recognition. Recall involves actively reconstructing the information, whereas recognition only requires a decision as to whether one thing among others has been encountered before. Recall is more difficult, because it requires the activation of all the neurons involved in the memory in question. In contrast, in recognition, even if a part of an object initially activates only a part of the neural network concerned, that may then suffice to activate the entire network.

Long–term memory can be further divided into explicit memory (which involves the subjects' conscious recollection of things and facts) and implicit memory (from which things can be recalled automatically, without the conscious effort needed to recall things from explicit memory, see Figure 7.2). Episodic (or, autobiographic) memory lets you remember events that you personally experienced at a specific time and place. Semantic memory is the system that you use to store your knowledge of the world; its content is thus abstract and relational and is associated with the meaning of verbal symbols.

Procedural memory, which is unconscious, enables people to acquire motor skills and gradually improve them. Implicit memory is also where many of our conditioned reflexes and conditioned emotional responses are stored. It can take place without the intervention of the conscious mind.

7.4 Genetically Coded Behavior

Whatever behavior one initiates, be it drinking, playing, reading, making strategic alliances, or making eyes at someone, it is always because one is subjectively feeling certain needs. A specific situation leads to a specific neural activity pattern in a person's brain that in turn leads to a specific behavior.

Each individual's genes activate a unique program for the development of that person's nervous system. But how this nervous system actually develops depends on each person's interactions with the environment, that is on unique personal experiences. The behaviors that this person is capable of are determined by the unique activity patterns of his/her nervous system, some of which are experienced as thoughts, emotions, memories, etc. Any given behavior by this person results from the interaction between his/her neural activity at that specific moment and his/her perception of this specific behavior.

Firstly, our most primitive behaviors (like our reflexes) are concerned with the present. These are incapable of adaptation. They make us react to external or internal stimuli automatically. Neurophysiologically speaking, these behaviors represent the activation of the "reptilian" structures of the brain, bringing the *hypothalamus* and the *brain stem* into play.

Secondly, our learned behaviors add our past experience to our present actions. These more sophisticated behaviors involve remembering pleasant or unpleasant sensations that one experienced in the past and the actions that caused us to experience them at the time. These behaviors represent most of the social and cultural knowledge that one acquires. In connection with these behaviors emotions arise, the awareness of the cardiovascular adjustments necessary for action. Neurophysiologically speaking, these behaviors represent the activation of the "mammalian" structures of the brain, bringing the *limbic system* into play.

Thirdly, our imagined behaviors respond to the present through past experience, by anticipation of the future result. They involve more elaborate planning. They call on the *creative imagination*, and hence on the *associative cortex*, to develop strategies for ensuring that our actions will be gratifying rather than painful. They represent the creative and innovative abilities of the human mind. Neurophysiologically speaking, these behaviors represent the activation of the "neocortical" structures of the brain, bringing the associative areas of the cerebral cortex into play.

7.5 Fight-or-Flight Reaction

A *human and animal behavior* is a set of movements coordinated by the nervous system to preserve the structure of the organism. The basic behavior is therefore to approach or explore the resources available in the environment. When an action to acquire one of these resources is rewarded, this gratifying behavior is positively reinforced, and the strategy through which the need was satisfied is memorized.

The other main basic behavior is to avoid pain, and hence to avoid situations that might lead to the organism's premature death. Fight, flight, and inhibition

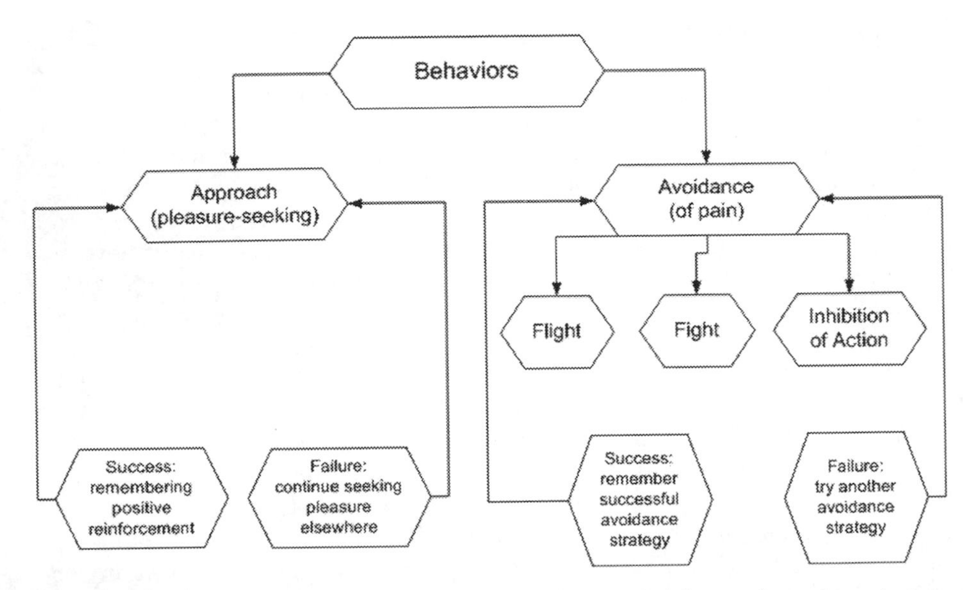

Fig. 7.3. Behavior is memorized if it succeeds and discarded for an alternative strategy if it fails.

of behavior are the three possible behavioral responses to a nociceptive stimulus. Just like gratifying behaviors, the response of fighting, fleeing, or inhibiting behavior in response to a threatening stimulus can be either effective or ineffective. The associated behavior is then memorized as either a winning or a losing strategy (see Figure 7.3).

Our environment is full of potentially gratifying objects that stimulate our approach behaviors. But it is also full of other people who also want to use these resources to ensure their own well-being. Each individual must therefore learn to decode other people's intentions in order to choose the attitude that will make his or her own actions the most effective.

For example, if you resisted somebody who was trying to take some resource away from you, and you ended up not only losing the resource but also getting injured in the struggle, then the next time, the memory of this failure might very well cause you to simply run away.

And if this person were someone from whom you could not flee (for example, because you were economically dependent on them), then you would learn that the best thing to do would be to inhibit your own behavior and accept your subordinate status. This is the way how social hierarchies are established.

The three reactions that let you avoid pain trigger hormonal and vasomotor adjustments that are controlled by the sympathetic nervous system. The activation of this system provides increased motor autonomy, mainly by increasing the oxygenation of the skeletal muscles. But these changes cannot last indefinitely, or they would interfere with the healthy functioning of those parts of the body, such as the internal organs, that were left temporarily short of blood. Once the source of the threat has disappeared, the body's equilibrium must therefore be rapidly restored.

7.6 The Importance of *Creative Visualization* in Motor Learning

Mental imagery, or *creative visualization* plays an important role in the learning of sport movements and the improvement of motor performance. The latest research is attempting to identify the determinants of efficient mental imaging in sport. Studies have already shown that people who are better at generating mental images make faster progress in motor learning, as do people who practice a movement physically before imagining it mentally. Research on mental imagery has shown that the processes for actually producing a movement and for representing it mentally are identical. Various experimental approaches have been used to show, for example, that the mental representation of an action seems to be based on the same mechanisms as the motor preparation for it.

With mental chronometry, for example, it has been shown that visual mental images preserve the spatial and structural characteristics of the object or scene that they represent. For example, it has been shown that the visual travel time between two points in a mental image of an object is proportional to the distance between these two points on the actual object.

Researchers have also found some physiological indicators whose activation resulted solely from mental imaging of actions. For example, a group of researchers measured how physical training and training by mental imaging affected the strength of finger muscles. Physical training increased finger muscle strength by 30%, but mental imaging alone still increased it by 22%. Since the subjects did not make any muscle contractions during their mental imagery training, the observed changes did not come from the peripheral motor system but actually came from the activation of circuits in the central motor system.

Here is yet another example. When subjects were asked to imagine walking or running at various speeds on a treadmill, their heart rate and total ventilation increased in proportion to the speed imagined in the course of this mental exercise, even though their oxygen consumption remained steady.

According to the theory for which Swedish neurobiologist David Ingvar coined the clever term "future memory", the *parietal cortex* is capable of producing internal models of movements to be performed, prior to any processing in the premotor and motor cortices. According to this theory, the brain is constantly simulating movements, only some of which are eventually externalized. This theory could provide a conceptual foundation for the mental training done by athletes and musicians, as well as for re-education through motor imagery.

The concept of motor imagery extends to the sensory modalities as well. When someone has any given sensory experience, then re-imagines its later, the brain activity in the two cases will be similar in both location and intensity. Because brain activity continuously influences the body, and vice versa, any experience has a given effect on the body, and re–visualizing it would generate a similar pattern of brain activity with similar effects on the endocrine system, the immune system, and so on. Speaking very generally, visualization can therefore be regarded as a form of autosuggestion or self–hypnosis which, by generating emotions, may have a beneficial physiological effect on the body.

Question: Do you think that top tennis players, like Federer, Nadal, Djokovic, and Roddick, can improve their performance with an appropriate sport–science knowledge?

Answer: Absolutely! Modern sports science can improve performance of any athlete! Very briefly, Roger Federer needs only *superb tennis weapons* (explained below). Rafael Nadal needs the same plus Federer's tennis knowledge. Andy Roddick mostly needs Federer's anticipation and mental strength. Novak Djokovic needs superb tennis weapons plus Federer's tennis knowledge.

8

The Law of High Performance

8.1 The Current Tennis Performance Criteria

The overriding principle governing *general sports performance* is the attempt of an individual, or a group of individuals, to perform a given task "in the best possible way." In this chapter we will focus on biomechanical and physiological principles of the *performance optimization* in the future tennis game. For the *tennis performance criteria* one can use the 10 points of the standard tennis game statistics (in brackets are the current ranks of Roger Federer, the greatest tennis champion, on October 22nd 2007, as given by *ATPtennis.com*):[1]

- *Service game:* (i) number of aces (4), (ii) 1st serve percentage (29), (iii) 1st serve points won (6), (iv) 2nd serve points won (1), (v) service games won (3), and (vi) break points saved (8).
- *Return of service:* (vii) points won returning 1st serve (4), (viii) points won returning 2nd serve (17), (ix) break points converted (36), and (x) return games won (10).

8.2 The Main Purpose of Sports Science

The very purpose of sports science is to provide solutions for two essential sport problems:

1. A *direct–training* problem: given the set of empirically proclaimed *talents*, develop the *champion model*.
2. An *inverse–selection* problem: given the *champion model*, develop the *talent model*.

Thus, sports science is all about *training methods* (directed to make a champion) and *selection methods* (directed to finding talents). In sports science all the

[1] Although one can immediately see that the standard statistics does not give us the full picture of the current tennis, as Federer is here ranked (1) only once, while, e.g., Ivo Karlovic, currently world number 25, is here ranked (1) four times.

T.T. Ivancevic et al.: Paradigm Shift for Future Tennis, COSMOS 12, pp. 49–50.
springerlink.com

champions are represented by the *champion model* for a particular discipline (e.g., tennis), and all the talents are represented by the *talent model* for the same discipline. In statistical language, both the champion and the talent have *the same factor structure* – only it is fully developed in case of the champion and yet undeveloped in case of the talent. For example, Nadal, Roddick, Federer, and Djokovic had all been talents. However, so far only one of them has proved to be a real champion – Roger Federer, the man who apparently defies all tennis statistics. Today, in my opinion, the highest chances to become future tennis champions have Nadal, Djokovic and Andy Murray.

9

The Law of Athleticism

9.1 Fast Legs of a Tennis Player

Running fast is the direct result of the athletes stride rate and stride length. Now, the question is how do one maximizes both of these to achieve top–level performances in the sprints, or fastest runs on the court. One cannot have a maximum stride length and stride rate and be their fastest; what is needed is a maximum stride rate with an optimal stride length. Maximal stride rate is how fast one can produce one stride, or about 10 of them in 20 meters. Stride rate is dependent upon a number of factors including, strength and mechanics. In order to produce greater stride rates one must be able to execute the correct stride cycle as fast as possible and with optimal length. Optimal stride length is one that allows the athlete to execute the correct stride pattern in as short a time frame as possible.

On the other hand, ground time is the largest contributor to stride rate. It is known that almost all athletes spend approximately the same amount of time in the air during the sprint stride. The big difference comes in the amount of time spent on the ground. The goal of all sprinters and fast-legged tennis players should be to spend as little time on the ground as possible. In order to achieve this, they need the necessary plyometric strength (explained later) to get them through the correct cycle.

Also, during the short sprints on the court, at each leg joint the musculo-tendinous units absorb force by stretching (eccentric) just before they shorten (concentric) to generate the take–off force.

9.2 Jumping Ability of Tennis Players

Jumping ability is defined as *leg power*. Recall that the muscular power is the product of muscular force (**F**) and movement velocity (**v**). It is represented by the area under the *force–velocity curve* (see Figure 2.4 above). In other words, it is the ability to generate muscular force quickly.

T.T. Ivancevic et al.: Paradigm Shift for Future Tennis, COSMOS 12, pp. 51–56.
springerlink.com

If you look at the force–velocity curve, you will see that high levels of power occur in the mid–range of either force or velocity. If an athlete develops greater power, this, in turn, enhances his ability to generate both force (strength) and velocity (speed). This amalgam of speed and strength may be more useful for athletic performance than strength alone.

Therefore, the leg power is the expression of the leg strength at speed. In the future tennis game, players at all levels will require certain levels of leg power to be successful, the higher the level, the higher the power needed.

The velocity (speed) component of the jumping ability is defined as the *stretch–shortening cycle* (to be explained later). Before one can focus on the stretch–reflex and its derivative, stretch–shortening cycle, one needs to know how to develop the strength (force) component of the jumping ability. There are several common methods for this, including:

1. General leg–strength exercises. These exercises are necessary to develop the force component of power. Squat exercises such as full squats, front squats, half squats, and split squats, develop the jumping musculature to a large degree. Very few, if any, successful high level field event athletes would not have good squatting ability. Even in novice athletes, some form of squatting could be performed with medicine balls or shot puts to challenge the jumping musculature. Although there is great benefit in performing general strength exercises such as squats and its variants, one must be careful not to use the squat as an end in itself as these exercises are slow in nature thus do not replicate the exact demands of the power events.
2. Special leg–strength exercises. These exercises attempt to convert general strength to specific jumping strength. Some examples may include Olympic lifts and their variants and jump squats. Some researchers suggest that the Olympic lifts are very similar in structure to a vertical jump. As jumping and throwing events have some component in the vertical direction the Olympic lifts are excellent exercises for these events.
3. Plyometric leg–strength exercises. These exercises are necessary to develop the velocity component of power and attempt to provide power improvement in a way which is specific to the required technique of an athlete. Examples of such exercises would include bounding and hopping, weighted bounding and hopping in addition to single and double legged box jumps.

9.3 More Dynamic Performance of Traditional Strength Exercises

Take for example the popular bench–press exercise. You can perform it in a traditional, bodybuilding–type form, where you remain tight throughout your entire body, and throughout the entire repetition. Alternatively, you can do it in a Russian plyometric form, which is more relaxed. This later one involves a faster negative movement (eccentric contraction), and a ballistic pressing action (concentric contraction), in reality more like a push press than a bench press (this is technically termed the stretch–shortening cycle). It is clear that the traditional exercise gives you a slow strength only, while the plyometric one gives you both strength and speed, that is the dynamical power. And this dynamical power is what one really needs for future tennis.

In simple mechanical words, given that force (**F**) equals mass (**m**) × acceleration (**a**), and mass is basically always constant in tennis, what one really needs is a huge acceleration. This huge acceleration will produce a huge racket-head speed, which will generate the lightning–speed of the ball.

9.4 A General Example of True Athleticism

In contemporary tennis, the most prominent example of *athleticism* is Rafael Nadal. Other examples include the current champion, Roger Federer, and American James Blake.

Notwithstanding the common public sympathy towards athleticism of the popular players, we need to remark that every serious 9-year-old (or older) gymnast can do a somersault on the street; or, equally, on the tennis court during the tennis game! This shows that the very idea of "athleticism" has historically been a little-bit misunderstood within the tennis community.

Let us have a look at the real athleticism of a person with almost the same body posture and weight as Federer's. We are naturally talking about the track-and-field hero, Jan Zelezny, the javelin world record holder, winner of three Olympic gold medals (Barcelona, Atlanta, Sydney) and three World Championships gold medals (Stuttgart, Gothenburg, Edmonton). He has five world javelin records, including the actual one of 98.48 m (set in 1996, Germany), as well as 34 performances over 90.00 m (which is more than all other javelin throwers combined).

This guy is not "a big boy". He was not over 2 m in height and not over 130 kg (as some "experts" might suggest, by considering his super–human throws). He did not look at all like Arnold Schwarzenegger, but rather as a younger and bigger brother of Bruce Lee. Zelezny's body height was only 1.86 m, (that was exactly between Roddick's and Federer's), while his body weight (at the time of his world-record throw) was only 80 kg. This is the same weight as Federer and the same as 1.87 m tall Novak Djokovic. At the same time, he is 5 kg less than Nadal, or 8 kg less than Roddick, or 8 kg less than Tommy Haas, even one kg less then the 1.82 m tall Fernando Gonzalez. That is to say, except for Federer and Djokovic, all these current elite tennis players are simply *too heavy*. Therefore, you do not need to be heavy and bulky to be able to hit hard! Tennis is not shot–put! The mass of the tennis racket is half the mass of the javelin. Therefore, common sense would say that the best javelin thrower of all times had to be heavier than top tennis players. Yet, at the time of his fantastic world record (as well as his 34 throws over 90.00 m), Zelezny looked pretty skinny. What he had, though, were muscles made of steel and lightning–fast reflexes. That is, Zelezny's body was an efficient slingshot!

In other words, a man with a body very similar to the current champion, Roger Federer, was able to throw an 800 g javelin (purposely reconstructed that it cannot fly far), almost 100 meters. Could you imagine the speed of a tennis serve or forehand with muscles similar to Zelezny's. With muscles trained (for years) in utilizing the stretch–reflex. Zelezny would not only break the strings, if he attempted the tennis serve, but would also brake the *frame* of the "modern" tennis racket. A totally new racket technology would be required for players with muscles similar to Zelezny's. This example shows the real *athleticism:* after years of proper stretch–reflex training, the body becomes a big slingshot, composed of a number of small slingshots: legs, hips, torso, shoulders, elbow and wrist – all working as a kinetic chain, in a dynamical power–sequence.

Now, we do not suggest that a tennis player will ever need to be able to throw a javelin close to 100 meters, like Zelezny. No, a 70 m throw of the current (reconstructed) javelin, quite achievable for junior javelin throwers, would give a 300+ km/h serve with a technique similar to Roddick's, as well as 240+ km/h forehand and a 200+ km/h (single–handed) backhand with a technique similar to Federer's.

What we are suggesting is that the future tennis champion will necesarily have something similar to Zelezny's body height and weight (which was, at the time of his record throws, quite similar to Novak Djokovic body height and weight), but more importantly, Zelezny-like slingshot muscles. It is just a matter of proper stretch–reflex based muscular conditioning. This conditioning would give the future tennis champion the ability to serve consistently over 300 km/h,[1] forehand consistently over 240 km/h and backhand consistently over 200 km/h. Just imagine the current top tennis players playing against someone with these three superb tennis weapons. Would they be able to win a single game?

9.5 The Efficiency Technique

Basically, for dealing with any sport–science issue, one has two possible approaches: sport physiology and sport biomechanics.

Sport and exercise physiology gives us a valid description of the energy systems involved in any kind of cyclic sport activity (like running, cycling, swimming, rowing. . .) of either short– , middle– or long–distance.

On the other hand, the current popular scientific discipline designed for analyzing sport techniques is biomechanics. Anatomical biomechanics describes various techniques using mostly methods from functional anatomy, while *Newton–Euler dynamics* (see Appendix 2: The Newton-Euler Law in Tennis Biomechanics) provides their mathematical modelling (and associated computer simulations), for the purpose of answering the What–If questions (by varying

[1] You don't have to be 9 feet tall nor does your combined height including your extended arm, racquet, and jump height need to be 9 feet or more. This is because both gravity and aerodynamic drag act on the ball during its flight. Gravity accelerates the ball downwards while drag (air resistance) creates a retarding force slowing the ball's forward motion. This causes a curved trajectory.

the athlete's parameters). Biomechanics can confidently do two things: (i) give us all Newtonian mechanical principles as guidance for development of any sport technique, and (ii) roughly describe the technique of a current champion (using the concept of bio-kinetic chains).

Both sport physiology and biomechanics are legitimate scientific tools for understanding standard sport activities and/or human movements. However, neither of them can give us the most precious answers to the crucial question: What is the *most efficient technique* for any particular human movement, both cyclic and non-cyclic? When it comes to understanding, prediction and control of the most efficient sport movements as well as training methods, both sport physiology and biomechanics fail.

For example, neither of them can help us understand the fastest ever tennis serve of American Andy Roddick (153 m/h), or the fastest tennis forehand of French Gael Monfils (120 m/h). These two guys are neither the tallest (which would give them a huge leverage), nor the strongest (which would give them a huge force), nor the fastest (which would give them a huge speed– arguably, Rafael Nadal is both stronger and faster then both of them) in the game. They are record holders for the serve and the forehand, and it wouldn't hurt even Federer to have a stronger serve and forehand – he would not need to go through so many tie–breaks.

To answer this, in our opinion, the most significant question in elite sport is the question that explains the secret technique of champions and to do this one needs a completely different scientific approach. What is the secret behind Roddick's serve and Monfils' forehand? The secret is the *stretch–reflex*. Moreover, the stretch–reflex is the unique secret behind all highly–efficient movements in sport, including all athletic throws, jumps, sprints, all weight lifts, fast gymnastic movements and so on.

For his fastest serve, Roddick used the stretch–reflex in (the prime mover muscles of) all major joints: right shoulder, right elbow, right wrist, both hips, both knees and both ankles. For his record forehand, Monfils used the stretch–reflex in the same joints. The two techniques look completely different, but the neuro-physiological basis is the same; the stretch–reflex utilized in (the prime mover muscles of) the same joints.

The underlying similarity between these two apparently different sport movements also teaches us another important lesson: the "monkey see, monkey do" approach of copying others' technique (without even an attempt to understand what's really happening), used by almost all coaches and athletes, sometimes even sport scientists – is blind. It shows only trivial and superficial similarities, which are good enough only for kindergarten sport, without any ability to understand, predict or control the movement. Surely, sport science should be able to do better than just "monkey see, monkey do." It should be able to provide the means for understanding, predicting, controlling and developing the most efficient human movements.

The stretch–reflex is the most efficient feedback-control mechanism in the human body. Therefore, any human movement that uses it simultaneously in several major joints – is immediately highly efficient.

The stretch–reflex recipe reads: make efficient slingshots out of all your major muscles: quadriceps, gluteus, soleus, pectoralis, deltoideus, biceps, triceps, palmar flexors, and all abdominal muscles. Make your major joints flexible and all your major muscles at the same time strong, fast and elastic, so that you can safely stretch and fire them like slingshots. Train them so that you make them efficient slingshots. That is, *base your strength and speed training on utilizing the stretch–reflex of all major muscles*, previously conditioned by the flexibility training. At the same time, base your sport technique (e.g., serve, forehand, backhand, etc) on the stretch–reflex, understanding that your muscles truly are natural slingshots ready to fire. Therefore, you need to develop your body full of slingshots, and your tennis weapons (e.g., serve, forehand, backhand, etc) need to be based on those slingshots, so that in a real game situation you can efficiently fire them. That is all! It is as simple as that!

You can easily see that popular plyometrics training is a part of this new science, but only a minor part. The concept of plyometrics training lacks the underpinning knowledge of the stretch–reflex, so naturally it proposes only a very limited number of plyometric exercises. Once one fully understands the applied stretch–reflex, one can design any number of plyometric exercises for any part of the body. One just needs to make his/her body full of slingshots, like Bruce Lee did three and a half decades ago, and like all athletics jumpers, throwers, sprinters and weightlifters do today.

10

The Law of Muscular Slingshots

10.1 Busting the Myth of Loops and Swings in Tennis Strokes

The basic idea of coaches and players, as well as the simplistic biomechanics of loops and swings in tennis shots (serve, forehand and backhand) are derived from the concept of a simple physical pendulum. An arm with a hand holding a racket has been seen as a more-or-less single rigid body, with no more than 3 *degrees of freedom* (DOF). And really, if you have a robotic arm with only 3 DOF, and you want to hit a ball with it, then you need to have a loop; and, even more, if you want to hit the ball hard, you need to have a big swing. That is absolutely true. In a language of modern biomechanics, "its phase space is a simple circle."[1] That is how we originally got our current loops – to play "nice tennis", and swings – to be able to "hit hard". In particular, you need a big swing as you want to use all the potential energy of the racket's weight, so, as a preparation for the shot (say, forehand), you lift the racket–head as high as possible along the circle (that is, above your head). This is the common picture behind all loops and swings in tennis. Although simple to understand, it is not as easily implemented, which is where the many expensive lessons on the court come in. This system has produced thousands of young tennis players, hitting the ball in virtually the same way, with the same distinguished tennis movements: loops and swings.

However, this picture is wrong! Our reality is much more complex than this simplistic model of a loop (circle) and a swing (potential energy along the circle), which produces only "tennis ballet".

[1] For technical details on modern geometrical biomechanics with hundreds of degrees of freedom, see one of my advanced scientific books: Human–like Biomechanics (Springer), Natural Biodynamics (World Scientific), Geometrical Dynamics of Complex Systems (Springer), High–Dimensional Chaotic and Attractor Systems (Springer), Neuro–Fuzzy Associative Machinery for Comprehensive Brain and Cognition Modelling (Springer), Applied Differential Geometry (World Scientific).

T.T. Ivancevic et al.: Paradigm Shift for Future Tennis, COSMOS 12, pp. 57–65.
springerlink.com

Firstly, the human arm has 9 DOF (not including fingers). Just the shoulder, elbow and wrist together have 9 DOF. This is a *redundant system*, because the racket itself has only 6 DOF (three translations and three rotations).[2] This means that there are infinite number of possible ways to hit the ball, and one can choose the best way in that situation. Therefore, a real "*phase space*" for the human arm with a racket is much more complicated than the simplistic circle–loop model. Besides, every coach advises, quite correctly, not to use the arm only, but the whole body. So one actually has several hundred DOF at his/her disposal to perform a tennis stroke.[3] Do you still believe that a simple circle and its associated loop is an appropriate model for the serve and ground-strokes?

Secondly, the question of utilizing the racket's own potential energy, that is, rising a racket–head up high as a preparation for the shot. This would make sense only for very weak players and very heavy rackets (in the same way as a weak player needs a long swing to accelerate the massive racket). Otherwise, in case of an athlete, and every elite tennis player needs to be an athlete, the current 400 gram racket is simply not heavy enough to make this potential–energy contribution to the shot significant. It would be significant with a 4 kg racket, but not with the current 400 gram one.

Put simply, loops are totally useless: it does not matter at all how you get into the position for the shot. Swings are almost useless. They could be used say in the fifth hour of a five–set tennis match, to compensate for the lack of strength in shots, but the same fatigue would make it hard to raise the racket–head high above your head.

Also, one can see that simplistic (bio)mechanics does not work with the human body. Robotics has already learned this lesson in the last three decades. Tennis has yet to learn it.

---

[2] Common robotic–tennis arm would have exactly 6 DOF – to match the racquet's own 6 DOF; it wouldn't have any excessive DOF; because of this 6–6 correspondence, a robotic arm would perform every racquet movement using the exactly prescribed trajectories. However, human arm (and human body as a whole) is a highly redundant system; this *mechanical redundancy* allows every tennis player to execute any movement in an infinite number of ways and to choose the one which is *optimal* with respect to racket-head speed, or spin, or energy efficiency.

[3] In the wooden era, the player really needed the long, flowing swing to accelerate the old, heavy racket up to hitting speed, unless he was very strong and fast. Remember, muscular force needs to overcome inertial one, or $force = mass\ times\ acceleration$, and *acceleration* is the cause of the racket–head speed. Therefore, if one has a massive racket and a weak muscular force, one needs a long movement to gradually accelerate and eventually get some racket–head speed. In addition, this gradual acceleration gave the player much more control of the racket head and allowed the player to hit the ball at approximately the same location on the racket face each time it was swung, because the "sweet spot" was small.

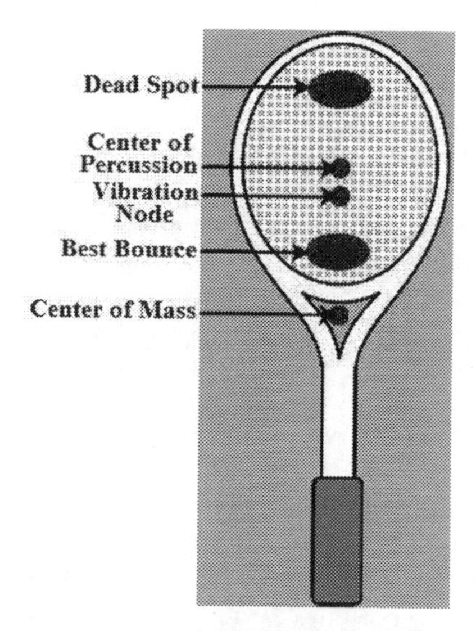

Fig. 10.1. Important points on a tennis racket. The common "sweet spot" is technically called the center of percussion (the point along the racquet's length where an impact produces no impulse reaction at the axis of rotation). Just below it is the vibration node. Note that the center of mass is below the racket–head.

The only efficient type of movements in tennis shots (serve, forehand and backhand) are whip–like movements,[4] based on muscular slingshots.[5]

10.2 The *Muscular Slingshot*

The stretch–reflex causes a stretched muscle to contract stronger and at the same time inhibits the antagonist muscle from contracting (that is, slowing the movement). Because this is an involuntary reflex response the rate of contraction is significantly (several times) faster and more powerful than a completely voluntary muscular contraction. In fact, the faster the muscle is stretched eccentrically, the greater the force will be on the following concentric contraction (see Appendix 1, for technical details).

Closely related to the muscular stretch–reflex is the stretch–shortening cycle, which occurs when elastic loading, through an eccentric muscular contraction, is immediately followed by an explosive concentric muscular contraction. The tension developed in the musculo-tendinous junction by the eccentric loading

[4] With the whip–like movement it is more difficult to hit the ball at exactly the same location on the racket–head each time. However, due to the characteristics of the modern racket and the heavy topspin strokes used, the resulting ball trajectory is much less sensitive to the exact location of the ball impact on the strings. New rackets appear to have larger "sweet spots", technically called the centers of percussion, see Figure 10.1.

[5] General experience from all other hitting, kicking and throwing sports or games.

of the muscle causes it to act in a similar manner to a rubber band. When this stored energy is released, it helps to increase the strength of the following concentric contraction. These neuromuscular considerations have huge ramifications for both the composition and training of the tennis serve, forehand and backhand.

When one trains his/her skeletal muscles so that they efficiently utilize both the stretch–reflex and the related stretch–shortening cycle, then one effectively makes powerful slingshots out of them.

The prototype of such a muscular slingshot is the javelin throw.

10.3 The Stretch–Reflex

The *stretch–reflex* (or, myotatic reflex) is the **secret** behind both speed and strength in sport. It is the quickest reflex (see Figure 6.5) in the human body, in which the *reflex arc* is a closed loop: both the receptor and the effector are in the same skeletal muscle (see Figure 10.2).

When a person walks and accidentally steps into a hole, the stretch–reflex of the stepping knee extensor (quadriceps muscle) is activated, generating additional force for knee extension, to re–establish the lost balance. What happens is that a muscle spindle (a length sensor within the quadriceps muscle) is stretched, causing a reflex arc to fire, generating a higher force than the voluntary muscular contraction alone. If the muscle is sharply stretched prior to its contraction, it generates a stronger force than without previous stretching. That is, the muscle behaves like a nonlinear spring (or, "elastic muscular component"). As an anatomical consequence, the attached body segment moves faster.

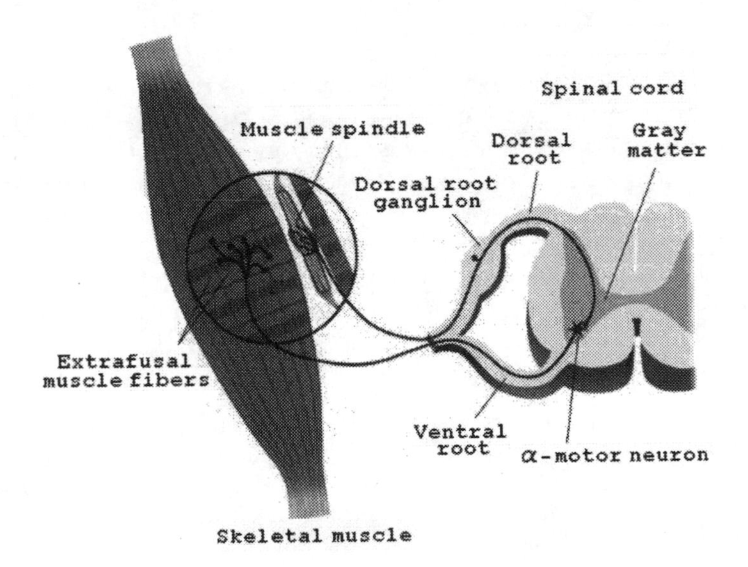

Fig. 10.2. Reflex arc of a myotatic stretch–reflex.

The stretch–reflex is the basis for all fast cyclic movements (such as sprinting) as well as for any combination of fast individual movements. It is also the basis for efficient weight lifting and athletic throws. In other words, speed and strength might have different muscular characteristics, but they have the same neural support: the stretch–reflex. Finally, the stretch–reflex is the only proper physiological basis behind the popular *plyometrics* muscular training.

Biomechanically speaking, the stretch–reflex (or, if you insist, the stretch–shortening cycle) involves a sharp eccentric contraction immediately followed by a strong reflex concentric contraction. There is the stretch–reflex in action in all, mostly cyclic, fast movements in full joint amplitude (e.g., in sprinting, one has a maximal knee flexion immediately followed by a sharp knee extension, followed by a maximal knee flexion. . .). Whenever one of the two antagonistic muscles is stretched, either by his antagonist, or by gravity, it responds by generating a stronger force than it would be able to generate voluntarily without the previous stretching. As an anatomical result, one also has the faster movement of the attached body segment (a calf in the sprinting case).

10.4 The Plyometrics

The modern–day method for developing the jumping ability called "plyometrics" is actually an almost half a century old Russian explosive strength method called the "shock method," firstly proposed in 1959 by Professor Yuri Verkhoshansky (see bibliography at the end of this book). Plyometrics is generally used for improving the explosive strength and the muscular reactive capacity. It is a type of exercise that utilizes a rapid eccentric movement, followed by a short amortization phase, and then followed by an explosive concentric movement, which enables the synergistic muscles to engage in the (myotatic) stretch–reflex during the stretch–shortening cycle. Plyometric exercises use explosive movements to develop muscular power, the ability to generate a large amount of force quickly. Plyometric training acts on both the musculotendinous and neurological levels to increase an athlete's power output without necessarily increasing their maximum strength output. Plyometrics are used to increase the speed or force of muscular contractions, often with goals of increasing the height of a jump or speed of a punch or throw.

Recall that for a muscle to cause movement, it must shorten; this is known as a concentric contraction. There is a maximum amount of force with which a certain muscle can concentrically contract. However, if the muscle is lengthened (eccentric contraction) just prior to the contraction, it will produce greater force through the storage of elastic energy. This effect requires that the transition time between eccentric contraction and concentric contraction (amortization phase) be very short. This energy dissipates rapidly, so the following concentric contraction must follow the eccentric stretch before this occurs. The process is frequently referred to as the *stretch–shortening cycle*, and is one of the underlying mechanisms of plyometric training.

In addition to the elastic–recoil of the musculo–tendonous system there is a neurological component. The stretch–shortening cycle affects the sensory response of the muscle spindles (stretch–reflex) and golgi tendon organs (GTO). It is believed that during plyometric exercise, the excitatory threshold of the GTO's is increased, meaning they become less likely to send signals to limit force production when the muscle has increased tension. This facilitates greater contraction force than normal strength or power exercise, and thus greater training ability.

The muscle spindles are involved in the stretch–reflex and are triggered by rapid lengthening of the muscle as well as absolute length. At the end of the rapid eccentric contraction, the muscle has reached a great length at a high velocity. This may cause the muscle spindle to enact a powerful stretch–reflex, further enhancing the power of the following concentric contraction. The muscle spindle's sensitivity to velocity is another reason why the amortization phase must be brief for a plyometric effect.

A longer term neurological component involves training the muscles to contract more quickly and powerfully by altering the timing and firing rates of the motor units. During a normal contraction, motor units peak in a de-synchronized fashion until tetanus is reached. Plyometric training conditions the neurons to contract with a single powerful surge rather than several disorganized contractions. The result is a stronger, faster contraction allowing a heavy load (such as the body) to be moved quickly and forcefully.

Some animals also take advantage of this effect; one is the kangaroo. If a kangaroo needed to use 100% new energy to contract its leg muscles every time it jumped, it would not be able to jump very far consistently. However, because of the muscles' ability to store energy from its previous jump like a spring, the kangaroo only needs to use a fraction of the total energy in the jump.

In everyday sport, plyometrics commonly refers to jumps and similar movements that involve eccentric (stretching) muscular contraction, immediately and rapidly followed by a concentric (shortening) contraction, that is the stretch–shortening cycle. The phase between these two contractions is referred to as the amortization phase. Energy stored during the eccentric phase is partially recovered during the concentric phase. In order to best use this stored energy the eccentric phase must be rapidly followed by the concentric.

One plyometric exercise that is very useful for tennis serve, involves catching and tossing a medicine ball to an assistant while the exerciser lies on their back. The triceps and chest muscles work both while they are lengthening (catch phase) and while contracting (toss phase).

As another example, a typical long jumper may spend as little as 0.1 seconds on the ground at take-off. A great deal of muscular force must be generated in this brief period of time. Through the correct use of plyometric exercises, this rate of force development can be enhanced. Note that high volume plyometric workouts will not enhance speed development. Therefore, by using plyometric jumping exercises one is trying to accomplish the following: (i) to shorten the time spent in the amortization phase, and (ii) to decrease the time spent on the ground yet generate maximum force.

Conventional wisdom dictates that in order to begin plyometric training there are prerequisite strength levels which are necessary. To begin to incorporate plyometric training in a program the prime concern is strength in the stabilizing muscles in order to prevent injury. The next concern after stabilization strength is eccentric strength. Eccentric strength is the limiting factor especially in more complex high volume and high intensity plyometric training. Without adequate levels of eccentric strength, rapid switching from eccentric to concentric work becomes very inefficient.

The following table gives several examples of plyometric jumping exercises:

Example	**Stress**	**Recovery time**
jump rope or ankle bounds or low amplitude jumps	very low	few hours
tuck jump or similar	low	one day
stair jumps or other similar short jumps	moderate	one to two days
hops or bounds for distance or similar	high	two days
depth jumps or other similar shock-type jumps	very high	three days

Proper execution of the exercises must be continually stressed regardless of the proficiency level. For the beginner, it is especially important to establish a sound technical base upon which to build the higher intensity work. Jumping is a constant interchange between force production and force reduction leading to a summation of forces utilizing all three joints of the lower body: the hip, knee, and ankle. The timing and coordination of all limb segments will yield a positive ground reaction force which results in a high rate of force production.

A key element in the execution of a proper technique is the landing. The shock of landing is not absorbed exclusively with the foot, rather it is a combination of the ankle, knee, and hip joints working together to absorb the shock of landing and then to transfer that force. The proper utilization of all three joints will allow the body to use the elasticity of the muscles to absorb the force of landing and then utilize that force in the subsequent movement.

Thus, plyometric exercises promote high movement speed, fast twitch fibre recruitment and elastic tendon energy release. They always involve: (i) an eccentric contraction; (ii) a brief amortization phase (with no change in muscle length); and (iii) a short concentric contraction delivering maximum force in a short period of time.

10.5 The Role of Flexibility

The capacity of muscles, tendons, ligaments and fascia to stretch, the range of motion in the joints and the ability of the muscles to contract and coordinate all define how one moves. Unlike strength, speed and other motor abilities, flexibility develops and determines efficient fluid movement. Heavy training and competition schedules place great stresses on our capacity for movement. One requires a systematic approach to training for full recovery both mentally and physically. The implementation of a proper flexibility program is imperative for this recovery, but more importantly for an increase in performance.

Flexibility training should be integrated on a continual basis within a yearly training plan. When practiced regularly it provides immediate relief from fatigue and muscle soreness. This is very important in accommodating increases in training volume and intensity as the year progress. Additionally, the accumulation of flexibility training attained by the athlete over an extended period of time will allow for an increased capacity to maintain these gains with less work.

Stretching should never be forced, but should be done with special care. It is important that the athlete/player focuses on the muscle group that is being stretched. This fosters greater body awareness, an overlooked attribute of a champion. Flexibility has to be both sequential and rhythmical in order to accommodate the effects of high training loads that have placed a greater load on the CNS. The constant stimulation of the nerve cells, whose high working capacity cannot be maintained for long, affects muscular action and ultimately athletic performance. When the period of competition and training has very high fuel consumption, fatigue sets in. Since blood glucose is depleted from the system, the CNS becomes fatigued. Properly applied stretching techniques promote enhanced blood circulation within the athlete. With this increase in circulation, there is removal of metabolic waste (lactic acid) products, as well as an increase in the transport of oxygen and nutrient to the muscle and tendon regions.

A properly designed flexibility program does not cause injury to the tissues but aids in their recovery and regeneration. Its greatest influence is at the myo-tendon junction, the transition zone between the muscle and tendon. The function of the tendon is to transmit the mechanical impulses that derive from muscular contraction to the joints. The myo-tendon junction must adapt each time to the functional needs of the musculoskeletal system. This region also helps in cushioning abrupt and violent movement. Most injuries occurring in this region are a result of micro-tears, referred to as micro-injuries. Micro-tears result in the development of scar tissue. As time progress and these injuries go untreated, a common result is chronic pain. Associated with chronic pain at the physical level are muscle imbalances and compensation shifts. If an athlete does not adhere to a proper flexibility program they may slowly develop into an involuntary contraction machine (due to a stretch-reflex overload).[6] Their higher muscle tone, a result of greater tightness, affects the nervous system. A possible outcome of this neural fatigue may be muscle atrophy, as well as poor muscle coordination. This affects the development of power (that is, both strength and speed).

10.6 How to Prevent Injuries

Firstly, it is very important for an athlete to be both physically and mentally fit for the competition and/or training. Besides, an athlete should be able to identify

[6] This is an example of possible misuse of the stretch reflex if muscles are not flexible enough.

the presence of a minor injury and distinguish it from fatigue. Here is a list of simple ways by which you can identify injury: (i) pain is sustained steadily and does not subside; (ii) an extent of tightness can be felt that restricts full–range motion; (iii) you feel light–headed or nauseous; (iv) what initially appeared to be a minor injury does not heal promptly.

Highly motivated athletes always strive to push beyond their limits, both in training and in the competitions. Unfortunately, if an athlete does not take sufficient rest when needed, their muscles can actually be injured, rather than become stronger. Such injuries include: cramps (muscles become excessively tight in contractions), contusion (internal bleeding, swelling, pain and stiffness are caused by a serious bruise), sprains (a ligament has been overstretched and torn) and strains (a muscle or tendon attachment has been over-stretched or torn).

For example, when lifting weights in weight–training, the body's natural response to the heavy weights is have microscopic tears in the muscles' connective tissues. This explains the soreness that one would usually feel after each weight–training exercise. It is essential for an athlete to rest after weight–training. During the duration which the athlete rests, the microscopic tears are repaired. In the repair process, muscles become larger and stronger. However, if an athlete fails to obtain adequate rest after weight–training, the repair process may be delayed or a true injury may result. This is because if the athlete continues to strenuously use the muscles without allowing rest for repair, the microscopic tears may become larger tears.

11

The Law of Whip–Like Movements

11.1 Biomechanical Kinetic Chain

Biomechanical term for the whip–like movement is the *kinetic chain*: the sequential flow of energy and momentum from bigger segments to smaller ones. Tennis requires sequenced activation of muscles and movement of bones and joints to achieve the motions, positions, and velocities seen in a player. This sequencing is known as the kinetic chain. Kinetic energy and momentum, as well as muscular power, are developed from the legs, hips and trunk muscles and transferred to the arm muscles. This allows the energy, momentum and power to be transferred efficiently to the hands, moving the racquet-head with maximum speed to the ball.

More precisely, to achieve the highly–efficient technique of the tennis serve, forehand and backhand, a proper sequencing of muscle stretch–reflex based actions must take place. Two movement strategies are critical in this respect: (i) proximal-to-distal firing patterns, and (ii) active acceleration–deceleration of body segments.

Research evidence suggests that a proximal-to-distal firing pattern is the most effective for increasing the racket-head speed in the serve, forehand and backhand. In such a sequencing pattern, the stronger more heavily muscled proximal (close to the torso) joints should become activated before the weaker but faster distal joints. This firing pattern has proven the most efficient due to the fact that it takes advantage of each joints' linear and angular momentum generating characteristics. It suggests that power (that is, both speed and strength combined) for the serve, forehand and backhand is primarily generated with: (i) leg extension, (ii) hip rotation, and (iii) trunk rotation and flexion – before the actual arm action. The actions of these proximal joints account for more than 50% of the total forces in the serve, forehand and backhand.

A second characteristic of efficient movement coordination in the serve, forehand and backhand is consecutive acceleration and deceleration of the main body segments. When done well this permits the player to achieve racket-head speeds far greater than they would if they did not use an optimal acceleration-deceleration coordination pattern. The mechanism for this benefit is the transfer

T.T. Ivancevic et al.: Paradigm Shift for Future Tennis, COSMOS 12, pp. 67–70.
springerlink.com

of both linear and angular momentum. This movement strategy aids in the transfer of momentum from the lower extremity to the upper, and from the upper extremity to the racket-head.

These two kinetic-chain patterns combined, generate a whip–like motion: when the upper leg and trunk musculature are the first to contract, greater separation is developed between the shoulders and hips which results in a whip effect as the hips are decelerated and the shoulders accelerate as they uncoil and the shot is released.

As already said, each muscular contraction in a chain can be either voluntary or reflex. By now, we should already know that the stretch–reflex based contraction is several times more efficient (that is, both faster and stronger) then the voluntary one.

11.2 Biomechanics of the Power High-Forehand

Power High-Forehand (**PHF**), with the aim of generating the *maximal possible racket-head speed*, should be executed as follows (see Figures 11.1 and 11.2).

Let's assume that you start from the base-line (as usual). You play your normal game, but as soon as you **anticipate**[1] a **high ball** coming at a waist/chest level (or higher) while passing over the service-line, you don't wait behind the base-line for the ball to come to you, but immediately sprint as fast as you can towards the service-line and a little-bit left from the ball, while at the same time leaving the racket-head behind your right shoulder (preparing for the shot). When you get close to the ball, you make sure that you stop about half-a-meter left from the ball's direction, with the left foot forward. This automatically forms the left closed stance (as if you want to execute the right-hand karate gyaku-zuki reverse punch), with the racket-head far behind your right shoulder. Then you are ready to intercept the ball and execute the winning shot.

Once you are in the proper left stance (as explained above), the PHF-shot is properly executed as a **full biomechanical chain**, based on the **sequence of stretch reflexes** in all included muscular actuators, one by one as a cascade. It is represented by the following sequence of individual joint movements:

[1] Here, we promote the *interceptive tennis tactics*, as a superior alternative to the current static tennis in which both players most of the time just wait behind the base line for the ball to bounce from the ground and come to them. In our Power interceptive tennis, the players do not wait for the ball to bounce, but rather intercept as soon as the ball passes the service line and hit it in the air. This significantly shortens the time for opponent to react. If executed properly, this shot has the same effect as the smash. However, while an opportunity for a smash usually happens only about once in a tennis match, if you develop proper interceptive skills you will realise that most of the balls coming over the net can be intercepted. This is tactical basis for the future tennis game. Proper interception is heavily based on the *anticipation* of the ball coming (both direction and height while passing over the service-line). Besides, you can also intercept low balls, using your ordinary forehand and backhand as an approaching shot; you just need to be ready to *hit the ball in the air* (not waiting for it to bounce).

Fig. 11.1. PHF–training 1: a "3 o'clock" shadow execution behind the service line.

1. Sharp right knee extension (with the right foot/heel firmly on the ground), while stretching the right hip muscles (gluteus medius and maximus);
2. Sharp right hip forward rotation, while leaving the right shoulder back and thus stretching the right waist muscles (external and internal abdominal muscles);
3. Sharp torso rotation to the right, while leaving the arm back and thus stretching the pectoral and deltoid muscles;
4. Sharp right-elbow movement forward, while leaving the racket-head back, and thus stretching the right triceps;
5. Sharp right-elbow extension, while still leaving the racket-head back, and thus stretching the right wrist flexor muscles; and
6. Sharp right-wrist flexion, with simultaneous pronation to generate some top-spin for ball control.

This, in a nutshell, is how the PHF-shot should be executed.

Fig. 11.2. PHF–training 2: a "2 o'clock" execution – intercepting the high ball behind the service line.

Now, how do you know that you are really executing the proper PHF-shot?

The first criterion is this: *it cannot be done slowly.* It can be performed moderately fast or extremely fast, more tensed or more relaxed, but never slowly. This is the first criterion for any stretch-reflex based movement.

Here is a simple familiar example: teaching long jump to small kids. You have a short run-up and then you jump. You can run slowly, but you simply cannot jump slowly. Why? Because a single-legged jump (from running) is based on the stretch reflex in all leg extensor muscles (soleus, quadriceps and gluteus). No stretch reflex, no jump – as simple as that.

It is the same with all naturally fast movements, including our PHF: it can be done in a more tensed or a more relaxed way, but it cannot be executed slowly. The very nature of all stretch-reflex based movements is its inherent speed (as well as strength). It is not meant for slow and/or weak movements.

12

The Newton-Euler Biomechanics Law

Modern biomechanics is rotational (see Figure 12.1) and can be formulated using Newton-Euler, Lagrangian or Hamiltonian formalism (see Appendix 2).

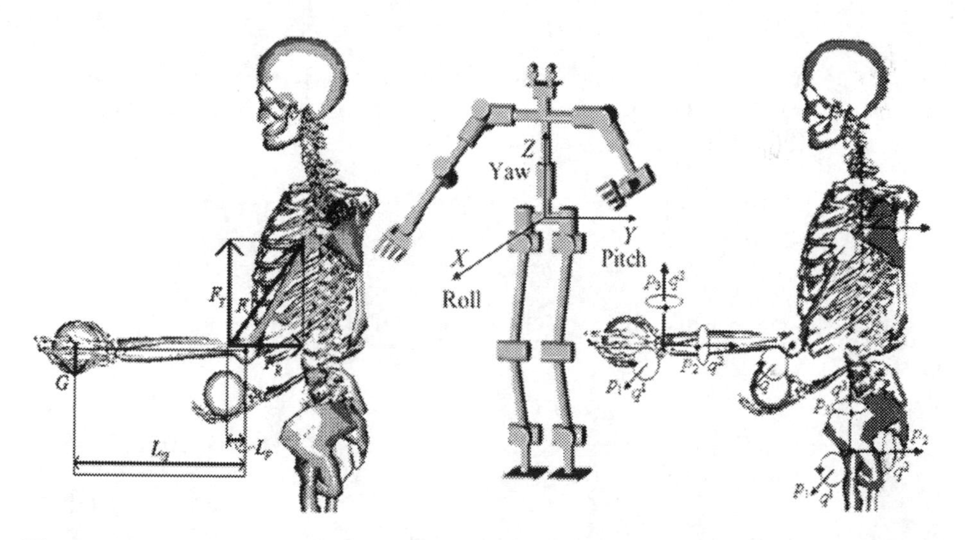

Fig. 12.1. Development of human biomechanics: Left – old translational approach; middle – humanoid robotics; and right – new rotational approach.

Here we give a brief on the *Newton-Euler dynamics* of the three-dimensional motion of a player's hand holding a tennis-racket, developed as a $Mathematica^{TM}$ code – ready for computer simulation. For more technical details, see [II06a, II06c].

We start by defining the three local *Cartesian translation coordinates*, sagittal, frontal and vertical, or $\{x_1, x_2, x_3\} = \{X, Y, Z\}$, placed at the center-of-mass of the racket, as well as the corresponding three *rotation (orientation) coordinates*, $\{\theta_1, \theta_2, \theta_3\} = \{roll, pitch, yaw\}$, around the Cartesian ones. In the traditional output form of *Mathematica v.7*, this reads:

$$\text{Vars} = \{x_1, x_2, x_3, \theta_1, \theta_2, \theta_3\}$$

T.T. Ivancevic et al.: Paradigm Shift for Future Tennis, COSMOS 12, pp. 71–72.
springerlink.com

Their time derivatives are given by:

$$v_i(t) = \{(x_1)'(t), (x_2)'(t), (x_3)'(t)\}$$

and

$$w_i(t) = \{(\theta_1)'(t), (\theta_2)'(t), (\theta_3)'(t)\}$$

Next, by introducing a racket mass $m = m_1 = m_2 = m_3$ and three grip/time-dependent inertia moments $\{J_1(t), J_2(t), J_3(t)\}$ around the Cartesian axes, the inertial definitions of both linear and angular momenta are given in *Mathematica* 7 by:

$$\begin{pmatrix} (p_1)'(t) \\ (p_2)'(t) \\ (p_3)'(t) \end{pmatrix} = \begin{pmatrix} m_1 (x_1)''(t) \\ m_2 (x_2)''(t) \\ m_3 (x_3)''(t) \end{pmatrix}$$

and

$$\begin{pmatrix} (\pi_1)'(t) \\ (\pi_2)'(t) \\ (\pi_3)'(t) \end{pmatrix} = \begin{pmatrix} J_1(t) (\theta_1)''(t) \\ J_2(t) (\theta_2)''(t) \\ J_3(t) (\theta_3)''(t) \end{pmatrix}$$

Now, if one introduces time-dependent muscular forces $\{F_1(t), F_2(t), F_3(t)\}$ and torques, $\{T_1(t), T_2(t), T_3(t)\}$, one can derive the three Newtonian differential equations of the translational racket motion as:

$$\begin{pmatrix} (p_1)'(t) = F_1(t) - m_3 v_3(t)\omega_2(t) + m_2 v_2(t)\omega_3(t) \\ (p_2)'(t) = F_2(t) - m_3 v_3(t)\omega_1(t) - m_1 v_1(t)\omega_3(t) \\ (p_3)'(t) = F_3(t) - m_2 v_2(t)\omega_1(t) + m_1 v_1(t)\omega_2(t) \end{pmatrix} \tag{12.1}$$

as well as the three Euler differential equations of the rotational racket motion as:

$$\begin{pmatrix} (\pi_1)'(t) = (J_2(t) - J_3(t))\,\omega_2(t)\omega_3(t) + (m_2 - m_3)\,v_2(t)v_3(t) + T_1(t) \\ (\pi_2)'(t) = (J_3(t) - J_1(t))\,\omega_1(t)\omega_3(t) + (m_3 - m_1)\,v_1(t)v_3(t) + T_2(t) \\ (\pi_3)'(t) = (J_1(t) - J_2(t))\,\omega_1(t)\omega_2(t) + (m_1 - m_2)\,v_1(t)v_2(t) + T_3(t) \end{pmatrix} \tag{12.2}$$

The combined system of differential equations (12.1)–(12.2) represents the total Newton-Euler dynamics of the three-dimensional motion of a player's hand holding a tennis-racket. If one can measure/provide all introduced parameters together with sample initial racket position and orientation (dependent on the initial body-configuration), one can readily simulate this system on any computer.

13

The Law of *Superb* Weapons

Each effective tennis shot, be it a serve, a forehand or a backhand, is a whip–like movement performed by a complex coordination of all the body's segments working to place the racquet in the correct position at the right time and apply the maximal summed force to the tennis ball.

As we already emphasized several times, the best power (strength + speed) exercise for both serve and forehand (and even for single-handed backhand) is a javelin throw. Not only that, but all speed and strength exercises practiced by elite javelin throwers are perfectly suited for future champion tennis players.

In short, a *superb serve, forehand and backhand are whip–like movements, each composed as a cascade of stretch–reflexes in all major joints, starting from the feet and ending with the hitting hand.*

13.1 Roddick's Serve Phenomenon

Recall that former world number one, Andy Roddick, holds the record for the world's fastest tennis serve: 153 m/h (or, 246 km/h) fired at Queen's Club, UK, in 2004. When he first met Patrick McEnroe, his Davis Cup coach, he said: "Whatever you do, don't say anything to me about my serve. If I think about it, I'm in trouble." Why? Because it is all reflex, more precisely stretch–reflex. If you think about something that is performed reflexively, you simply mess it up. Therefore, it is crucial that the elite player develops a fully reflex–based technique. This will generate the highest possible racket–head speed of an elite athlete and thus maximize their performance/efficiency.[1]

However, coaches and sports scientists should analyze the most efficient movements to be able to teach the model techniques. For example, Professor Bruce Elliott from the University of Western Australia, has extrapolated the contributions of the body segments to racket-head speed using 3D video– and computer

[1] Assuming that the player is already capable of consistently getting the serve in the square, keep their serve deep, able to serve to the opponent's backhand, body and/or forehand at will, and effectively use slice and/or topspin kick.

T.T. Ivancevic et al.: Paradigm Shift for Future Tennis, COSMOS 12, pp. 73–76.
springerlink.com

analysis. "These contributions vary from person to person," Elliott says, "but the data shows the clear importance of the trunk, shoulder internal rotation and wrist flexion in the swing to impact."

13.2 The Best Strength Exercise for the Serve

Apart from the javelin throw, the best strength exercise for the serve is the axe chop/sledgehammer. Its purpose is to simultaneously develop strength, speed and flexibility in the shoulder girdle, pectoral and upper back musculature. This exercise requires a sledgehammer or lumber axe, approximately of the mass and handle–length of the tennis racket, as well as a stable knee–height hitting surface such as a tractor tire, mound of dirt, or a large log. The athlete stands in front of a knee–height hitting surface. After a proper warm-up, the axe / hammer is brought over the head and swung violently down onto the hitting surface. The key for this exercise is to let the weight of the axe or hammer pull the arms back so that a stretch is felt through the shoulders and upper back. This will initiate a stretch–reflex contraction, while developing strength, speed and flexibility for throwing. It is important that: (i) the athlete initiate the movement with the whole body rather than just the arm, which will create a whip–like effect on the axe or hammer; and (ii) to imitate the serve movement as close as possible.

Other strength exercises include various overhead throws, like medicine ball throw and weighted ball throw. In particular, catching and tossing a medicine ball to an assistant while the exerciser lies on their back.

13.3 The Best Speed Exercise for the Serve

The best speed exercise for the serve is simulating the serve movement (without the ball) with a badminton racket instead of a tennis racket. This should be performed in a double series: 10 repetitions of a shadow serve with a tennis racket, then 10 repetitions with a badminton racket, then a 2 minute pause; in such a way to make 100 of each, performed with a maximal speed (after a proper warm-up).

13.4 Common Technical Misconceptions about Tennis Forehand

The wooden racket era, characterized by the heavy rackets and the weaker players, was dominated by the following 5 classical postulates for the proper technique for the tennis forehand:

1. You must have a loop in the stroke, either big or small, but the bigger the better;
2. If you want to "hit hard," you need to have a huge swing;
3. You need to rotate hips and shoulders together, that is simultaneously;

4. With a series of little steps you need to put yourself into the proper position for the effective forehand; and
5. The wrist must be fixed totally rigid.

The well–known result of this 5-point approach to the forehand technique was two–fold: (a) nice and slow “ballet on the court” and (b) tennis elbow injury. Today, with the light-weight metal rackets and stronger players, the 5th postulate has been dropped: one can see now plenty of wrist slaps.

However, the other 4 postulates are still assumed valid. Well, from the perspective of high efficiency/maximum performance in tennis of the future – all four postulates are wrong. They are all pure “cosmetics” without any substance.

To dispel this “myth of a proper tennis forehand,” in this chapter we have given a biomechanical description of a whip–like tennis forehand movement, composed as a series of stretch–reflexes in all major joints.

13.5 A Complex Reaction

A *complex reaction* has the highest importance in every sport's duel, including tennis. In general, a complex reaction has two components: smart anticipation and lightning–fast reaction. You first anticipate the opponent's movement and then you react by making your own movement.

Good anticipation of the opponent's actions is the essential characteristic of a master in any sports duel. The best anticipation is actually called “mind reading”. It is the difference between master and disciple. It can be learned, and it should be learned if you have high expectations from tennis. Of the current top tennis players, Federer, Nadal and Djokovic have the best anticipation – and coincidentally they are world number one, two and three. Before them, Sampras and Agassi had the best anticipation, and so on. Apart from getting the experience from many tennis tournaments (at different competition levels), the best tool for developing anticipation would be the bullet/blitz chess (see The Law of Blitz–Chess in Tennis).

On the other hand, the basic speed of reaction is largely genetically predetermined and cannot be significantly improved. The best means for its developing is both technique and training based on the stretch–reflex.

13.6 The Main Characteristics of a Good Return

For the current first serve in the range of about 200 $km/h - 220\, km/h$, the available time budget for the return player is approximately 0.7 s on the slowest clay courts, approximately 0.6 s on faster hard courts, and approximately 0.5 s on the fastest lawn courts. These are average numbers, with about 10% deviations due to differences in the conditions of ball flight and individual returning strategies (e.g., the position of the return player on the court etc.).

The whole return can be roughly divided into three phases: (i) anticipating the ball, (ii) movement regulation and reprogramming, and (iii) hitting the ball.

During the time of movement regulation the players adapt their racket movement to the demands of the approaching ball. The sequencing of maximum segment velocities (i.e., first hips, then shoulder, then elbow, and finally wrist), which is a common feature of both the ground-strokes and serve, cannot be clearly seen during the return. This is due to: (a) the unpredictable nature of the return and the high time pressure under which players are placed; and (b) the very high demand of precision on the return. Currently, on the 1st serve return, precision–orientation is dominant, while on the 2nd serve return, ball–speed generation is dominant (the racket–head speed at impact is significantly lower during 1st serve returns). Also, the ability to reprogram an incorrect anticipation/decision (e.g., shifting from backhand to forehand return) within a fraction of a second is a very important factor on the return.

Therefore, it is highly recommended to: (i) train anticipatory abilities; (ii) train returns by varying the time pressure imposed on the players in order to improve their time management; and (iii) offer tasks for reprogramming.

14

The Law of Blitz–Chess

Like in any other sport game, the important part of the tennis game is tactics. Current tennis tactics will be slightly simplified in the future tennis game, as, due to the highly–increased speed of the ball, long rallies will rarely exist. Therefore, this "tennis chess" will consist of up to 3–4 movements at any one time.

Question: Assuming that both players are right-handed players, could you show us the optimal serve tactics?

A1. If you serve from the right side of the baseline centermark, then Figure 14.1 shows the optimal serve tactics.

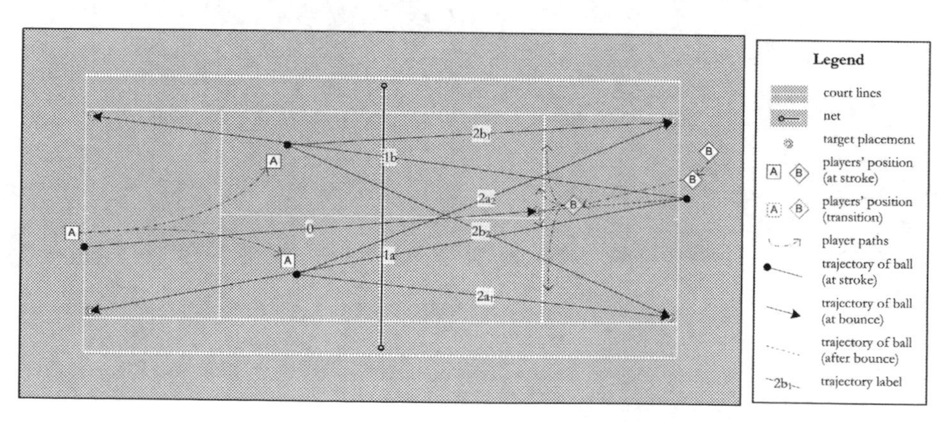

Fig. 14.1. Right–handed player targeting "T", serving to the right–handed player's backhand.

A2. If you serve from the left side of the centermark, then Figure 14.2 shows the optimal serve tactics.

T.T. Ivancevic et al.: Paradigm Shift for Future Tennis, COSMOS 12, pp. 77–79.
springerlink.com

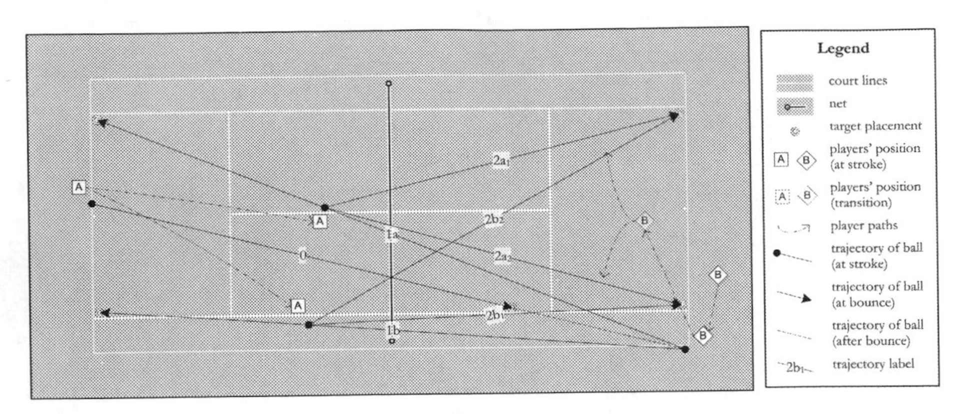

Fig. 14.2. Right–handed player targeting wide, serving to the left–handed player's backhand.

Question: Assuming that the right–handed player is serving and left–handed player receiving, could you show us the optimal serve tactics?

A1. If you serve from the right side of the centermark, then Figure 14.3 shows the optimal serve tactics.

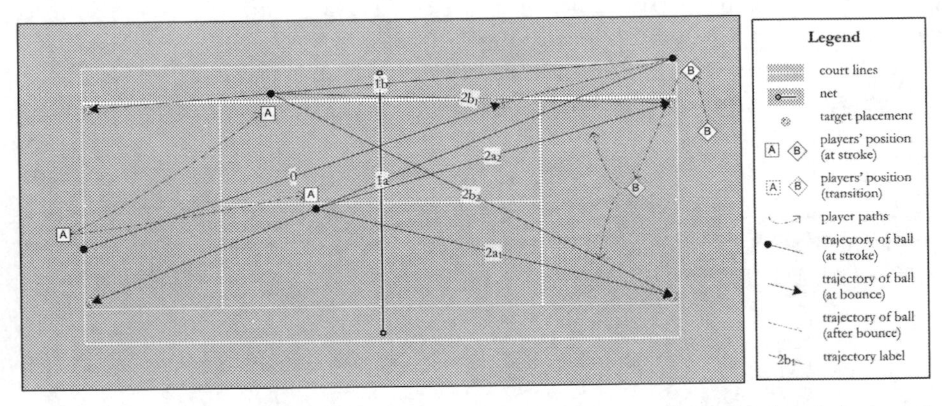

Fig. 14.3. Right–handed player targeting wide, serving to the right–handed player's backhand.

A2. If you serve from the left side of the centermark, then Figure 14.4 shows the optimal serve tactics.

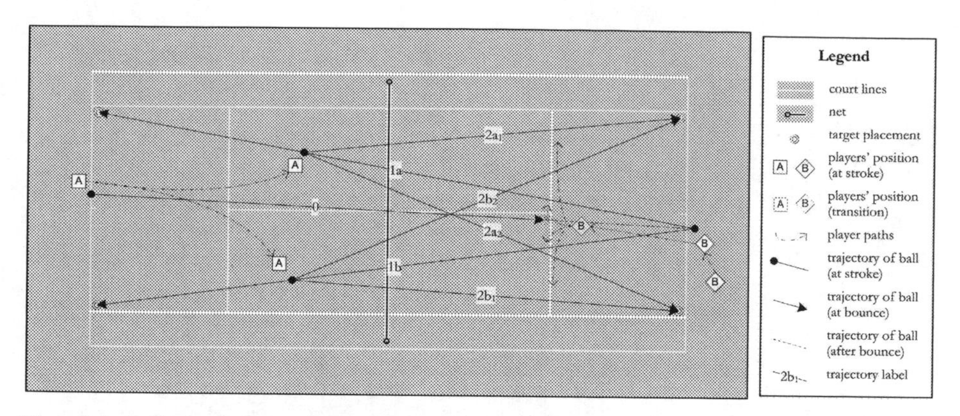

Fig. 14.4. Right–handed player targeting "T", serving to the left–handed player's backhand.

15

The Law of Artificial Intelligence

In this chapter we formulate a fuzzy–logic,[1] *attack* (AT) and *counter–attack* (CA) model for the tennis game.

Attack Model: Tennis Serve

A. Simple Attack: Serve Only. The simple AT–dynamics is represented by a single fuzzy associative memory (FAM) map

$$\underset{CAT}{TARGET} \xrightarrow[FAM]{\mathcal{F}^{AT}} \underset{CAT}{ATTACK}$$

[1] Fuzzy logic is a powerful problem–solving methodology with a myriad of applications in embedded control and information processing (e.g., train controllers, aircraft autopilots, air conditioning, control of nuclear reactors, etc). Fuzzy provides a remarkably simple way to draw definite conclusions from vague, ambiguous or imprecise information. In a sense, fuzzy logic resembles human decision making with its ability to work from approximate data and find precise solutions. Fuzzy logic usually works as a set of fuzzy IF–THEN rules. For example, in case of a heater controller with two inputs (temperature and humidity) and one output (fan speed) one has the following set of fuzzy rules:

IF temperature IS cold AND humidity IS high THEN fan-spd IS high
IF temperature IS cool AND humidity IS high THEN fan-spd IS medium
IF temperature IS warm AND humidity IS high THEN fan-spd IS low
IF temperature IS hot AND humidity IS high THEN fan-spd IS zero

IF temperature IS cold AND humidity IS med THEN fan-spd IS medium
IF temperature IS cool AND humidity IS med THEN fan-spd IS low
IF temperature IS war m AND humidity IS med THEN fan-spd IS zero
IF temperature IS hot AND humidity IS med THEN fan-spd IS zero

IF temperature IS cold AND humidity IS low THEN fan-spd IS medium
IF temperature IS cool AND humidity IS low THEN fan-spd IS low
IF temperature IS warm AND humidity IS low THEN fan-spd IS zero
IF temperature IS hot AND humidity IS low THEN fan-spd IS zero.

T.T. Ivancevic et al.: Paradigm Shift for Future Tennis, COSMOS 12, pp. 81–85.
springerlink.com

In the case of simple tennis serve, this AT–scenario reads

$$\underset{OPPONENT-IN}{O \ni o_m} \xrightarrow{\mathcal{F}^{AT}} \underset{SERVE-OUT}{SR \ni sr_n}$$

where the two sets, $O_{\dim=2} \ni o_m$ and $SR_{\dim=3} \ni sr_n$, contain the temporal fuzzy variables $\{o_m = o_m(t)\}$ and $\{sr_n = sr_n(t)\}$, respectively opponent–related (target information) and serve–related, partitioned by overlapping Gaussians, $\mu(z) =$ $exp\left[\frac{-(z-m)^2}{2\sigma^2}\right]$, and defined as:

$$\underset{OPPONENT-IN}{O} : \begin{array}{l} o_1 = Opp.Posit.Left.Right : (center, medium, wide), \\ o_2 = Opp.Antcp.Left.Rght : (runCenter, stay, runWide), \end{array}$$

$$\underset{SERVE-OUT}{SR} : \begin{array}{l} sr_1 = 1.Serve.Speed : (low, medium, high) \\ sr_2 = 2.Serve.Spin : (low, medium, high) \\ sr_3 = 3.Serve.Placement : (center, medium, wide) \end{array}$$

In the fuzzy–matrix form this simple serve reads

$$\overset{O:OPPONENT-IN}{\begin{bmatrix} o_1 = Opp.Posit.Left.Right \\ o_2 = Opp.Anticip.Left.Right \end{bmatrix}} \xrightarrow{\mathcal{F}^{AT}} \overset{SR:\ SERVE-OUT}{\begin{bmatrix} sr_1 = 1.Serve.Speed \\ sr_2 = 2.Serve.Spin \\ sr_3 = 3.Serve.Place \end{bmatrix}}$$

B. Attack–Maneuver: Serve–Volley. The generic advanced AT–dynamics is given by a composition of FAM functors

$$\underset{CAT}{TARGET} \xrightarrow[FAM]{\mathcal{F}^{AT}} \underset{CAT}{ATTACK} \xrightarrow[FAM]{\mathcal{G}^{AT}} \underset{CAT}{MANEUVER}$$

In the case of advanced tennis serve, this AT–scenario reads

$$\underset{OPPONENT-IN}{O \ni o_m} \xrightarrow{\mathcal{F}^{AT}} \underset{SERVE-OUT}{SR \ni sr_n} \xrightarrow{\mathcal{G}^{AT}} \underset{RUN-VOLLEY}{RV \ni rv_p}$$

where the new n–category, $RV_{\dim=2} \ni rv_p$, contains the opponent–anticipation driven volley–maneuver, expressed by fuzzy variables $\{rv_p = rv_p(t)\}$, partitioned by overlapping Gaussians and given by:

$$\underset{RUN-VOLLEY}{RV} : \begin{array}{l} rv_1 = RV.For : (baseLine, center, netClose) \\ rv_2 = RV.L.R. : (left, center, right) \end{array}$$

In the fuzzy–matrix form this advanced serve reads

$$\overset{O:OPPONENT-IN}{\begin{bmatrix} o_1 = Opp.Posit.L.R. \\ o_2 = Opp.Anticip.L.R. \end{bmatrix}} \xrightarrow{\mathcal{F}^{AT}} \overset{SR:\ SERVE-OUT}{\begin{bmatrix} sr_1 = 1.Serve.Speed \\ sr_2 = 2.Serve.Spin \\ sr_3 = 3.Serve.Place \end{bmatrix}} \xrightarrow{\mathcal{G}^{AT}} \overset{RV:\ RUN-VOLEY}{\begin{bmatrix} rv_1 = RV.For \\ rv_2 = RV.L.R \end{bmatrix}}$$

Counter–Attack Model: Tennis Return

A. Simple Return. The simple CA–dynamics reads:

$$\underset{CAT}{ATTACK} \xrightarrow[FAM]{\mathcal{F}^{CA}} \underset{CAT}{MANEUVER} \xrightarrow[FAM]{\mathcal{G}^{CA}} \underset{CAT}{RESPONSE}$$

In the case of a simple tennis return, this CA–scenario consists purely of conditioned–reflex reaction, no decision process is involved, so it reads:

$$\underset{BALL-IN}{B \ni b_{\mathcal{K}}} \xrightarrow{\mathcal{F}^{CA}} \underset{RUNNING}{R \ni r_{\mathcal{J}}} \xrightarrow{\mathcal{G}^{CA}} \underset{SHOT-OUT}{S \ni s_k}$$

where the n–categories $B_{\mathrm{dim}=5} \ni b_{\mathcal{K}}$, $R_{\mathrm{dim}=3} \ni r_{\mathcal{J}}$, $S_{\mathrm{dim}=4} \ni s_k$, contain the fuzzy variables $\{b_{\mathcal{K}} = b_{\mathcal{K}}(t)\}$, $\{r_{\mathcal{J}} = r_{\mathcal{J}}(t)\}$ and $\{s_k = s_k(t)\}$, respectively defining the ball inputs, our player's running maneuver and his shot–response, K.e.,

$$\overset{B:\,BALL-IN}{\begin{bmatrix} b_1 = Dist.L.R. \\ b_2 = Dist.\mathcal{F}.B. \\ b_3 = Dist.Vert \\ b_4 = Speed \\ b_5 = Spin \end{bmatrix}} \xrightarrow{\mathcal{F}^{CA}} \overset{R:\,RUNNING}{\begin{bmatrix} r_1 = Run.L.R. \\ r_2 = Run.\mathcal{F}.B. \\ r_3 = Run.Vert \end{bmatrix}} \xrightarrow{\mathcal{G}^{CA}} \overset{S:\,SHOT-OUT}{\begin{bmatrix} s_1 = Backhand \\ s_2 = Forehand \\ s_3 = Volley \\ s_4 = Smash \end{bmatrix}}$$

Here, the existence of efficient weapons within the $\underset{SHOT-OUT}{S}$ arsenal–space, namely $s_k(t) : s_1 = Backhand, s_2 = Forehand, s_3 = Voley$ and $s_4 = Smash$, is assumed.

The universes of discourse for the fuzzy variables $\{b_{\mathcal{K}}(t)\}$, $\{r_{\mathcal{J}}(t)\}$ and $\{s_k(t)\}$, partitioned by overlapping Gaussians, are defined respectively as:

$$\underset{BALL-IN}{B} : \begin{array}{l} b_1 = Dist.L.R. : (veryLeft, left, center, right, veryRight), \\ b_2 = Dist.\mathcal{F}.B. : (baseLine, center, netClose), \\ b_3 = Dist.Vert : (low, medium, high), \\ b_4 = Speed : (low, medium, high), \\ b_5 = Spin : (highTopSpin, lowTopSpin, flat, \\ \qquad lowBackSpin, highBackSpin). \end{array}$$

$$\underset{RUNNING}{R} :$$
$$\begin{array}{l} r_1 = Run.L.R. : (veryLeft, left, center, right, veryRight), \\ r_2 = Run.\mathcal{F}.B. : (closeFront, front, center, back, farBack), \\ r_3 = Run.Vert : (squat, normal, jump). \end{array}$$

$$\underset{SHOT-OUT}{S} : \begin{array}{l} s_1 = Backhand : (low, medium, high), \\ s_2 = Forehand : (low, medium, high), \\ s_3 = Volley : (backhand, block, forehand), \\ s_4 = Smash : (low, medium, high). \end{array}$$

B. Advanced Return. The advanced CA–dynamics includes both the information about the opponent and (either conscious or subconscious) decision making. This generic CA–scenario is formulated as the following composition + fusion of FAM functors:

$$\begin{array}{ccccccc}
\underset{CAT}{ATTACK} & \xrightarrow[FAM]{\mathcal{F}^{CA}} & \underset{CAT}{MANEUV} & \xrightarrow[FAM]{\mathcal{G}^{CA}} & \underset{CAT}{DECISION} & \xrightarrow[FAM]{\mathcal{H}^{CA}} & \underset{CAT}{RESP} \\
 & & & & \uparrow {\scriptstyle \mathcal{K}^{CA}\ FAM} & & \\
 & & & & \underset{CAT}{TARGET} & &
\end{array}$$

where we have added two new $n-$categories, $\underset{CAT}{TARGET}$ and $\underset{CAT}{DECISION}$, respectively containing information about the opponent as a target, as well as our own aiming decision processes. In the case of advanced tennis return, this reads:

$$\begin{array}{ccccccc}
\underset{BALL-IN}{B \ni b_{\mathcal{K}}} & \xrightarrow{\mathcal{F}^{CA}} & \underset{RUNNING}{R \ni r_{\mathcal{J}}} & \xrightarrow{\mathcal{G}^{CA}} & \underset{DECISION}{D \ni d_l} & \xrightarrow{\mathcal{H}^{CA}} & \underset{SHOT-OUT}{S \ni s_k} \\
 & & & & \uparrow {\scriptstyle \mathcal{K}^{CA}} & & \\
 & & & & \underset{OPPONENT-IN}{O \ni o_m} & &
\end{array}$$

where the two additional $n-$categories, $O_{\dim=4} \ni o_m$ and $D_{\dim=5} \ni d_l$, contain the fuzzy variables $\{o_m = o_m(t)\}$ and $\{d_l = d_l(t)\}$, respectively defining the opponent–related target information and the aim–related decision processes, both partitioned by overlapping Gaussians and defined as:

$$\underset{OPPONENT-IN}{O} : \begin{array}{l}
o_1 = Opp.Posit.L.R. : (left, center, right), \\
o_2 = Opp.Posit.\mathcal{F}.B. : (netClose, center, baseLine), \\
o_3 = Opp.Anticip.L.R. : (runLeft, stay, runRight), \\
o_4 = Opp.Anticip.\mathcal{F}.B. : (runNet, stay, runBase).
\end{array}$$

$$\underset{DECISION}{D} : \begin{array}{l}
d_1 = Aim.L.R. : (left, center, right), \\
d_2 = Aim.\mathcal{F}.B. : (netClose, center, baseLine), \\
d_3 = Aim.Vert : (low, medium, high), \\
d_4 = Aim.Speed : (low, medium, high), \\
d_5 = Aim.Spin : (highTopSpin, lowTopSpin, noSpin, \\
\qquad\qquad\qquad lowBackSpin, highBackSpin).
\end{array}$$

The corresponding fuzzy–matrices read:

$$\overset{B:\,BALL-IN}{\begin{bmatrix} b_1 = Dist.L.R. \\ b_2 = Dist.\mathcal{F}.B. \\ b_3 = Dist.Vert \\ b_4 = Speed \\ b_5 = Spin \end{bmatrix}}, \quad \overset{R:\,RUNNING}{\begin{bmatrix} r_1 = Run.L.R. \\ r_2 = Run.\mathcal{F}.B. \\ r_3 = Run.Vert \end{bmatrix}}, \quad \overset{D:\,DECISION}{\begin{bmatrix} d_1 = Aim.L.R. \\ d_2 = Aim.\mathcal{F}.B. \\ d_3 = Aim.Vert \\ d_4 = Aim.Speed \\ d_5 = Aim.Spin \end{bmatrix}},$$

$$\begin{bmatrix} O{:}OPPONENT-IN \\ o_1 = Opp.Posit.L.R. \\ o_2 = Opp.Posit.\mathcal{F}.B. \\ o_3 = Opp.Anticip.L.R. \\ o_4 = Opp.Anticip.\mathcal{F}.B. \end{bmatrix}, \quad \overset{S:\,SHOT-OUT}{\begin{bmatrix} s_1 = Backhand \\ s_2 = Forehand \\ s_3 = Volley \\ s_4 = Smash \end{bmatrix}}.$$

16

The Law of Mental Training

16.1 The Best Preparation for *Mental Speed*

The best preparation for the *mental speed* in tennis is the so–called *lightning chess*, or *bullet chess*. It is the faster version of the *blitz chess* game, where each side has less than 3 minutes to complete all of their moves. Often bullet chess is so fast that tactics and skill are secondary to quick moves. Under United States Chess Federation (USCF) rules, bullet games are considered blitz. Every year, an over-the-board lightning chess tournament is held in Apeldoorn in the Netherlands. The time control is 2 minutes per player per game. It is the only official championship in bullet chess. The winner can claim the title Open Dutch Champion in Lightning Chess.

16.2 The Most Effective Visualization Exercise for Tennis

The most effective visualization exercise for tennis champions has the following seven phases:

1. You need to decide what level in tennis you want to achieve (e.g., "world number one," or "in the top 10," etc.). This needs to be something that you believe that you can do and that you deserve.
2. Imagine yourself as a winner at that level.
3. Visualize that your every first serve results in an ace.
4. Visualize that you can efficiently return every serve.
5. Visualize that your every ground-stroke is a winner.
6. Visualize that you are full of energy.
7. Visualize the enjoyment of all your aces and winners.

16.3 The Common Problems with a Player's Mental Game

The common problems are: (i) confusion about strategy; (ii) trying to do too many things at once; (iii) being easily distracted; (iv) too much concern about

T.T. Ivancevic et al.: Paradigm Shift for Future Tennis, COSMOS 12, pp. 87–91.
springerlink.com

winning and losing; (v) perfectionism; (vi) complacency; (vii) having no plan or clear goal; (viii) too much spontaneity and creativity; (ix) lack of humility; (x) inappropriate reaction to errors; and (xi) negative self talk (self–criticism).

To improve your mental game you need to: (i) set clear goals; (ii) create action steps that will take you closer to these goals (learn to visualize: try to "see" in your mind what you want to achieve); (iii) use positive self–talk (self–criticism); (iv) practice yoga breathing techniques; (v) always put forth 100% effort; (vi) be process oriented; (vii) stay detached; (viii) no future tripping, no past tripping – stay in the present; (ix) be non–reactive to the opinions of others; (x) tolerate your inability to be perfect; (xi) do what it takes to have fun and smile; (xii) define winning in a way that includes more than just the final score.

16.4 Focus on the Ball and Still Be Aware of the Opponent

Yes, this is a great problem for most players, and apparently Roger Federer solves it consistently and routinely. Aside from a great tolerance to competitive pressure, focus is one of the key characteristics of a champion. You really need to "fix your eyes on the ball," as every coach advises, and yet to have a full amount of the so–called *situation awareness* (SA), which is defined in modern psychology by M.R. Endsley (1995) as "the perception of the elements in the environment within a volume of time and space, the comprehension of their meaning, and projection of their status in the near future" (see Appendix). For example, SA is a key factor in the training of fighter pilots and *Formula* 1 racing drivers. The outcome of any complex situation critically depends on SA. Therefore, you need to develop: (i) strong focus on the moving tennis ball; (ii) ability for strong SA, as defined above; and (iii) to have both focus on the ball and SA at the same time. This is similar to learning to play a hard piece on a piano: first you learn the right–hand part, then you learn the left–hand part, then you learn how to combine them together. Clearly, both physical and mental fatigue (combined with other disturbances, like e.g., rain, wind, noisy audience, big overhead screen, etc.) will gradually reduce both focus and SA. Therefore, besides physical speed and strength, you obviously need mental strength and endurance, to be able to win.

16.5 The Mental Profile of a Future Tennis Champion

The future tennis champion will be a rare combination of talent, hard work and the right mental profile. Even in today's tennis, often the difference between the good and elite players is their mental qualities. In this respect, most important are the following three psychological characteristics: (i) confidence, (ii) anxiety, and (iii) motivation.

(i) Confidence is an emotion or state of mind commonly associated with athletic success. Indeed, the following quote from the former number one Jimmy Connors provides great insight into the confidence level of an elite athlete:

> The whole thing is never to get negative about yourself. Sure, it's possible that the other guy you're playing is tough, and that he may have beaten you the last time you played, and okay, maybe you haven't been playing all that well yourself. But the minute you start thinking about these things you're dead. I go out to every match convinced that I'm going to win. That is all there is to it.

In general, elite athletes tend to have very high levels of confidence and feel that these high levels are needed for the performances that they are looking for. About 90% of all elite athletes have a very high level of self-confidence. Confidence is usually a result of an athlete anticipating success in their upcoming event. An athlete's anticipated outcome is the greatest indicator of confidence. This expectation for success can be based on an athlete's confidence in themselves, emotional readiness, physical ability, knowledge of the opponent, goals, strategies, physical condition, or on the coach. To reach the very pinnacle of sport, an athlete must have high confidence in their abilities; and getting to that elite level and all the preceding successes that it took to get to that level must surely build the confidence levels of an athlete.

(ii) The link between anxiety and poor performance in sport has been known for a long time. Stories abound of athletes or teams that performed poorly because they underestimated their opponent (below optimum anxiety levels) or worried themselves out of the game (above optimum anxiety levels). Dealing with anxiety successfully is an important characteristic of a future tennis champion. The ability to cope with pressure and anxiety will be an integral part of a champion.

One of the earliest models that attempted to explain the relationship between arousal/anxiety and performance was the so–called *inverted–U hypothesis*, which stated that as arousal increased, performance would increase as well; but if arousal became too great performance would deteriorate. In other words, as stress began to build an individual still felt confident in their ability to control it and performance would improve. However, once a stressor became so great that the individual started to doubt their ability to cope with it, performance would decline.

An individualistic approach was added to this hypothesis when the concept of individualized zones of optimal functioning, or IZOFs, was developed. According to this theory, each individual has an optimal level of pre-performance anxiety. If the athlete is in this "zone," peak performances will be the result. However, if anxiety levels are too high or too low, the athlete will not see optimum results. IZOFs can be determined by repeatedly measuring anxiety and performance or through athlete's recall of anxiety levels prior to peak performances.

Depending on the individual, anxiety levels can have a variety of effects on athletic performance. Today it is well-known that anxiety can be reduced through mental imagery, relaxation, and cognitive intervention. These methods not only aim at reducing stress and anxiety levels but also aim to improve confidence levels. The goal is to help the athlete enter his or her IZOF.

(iii) To become a champion in any sport requires many hours, days and years of training. Often this training is rigorous, painful, or exhausting. However, the athletes who have reached the pinnacle of their sport have more than likely put in their time to get to achieve that high level of success. To do this, these athletes must have something that motivates them to continually push their bodies, and come back from whatever struggles or setbacks they may experience along the way. This motivation may come intrinsically or extrinsically. Intrinsic motivation is an athletes' personal drive to achieve their goal. This may be setting a school record, winning a race, or defeating a particular opponent. Extrinsic motivation is the resulting motivation from an outside source such as parents, coaches, or teammates.

There are many players who have the talent to succeed but very few who have the motivational drive to do what it takes to become a champion. In light of this, it appears that intrinsic motivation may be the greater determinant of achieving success in sports. To achieve at an elite level in sport, an athlete must have the motivation to train hard on a daily basis and to overcome any obstacles or setbacks that they might face in reaching or maintaining that level of performance.

Overall, it seems that the following traits would be common among elite athletes: extreme self-confidence, low performance anxiety, and high motivation. These three things are very closely related and would seem to form a cyclic pattern, positively influencing one another.

16.6 Mastering the Effective Visualization

In the 1980s and 90s, Dr. Denis Waitley implemented what he called "Visual Motor Rehearsal" (VMH) into the U.S. Olympics program. He and his researchers had found that when an athlete competed in an event only in their mind, the same nervous reaction in the body occurred as when they did their event in real life. This was another proof: The mind cannot tell the difference between an actual, "real–life" event and a vividly imagined one.[1] In this way, the VMH has shown that by merely thinking and visualizing their event, athletes can enhance their performance.[2]

How to apply the VMH to tennis? Before a tennis match, sit back, relax, close your eyes and play the game in your mind. Do the VMH every time you have an important match. The more you practice the more effective VMH will be for

[1] See "The Psychology of Winning," a famous tape program for self–improvement, by Dr. Denis Waitley.

[2] More technically speaking, relationships between sensory and motor events can be learned, although the events may have no prior association. For example, although most drivers know how to respond to a traffic signal, the sensory–motor rules governing these relationships require learning, because no intrinsic association exists between a traffic signal color or its spatial position and the appropriate movements to modify the speed of an automobile. These arbitrary cue–response associations become learned through experience, commonly by trial and error.

you. Do the same with learning a new technique: first watch the chosen ideal (a champion) and then visualize the same movement as performed by yourself. The more you visualize the less you need to practice physically.

16.7 Be Happy and Think Clearly All the Time

Yes, we believe one can be happy and still think clearly in big tournaments. We would like to quote from the book *Born to Believe*, by Andrew Newberg and Mark Waldman:

> The brain is very happy when you are focused on what you love doing. The more you focus on what you truly love and desire, the volume gets turned down in those parts of the limbic system where the destructive emotions of fear, anger, depression and anxiety are controlled. This allows you to think more clearly.
> You also turn up the volume in other parts of the limbic system that generate positive emotions. When this happens, you get a release of dopamine, endorphins, and a variety of stress-reducing hormones and neurotransmitters, which enhances clear thinking. The more you focus on what you truly love, the healthier you are likely to be, and the more you feel the positive effects of those stress-reducing neuro-chemicals in your body and mind.
> You can have a decrease in negative emotions and an increase in positive emotions when you align yourself with what you believe is most important to you.

16.8 Realistic Goals versus a Wish to Be a Champion

A focused desire to become a champion is a major driving mental force. Again, we quote from the book *Born to Believe*:

> When one focuses on the big questions, the really big questions, one is challenging their brains to think outside the box, and this causes the structure of our neurons to change, particularly in our frontal lobes, that part of the brain that controls logic, reason, language, consciousness, and compassion.
> New axons grow, reaching out to new dendrites to communicate in ways that our brains have never done before. When contemplating the *big* questions one uses their frontal lobes to alter the function of other parts of their brains.

17

The Quantum Intention Law

17.1 Quantum Brain and Mind

In this section we will introduce modern concepts of quantum brain and mind, mostly following a modern trend set-up in the journal *NeuroQuantology*. For example, according to some of the recent papers from this journal (see [Per03]), 'brain is classical' (i.e., governed by classical biophysics), while 'mind is quantum',[1] and human consciousness is generated by 'neuro-quantization', which takes place inside brain's microtubules (see, e.g. [Pen89, Pen94]).

17.1.1 Classical Neurodynamics

To give a brief introduction to classical neurodynamics, we start from the fully recurrent, $N-$dimensional, RC transient circuit, given by a nonlinear vector differential equation [Hay98, Kos92, II07b]:

$$C_j \dot{v}_j = I_j - \frac{v_j}{R_j} + w_{ij} f_i(v_i), \qquad (i, j = 1, ..., N), \tag{17.1}$$

where $v_j = v_j(t)$ represent the activation potentials in the jth neuron, C_j and R_j denote input capacitances and leakage resistances, synaptic weights w_{ij} represent conductances, I_j represent the total currents flowing toward the input nodes, and the functions f_i are sigmoidal.

Geometrically, equation (17.1) defines a smooth autonomous vector–field $X(t)$ in ND neurodynamical phase–space manifold M, and its (numerical) solution for the given initial potentials $v_j(0)$ defines the autonomous neurodynamical phase–flow $\Phi(t) : v_j(0) \to v_j(t)$ on M.

In AI parlance, equation (17.1) represents a generalization of three well–known recurrent NN models (see [Hay98, Kos92, II07b]):

---

[1] From Delight we all came into existence; in Delight we grow and play our respective roles; at the end of our journey's close we return to Delight – *Upanishads*.

T.T. Ivancevic et al.: Paradigm Shift for Future Tennis, COSMOS 12, pp. 93–109.
springerlink.com

(i) continuous Hopfield model [Hop84],
(ii) Grossberg ART–family cognitive system [CG83], and
(iii) Hecht–Nielsen counter–propagation network [Hec87, Hec90].

Physiologically, equation (17.1) is based on the Nobel–awarded *Hodgkin–Huxley equation* of the neural action potential (for the single squid giant axon membrane) as a function of the conductances g of sodium, potassium and leakage [HH52, Hod64]:

$$C\dot{v} = I(t) - g_{Na}(v - v_{Na}) - g_K(v - v_K) - g_L(v - v_L),$$

where bracket terms represent the electromotive forces acting on the ions.

The *continuous Hopfield circuit* model [Hop84]:

$$C_j\dot{v}_j = I_j - \frac{v_j}{R_j} + T_{ij}u_i, \qquad (i,j = 1, ..., N), \tag{17.2}$$

where u_i are output functions from processing elements, and T_{ij} is the inverse of the resistors connection–matrix becomes equation (17.1) if we put $T_{ij} = w_{ij}$ and $u_i = f_i[v_j(t)]$.

The Grossberg *analogous ART2 system* is governed by activation equation:

$$\varepsilon\dot{v}_j = -Av_j + (1 - Bv_j)I_j^+ - (C + Dv_j)I_j^-, \qquad (j = 1, ..., N),$$

where A, B, C, D are positive constants (A is dimensionally conductance), $0 \leq \varepsilon << 1$ is the fast–variable factor (dimensionally capacitance), and I_j^+, I_j^- are excitatory and inhibitory inputs to the jth processing unit, respectively.

General *Cohen–Grossberg activation equations* [CG83] have the form:

$$\dot{v}_j = -a_j(v_j)[b_j(v_j) - f_k(v_k)m_{jk}], \qquad (j = 1, ..., N), \tag{17.3}$$

and the *Cohen–Grossberg theorem* ensures the global stability of the system (17.3). If

$$a_j = 1/C_j, \qquad b_j = v_j/R_j - I_j, \qquad f_j(v_j) = u_j,$$

and constant $m_{ij} = m_{ji} = T_{ij}$, the system (17.3) reduces to the Hopfield circuit model (17.2).

The Hecht–Nielsen *counter–propagation network* is governed by the activation equation [Hec87, Hec90]:

$$\dot{v}_j = -Av_j + (B - v_j)I_j - v_j\sum_{k\neq j} I_k, \qquad (j = 1, ..., N),$$

where A, B are positive constants and I_j are input values for each processing unit.

Provided some simple conditions are satisfied, namely, say symmetry of weights $w_{ij} = w_{ij}$, non–negativity of activations v_j and monotonicity of transfer functions f_j, the system (17.1) is globally asymptotically stable (in the sense of Liapunov energy functions). The fixed–points (stable states) of the system correspond to the fundamental memories to be stored, so it works as content–addressable memory (AM). The initial state of the system (17.1) lies inside the basin of attraction of its fixed–points, so that its initial state is related to appropriate memory vector. Various variations on this basic model are reported in the literature [Hay98, Kos92], and more general form of the vector–field can be given, preserving the above stability conditions.

17.1.2 Computation at the Edge of Chaos

Depending on the connectivity, the so–called *recurrent neural networks* (see [II07b]), consisting of simple computational units, can show very different types of dynamics, ranging from totally ordered (linear–like behavior) to chaotic. Using the the *mean–field theory* approach with *evolving Hamming distance*, Bertschinger and Natschläger analyzed how the type of dynamics (ordered or chaotic) exhibited by randomly connected networks of threshold gates driven by a time-varying input signal depended on the parameters describing the distribution of the *connectivity matrix* (see [II08c]). In particular, the authors calculated the critical boundary in *parameter space* where the transition from ordered to chaotic dynamics takes place. Employing a recently developed framework for analyzing real-time computations, they showed that only near the critical boundary could such networks perform complex computations on time series. This result strongly supports conjectures that dynamical systems that are capable of doing complex computational tasks should operate near the *edge of chaos*, that is, the transition from ordered to chaotic dynamics (see [II08b]).

In particular, the authors pointed out the following: (i) Dynamics near the critical line are a general property of input–driven dynamical systems that support complex real-time computations; and (ii) Such systems can be used by training a simple readout function to approximate any arbitrary complex filter. This suggests that to exhibit sophisticated information processing capabilities, an adaptive system should stay close to *critical dynamics*. Since the dynamics is also influenced by the statistics of the input signal, these results indicate a new role for plasticity rules: stabilize the dynamics at the critical line. Such rules would then implement what could be called *dynamical homeostasis*, which could be achieved in an unsupervised learning manner.

Now, if critical dynamics at the edge of chaos of classical adaptive systems is essentially nonlinear, then what can we say about modern *adaptive quantum systems*? While quantum mechanics is based on superposition principle (see Chapter 3), and thus is essentially linear, adaptation introduces critical nonlinearity. In this way, we come to the nonlinear deformation/extension of the linear *Schrödinger equation*.

An impetus to hypothesize a quantum brain model comes from the brain's necessity to unify the neuronal response into a single percept. Anatomical, neurophysiological and neuropsychological evidence, as well as brain imaging using fMRI and PET scans, show that separate functional maps exist in the biological brain to code separate features such as direction of motion, location, color and orientation. How does the brain compute all this data to have a coherent perception?

To provide a partial answer to the above question, a *quantum neural network* (QNN) (see Figure 17.1) has been proposed (see [II08c] and references therein), in which a collective response of a neuronal lattice is modeled using the Schrödinger equation:

$$\mathrm{i}\partial_t \psi(x,t) = -\frac{1}{2}\Delta\psi(x,t) + V(x)\psi(x,t), \tag{17.4}$$

where $\psi(x,t)$ is the wave function, or probability amplitude, associated with the quantum–mechanical system at a space-time point (x,t) and Δ is the standard *Laplacian operator*. It is shown that an external stimulus reaches each neuron in a lattice with a probability amplitude function ψ_i. Such a hypothesis would suggest that the carrier of the stimulus performs quantum computation. The collective response of all the neurons is given by the *quantum superposition equation*,

$$\psi = \sum_{i=0}^{N} c_i \psi_i = c_i \psi_i.$$

The QNN hypothesis suggests that the time evolution of the collective response $\psi = \psi(x,t)$ is described by equation (17.4). A neuronal lattice sets up a spatial potential field $V(x)$. A quantum process described by a quantum state ψ which mediates the collective response of a neuronal lattice evolves in the spatial potential field according to equation (17.4). Thus the '*classical brain*' sets up a spatio-temporal potential field $V(x,t)$ and the '*quantum brain*' is excited by this potential field to provide a collective response $\psi = \psi(x,t)$ (see [II08c] and references therein).

Mathematical basis for the QNN presented in Figure 17.1 is the *nonlinear Schrödinger equation* (NLS), which gives a closed–loop feedback dynamics for

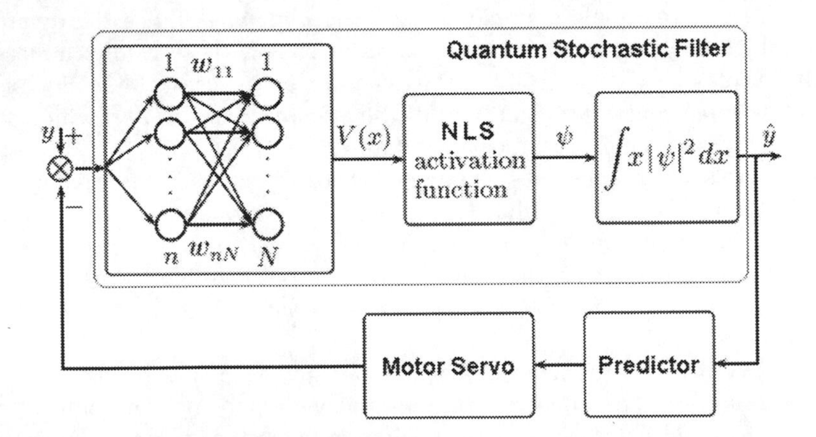

Fig. 17.1. Quantum neural network has three blocks: quantum stochastic filter based on the *nonlinear Schrödinger equation*, Kalman–like predictor and neural motor servo, corresponding to the Kalman–like corrector.

the plant defined by the linear Schrödinger equation (17.4). Informally, QNN is represented by *adaptive NLS*, that is NLS with 'synaptic weights' w_i replacing quantum superposition constants c_i. Formally, it is the Schrödinger equation (17.4) with added cubic nonlinearity and synaptic weights.

Consider a *classical perceptron* [II07b], i.e., the system with n input channels $x_1, \dots, x_n$ and one output channel y, given by

$$y = f(\sum_{j=1}^{n} w_j x_j), \tag{17.5}$$

where $f(\cdot)$ is the perceptron activation function and w_j are the weights tuning during learning process.

The *perceptron learning algorithm* works as follows:

1. The weights w_j are initialized to small random numbers.
2. A *pattern vector* $(x_1, \dots, x_n)$ is presented to the perceptron and the output y generated according to the rule (17.5)
3. The weights are updated according to the rule

$$w_j(t+1) = w_j(t) + \eta(d-y)x_j, \tag{17.6}$$

where t is discreet time, d is the desired output provided for training and $0 < \eta < 1$ is the step size.

It will be hardly possible to construct an exact analog of the nonlinear activation function f, like sigmoid and other functions of common use in neural networks, but we will show that the leaning rule of the type (17.6) is possible for a quantum system too.

Consider a quantum system with n inputs $|x_1\rangle, \dots, |x_n\rangle$ of the form (17.8), and the output $|y\rangle$ derived by the rule

$$|y\rangle = \hat{F} \sum_{j=1}^{n} \hat{w}_j |x_j\rangle, \tag{17.7}$$

where $\hat{w}_j$ become 2×2 matrices acting on the basis $(|0\rangle, |1\rangle)$, combined of phase shifters $\mathrm{e}^{\mathrm{i}\theta}$ and beam splitters, and possibly light attenuators, $\hat{F}$ is an unknown operator that can be implemented by the network of quantum gates.

Also, quantum computation on optical modes using only beam splitters, phase shifters, photon sources and photo detectors is possible. Following this concept, we can assume the existence of a qubit [II08c]

$$|x\rangle = \alpha|0\rangle + \beta|1\rangle, \tag{17.8}$$

where $|\alpha|^2 + |\beta|^2 = 1$, where the states $|0\rangle$ and $|1\rangle$ are understood as different polarization states of light.

Consider the simplistic case with $\hat{F} = 1$ being the identity operator. The output of the quantum perceptron at the time t will be [II08c]

$$|y(t)\rangle = \sum_{j=1}^{n} \hat{w}_j(t)|x_j\rangle.$$

In analogy with classical case (17.6), let us provide a learning rule

$$\hat{w}_j(t+1) = \hat{w}_j(t) + \eta(|d\rangle - |y(t)\rangle)\langle x_j|, \tag{17.9}$$

where $|d\rangle$ is the desired output.

It can be shown that the learning rule (17.9) drives the quantum perceptron into desired state $|d\rangle$ used for learning. Using the rule (17.9) and taking the module-square difference of the real and desired outputs, we yield

$$\| |d\rangle - |y(t+1)\rangle \|^2 = \| |d\rangle - \sum_{j=1}^{n} \hat{w}_j(t+1)|x_j\rangle \|^2 = (1 - n\eta)^2 \| |d\rangle - |y(t)\rangle \|^2.$$

For small η $(0 < \eta < 1/n)$ and normalized input states $\langle x_j|x_j\rangle = 1$ the result of iteration converges to the desired state $|d\rangle$. The whole network can be then composed from the primitive elements using the standard rules of ANN architecture. For more technical details, see [II08c].

17.1.3 Connectionism, Control Theory and Brain Theory

Let us start with the classical *connectionist brain theory*[2] (see [Arb98] for a general overview). A recent paper [Roy08] proposes a engineering new theory for the supposed internal mechanisms of the brain function. It postulates that there are controllers in the brain and that there are parts of the brain that control other parts. Thus, the new theory refutes the standard connectionist theory that there are no separate controllers in the brain for higher level functions and that all control is 'local and distributed' at the level of the cells. Connectionist algorithms themselves are used to prove this theory. The author claims that there is evidence in the neuroscience literature to support this theory. Thus, this paper proposes a control theoretic approach for understanding how the brain works and learns. That means that control theoretic principles should be applicable to developing systems similar to the brain.

Having established the connection between *connectionism* and the control theory, it is a small step to argue that if connectionist models are brain-like, then, given that they use executive controllers, the internal mechanisms of the brain must also use executive controllers. Hence, it can be generalized and claimed that there are parts of the brain that control other parts. A control theoretic approach to understanding how the brain works and learns could overcome many

[2] As already mentioned, *connectionism* is an approach in the fields of artificial intelligence, cognitive psychology/cognitive science, neuroscience and philosophy of mind, that models mental or behavioral phenomena as the emergent processes of interconnected networks of simple units. There are many forms of connectionism, but the most common forms use ANNs.

of the limitations of connectionism and lead to the development of more powerful models that can properly replicate the functions of the brain, such as autonomous learning without external intervention. None of the supervised connectionist algorithms can learn without human intervention that requires setting various learning parameters by a trial-and-error process. Thus, it is impossible to build autonomous robots (software or hardware), ones that can learn on their own, with these connectionist algorithms. The new proposed control theoretic paradigm, however, should allow the field to freely explore other means of learning in neural networks [Roy08].

17.1.4 Neocortical Biophysics

Basic brain physiology will be presented in Part II. In this section we focus on its latest and most important part, called *neocortex*[3] performs a large variety of complex computations in real time. It is conjectured that these computations are carried out by a network of cortical micro-circuits, where each micro-circuit is a rather stereotypical circuit of neurons within a cortical column. A characteristic property of these circuits and networks is an abundance of feedback

[3] The *neocortex* (Latin for 'new bark' or 'new rind') is a part of the brain of mammals. It is the outer layer of the cerebral hemispheres, and made up of six layers, labeled I to VI (with VI being the innermost and I being the outermost). The neocortex is part of the *cerebral cortex* (along with the archicortex and paleocortex, which are cortical parts of the limbic system). It is involved in higher functions such as sensory perception, generation of motor commands, spatial reasoning, conscious thought and, in humans, language.

The neocortex consists of the grey matter, or neuronal cell bodies and unmyelinated fibers, surrounding the deeper white matter (myelinated axons) in the cerebrum. Whereas the neocortex is smooth in rodents and other small mammals, it has deep grooves (sulci) and wrinkles (gyri) in primates and other larger mammals. These folds increase the surface area of the neocortex considerably without taking up too much more volume. This has allowed primates and especially humans to evolve new functional areas of neocortex that are responsible for enhanced cognitive skills such as working memory, speech, and language. The neocortex contains two primary types of neurons, excitatory pyramidal neurons (80% of neocortical neurons) and inhibitory interneurons (20%). The neurons of the neocortex are also arranged in vertical structures called neocortical columns.

The neocortex is divided into frontal, parietal, temporal, and occipital lobes, which perform different functions. For example, the occipital lobe contains the primary visual cortex, and the temporal lobe contains the primary auditory cortex. Further subdivisions or areas of neocortex are responsible for more specific cognitive processes. In humans, the frontal lobe contains areas devoted to abilities that are enhanced in or unique to our species, such as complex language processing localized to the ventrolateral prefrontal cortex (Broca's area) and social and emotional processing localized to the orbito-frontal cortex.

The female human neocortex contains approximately 19 billion neurons while the male human neocortex has 23 billion on average [PG97]. Additionally, the female neocortex has more white matter, while the male neocortex contains more grey matter. The implications of such differences are not fully known.

connections. But the computational function of these feedback connections is largely unknown. Two lines of research have been engaged to solve this problem. In one approach, which one might call the constructive approach, one builds hypothetical circuits of neurons and shows that (under some conditions on the response behavior of its neurons and synapses) such circuits can perform specific computations. In another research strategy, which one might call the analytical approach, one starts with data-based models for actual cortical micro-circuits, and analyzes which computational operations such "given" circuits can perform under the assumption that a learning process assigns suitable values to some of their parameters (e.g., synaptic efficacies of readout neurons). An underlying assumption of the analytical approach is that complex recurrent circuits, such as cortical micro-circuits, cannot be fully understood in terms of the usually considered properties of their components. Rather, system-level approaches that directly address the dynamics of the resulting recurrent neural circuits are needed to complement the bottom-up analysis [MJS07]. This line of research started with the identification and investigation of so-called *canonical micro-circuits* [DKM95]. Several issues related to cortical micro-circuits have also been addressed in the work of S. Grossberg and collaborators from Boston (see [Gro03] and the references therein). Subsequently it was shown that quite complex real–time computations on spike trains can be carried out by such 'given' models for cortical micro-circuits (see [DM04] for a review). A fundamental limitation of this approach was that only those computations could be modeled that can be carried out with a fading memory, more precisely only those computations that require integration of information over a time-span of 200 ms to 300 ms (its maximal length depends on the amount of noise in the circuit and the complexity of the input spike trains). In particular, computational tasks that require a representation of elapsed time between salient sensory events or motor actions, or an internal representation of expected rewards, working memory, accumulation of sensory evidence for decision making, the updating and holding of analog variables such as for example the desired eye position, and differential processing of sensory input streams according to attentional or other internal states of the neural system could not be modeled in this way. Previous work on concrete examples of artificial neural networks and cortical micro-circuit models had already indicated that these shortcomings of the model might arise only if one assumes that learning affects exclusively the synapses of readout neurons that project the results of computations to other circuits or areas, without giving feedback into the circuit from which they extract information. This scenario is unrealistic from a biological perspective, since pyramidal neurons in the cortex typically have in addition to their long projecting axon a large number of axon collaterals that provide feedback to the local circuit [MJS07]. Also, abundant feedback connections exist on the network level between different brain areas. For statistical mechanics of neocortical interactions (using path integral–related methods), see [Ing82, Ing97, Ing98].

Schematically, *neocortical dynamics* is depicted in Figure 17.2 (a), which illustrates the universally accepted six laminae of the neocortex [Sze78]. The pyramidal apical dendrites finish in a tuft-like branching in lamina I. It has been accepted (see, e.g. [Bec08]) that the apical bundles of dendrites are the basic anatomical units of the neocortex. They are observed in all areas of the cortex that have been investigated in all mammals, including humans. It has been proposed that these bundles are the cortical units for reception, which would give them a preeminent role. Since they are composed essentially of dendrites, the name *dendron* was adopted by John Eccles[4] [Ecc90, Ecc94]. Figure 17.2 (b) illustrates a typical *spine synapse* that makes an intimate contact with an apical dendrite of a pyramidal cell. The inner surface of a bouton confronting the synaptic cleft d forms the presynaptic vesicular grid (PVG). A nerve impulse propagating into a bouton causes a process called exocytosis. A nerve impulse evokes at most a single exocytosis from a PVG [Bec08]. Exocytosis is the basic unitary activity of the cerebral cortex. Each all-or-nothing *exocytosis* of synaptic transmitter substance results in a brief excitatory postsynaptic depolarization (EPSP). Summation by electrotonic transmission of many hundreds of these milli-EPSPs is required for an EPSP large enough (1020 mV) to generate the discharge of an impulse by a pyramidal cell. This impulse will travel along its axon to make effective excitation at its many synapses. This is the conventional macro-operation of a pyramidal cell of the neocortex, and it can be satisfactorily described by conventional neuroscience, even in the most complex design of neural network theory and neural group selection (see, e.g. [Sze78]).

Experimental analysis of transmitter release by spine synapses of hippocampal pyramidal cells has revealed a remarkably low exocytosis probability per excitatory impulse. This means that there must exist an activation barrier against opening of an ion channel in the PVG. Activation can either occur purely stochastically by thermal fluctuations, or by stimulation of a trigger process. Recently, F. Beck has proposed a two–state *quantum trigger* which is realized by *quasi-particle tunneling*.[5] This is motivated by the predominant role of exocytosis as the synaptic regulator of cortical activity which is certainly not completely at random [Bec08]. The quasi-particle assumption allows the treatment of the

---

[4] Sir John Carew Eccles (January 27, 1903 – May 2, 1997) was an Australian neurophysiologist who won the 1963 Nobel Prize in Physiology or Medicine for his work on the synapse. He shared the prize together with Alan Lloyd Hodgkin and Andrew Fielding Huxley, who formulated the celebrated *Hodgkin–Huxley neural model*.

[5] The so-called *quantum tunneling effect* is a nanoscopic phenomenon in which a quantum particle violates the principles of classical mechanics by penetrating a potential barrier or impedance higher than the kinetic energy of the particle. A barrier, in terms of quantum tunneling, may be a form of energy state analogous to a 'hill' or incline in classical mechanics, which classically suggests that passage through or over such a barrier would be impossible without sufficient energy. On the quantum scale, objects exhibit wave-like behavior; in quantum theory, quanta moving against a potential energy 'hill' can be described by their wave-function, which represents the probability amplitude of finding that particle in a certain location at either side of the 'hill'. If this function describes the particle as being on the other side of the 'hill', then there is the probability that it has moved through, rather than over it, and has thus 'tunneled'.

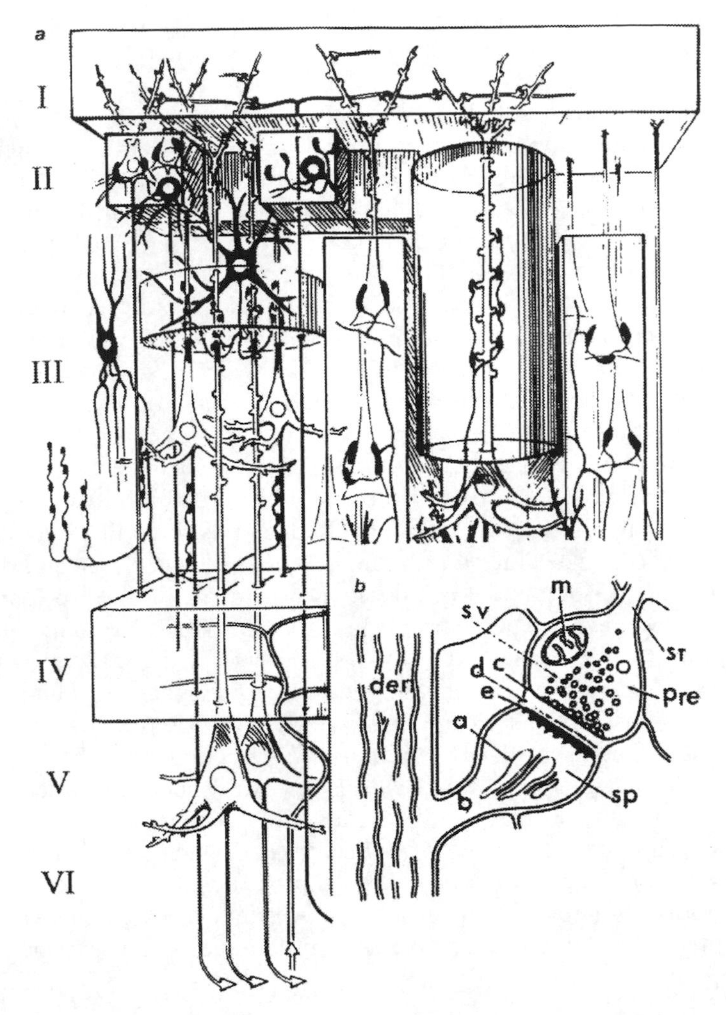

Fig. 17.2. Schematic of *neocortical biophysics* (adapted from [Bec08]): (a) 3D construct by [Sze78] showing cortical neurons of various types; there are two pyramidal cells in lamina V and three in lamina III, one being shown in detail in a column to the right, and two in lamina II. (b) Detailed structure of a spine (sp) synapse on a dendrite (den); st, axon terminating in the synaptic bouton or presynaptic terminal (Pre); sv, synaptic vesicles; c, presynaptic vesicular grid (PVG in text); d, synaptic cleft; e, postsynaptic membrane; a, spine apparatus; b, spine stalk; m, mitochondrion [Gra82].

complicated molecular transition as an effective one–body problem whose solution follows from the time dependent Schrödinger equation (in normal units, see Part II)

$$\mathrm{i}\partial_t\psi(q,t) = -\frac{1}{2}\Delta\psi(q,t) + V(q)\psi(q,t).$$

Here it is assumed that the activated state of the presynaptic cell lasts for a finite time period $[t_0, t_1]$ only before it recombines. Recombination of the activated state defines the measurement process, leading to von Neumann's *collapse of the tunneling state* [Bec08].

The total wave-function $\psi(q,t)$ can be separate at the final time t_1 into two components, representing left and right parts:

$$\psi(q,t_1) = \psi_{\text{left}}(q,t_1) + \psi_{\text{right}}(q,t_1),$$

which constitutes the two amplitudes for the alternative results of the same process: which, after collapse, determine: either exocytosis has happened (ψ_{right}), or exocytosis has not happened (ψ_{right}) (inhibition). State collapse transforms this into the probabilities [Bec08]:

$$\text{Exocytosis probability} : P_{\text{exocyt}}(t_1) = \int |\psi_{\text{right}}(q,t_1)|^2 dq,$$

$$\text{Inhibition probability} : P_{\text{inhibit}}(t_1) = \int |\psi_{\text{right}}(q,t_1)|^2 dq.$$

17.1.5 Quantum Neurodynamics

All theories that hold the neuron as the functional basis of consciousness must bridge a gap between a property of conscious experience and a fundamental tenet of neuronal processing. While evidence suggests that the brain subdivides perceptual processing into modality (e.g., the visual, the tactile) and submodality (e.g., color, temperature), our perceptions themselves are a unified experience [Mas04]. If the anatomic substrate of perceptual modality is functionally and spatially discrete neuronal subpopulations, how is information ultimately synthesized to create the manifest oneness of experience? The *cognitive binding problem*[6] is a central question in the study of consciousness: how does the brain synthesize its modal and submodal processing systems to generate a unity of conscious experience? *Binding* is thought to occur at virtually all levels of perceptual (and motor) processing, and is thought to be a crucial event for consciousness itself [CK90, Cri94]. Although a relatively recent question in neuroscience, the *binding problem* may have made its first appearance in Immanuel Kant's *Critique of Pure Reason*. Kant's principle of a *transcendental unity of apperception*[7] describes the synthesis of the *knowledge of the manifold* [Kan65]:

[6] The term *cognitive binding problem* is attributed to C. von der Mahlsburg [Mal81].

[7] *Apperception* (from Latin ad + percipere = to perceive) has the following meanings: (i) in epistemology, it is the introspective or reflective apprehension by the mind of its own inner states; (ii) in psychology, it is the process by which new experience is assimilated to and transformed by the residuum of past experience of an individual to form a new whole (in short, it is to perceive new experience in relation to past experience); (iii) in philosophy, Kant distinguished empirical apperception from transcendental apperception (the first is 'the consciousness of the concrete actual self with its changing states', the so-called 'inner sense'; the second is 'the pure, original, unchangeable consciousness which is the necessary condition of experience as such and the ultimate foundation of the synthetic unity of experience'.

> There can be in us no modes of knowledge, no connection or unity of consciousness of one mode of consciousness with another, without that unity of consciousness which precedes all data of intuitions, and by relation to which representation of objects is alone possible... For the mind could never think its identity in the manifoldness of its representations, and indeed think this identity *a priori*, if it did not have before its eyes the identity of its act, whereby it subordinates all synthesis of apprehension (which is empirical) to a transcendental unity, thereby rendering possible their interconnection according to a priori rules

There have been various solutions proposed for the binding problem, which can be summarized as *binding by convergence* (information is bound by higher order neurons that collect various responses and fire as a *binding unit* when the full set of inputs converge), *binding by assembly* (binding occurs as a result of self–organizing Hebbian cell assemblies), and *binding by synchrony* (binding results from the synchronized firing of neuronal sub-populations that are spatially discrete) [Mas04].

The discussion of distributed domains of reality and consciousness leads us into a recent view of the brain that does not regard the neuron as a unit of cognitive perception, but rather regards the brain as one indivisible entity. This view is based on the emerging field of quantum neurodynamics [Mas04]. The brain, instead of being a Newtonian object obeying classical laws, is posited to be a macroscopic quantum object obeying the same laws found at the Planck scale [Pen94]. The large–scale quantum coherence would render the brain a *Bose–Einstein condensate*, where properties of quantum wave-functions hold at a macroscopic level. These properties would include non-locality and quantum superposition. Such large- scale coherence can be found in superconductivity and superfluidity, where the environment is disentangled from the quantum events. Superconductivity and superfluidity occur at temperatures just above absolute zero, where it is difficult for the environment to interfere with the quantum coherence. H. Frölich, however, predicted that quantum coherence could also occur in the temperature of a biologic environment (e.g. the brain) with high metabolic energy and extreme dielectric properties of the material involved [FK83]. It has been posited by S. Hameroff that the cytoskeletal elements of *microtubules* could be just such a material [Ham87]. Microtubules are composed of tubulin dimers, which can exist in multiple conformations based on dielectric properties, leading to dipole oscillations that could be transmitted through the length of the tubule. These superconducting *Frölich waves* could be transmitted throughout the protein network and gap junctions: from cytoskeleton to membrane protein to extracellular matrix to adjacent membrane protein to cytoskeleton, and so on. It is of interest that the supporting glial cells, the cellular majority in the brain, would contribute to this network, recalling C. Golgi's conception of the *syncytium*. It is currently a point of theoretical contention whether quantum coherence in the brain would decohere from a waveform collapse of superposed states (termed objective reduction) [HP96], or from the 'warm, wet, and noisy' environment of the brain [Teg00]. It has been suggested that the hollow core of microtubules, ordered water, and actin gelation may buffer the quantum events from the environment. Magnetic resonance imaging of the brain by quantum coherence (albeit an induced artifact of the procedure) has also been suggested as proof of principle [HHT02].

Hameroff has suggested that quantum computation within the microtubules could be involved in consciousness, and proposes the activity of general anesthetics as support for this claim [HW83]. He has argued that the similar activity of diverse chemical structures within the class of anesthetics can be explained by a common action of binding to hydrophobic domains and modulating dipole moments in the tubulin components of microtubules. This is supported by the fact that anesthetics inhibit prokaryotic motility, a function mediated by microtubules. It is of interest that recent studies using quantitative EEG suggest that anesthetics of various pharmacologic properties may act by interrupting cognitive binding.

17.1.6 Bi-stable Perception and Consciousness

Recall from Gestalt psychology (see, e.g. [HHB01]) that *bi-stable perception* arises whenever a stimulus can be thought in two different alternatives ways. The essential points are here on the terms 'be aware of ambiguity' and 'manifest state of only one percept at time'. For example consider the ambiguous Gestalt picture called *Rubin face*[8] given in Figure 17.3. A quantum–mechanical counterpart

[8] Rubin's vase, also known as the *figure–ground vase*) is a famous set of cognitive optical illusions developed around 1915 by the Danish Gestalt psychologist Edgar Rubin. His fundamental *principle of cognitive optical illusions* reads: When two fields have a common border, and one is seen as figure and the other as ground, the immediate perceptual experience is characterized by a shaping effect which emerges from the common border of the fields and which operates only on one field or operates more strongly on one than on the other (see [HHB01]). The illusion generally presents the viewer with a mental choice of two interpretations, each of which is valid. Often, the viewer sees only one of them, and only realizes the second, valid, interpretation after some time or prompting. When they attempt to simultaneously see the second and first interpretations, they suddenly cannot see the first interpretation anymore, and no matter how they try, they simply cannot encompass both interpretations simultaneously – one occludes the other. The illusions are useful because they are an excellent and intuitive demonstration of the figure–ground distinction the brain makes during visual perception. Rubin's figure–ground distinction, since it involved higher–level cognitive pattern matching, in which the overall picture determines its mental interpretation, rather than the net effect of the individual pieces, influenced the Gestalt psychologists, who discovered many similar illusions themselves. Normally the brain classifies images by what surrounds what, establishing depth and relationships. If something surrounds another thing, the surrounded object is seen as figure, and the presumably further away (and hence background) object is the ground, and vice versa. This makes sense, since if a piece of fruit is lying on the ground, one would want to pay attention to the 'figure' and not the 'ground'. However, when the contours are not so unequal, ambiguity starts to creep into the previously simple inequality, and the brain must begin 'shaping' what it sees; it can

of such mind processes was proposed in [Con08]. In this quantum model of mental states, the state of the potential consciousness is represented by a quantum state–vector in Hilbert space. If we indicate e.g., a 2D case with potential states $|1\rangle$ and $|2\rangle$, the potential state $|\psi\rangle$ of bi–stable perception is given by

$$|\psi\rangle = a|1\rangle + b|2\rangle,$$

with the same *perception energy* E given by the *Schrödinger equation*,

$$\mathrm{i}\partial_t|\psi\rangle = E|\psi\rangle.$$

Here a and b represent the respective probability amplitudes of the two stable perception states, so that $|a|^2$ gives the probability that the state of consciousness, represented by the percept $|1\rangle$ will be finally actualized or manifested during perception; conversely, $|b|^2$ represents the probability that state (percept) $|2\rangle$ of consciousness will be actualized or manifested during perception. Closely–related is *binocular rivalry model* of [Man07], with a wave-function between perceptual events describing a state of potential consciousness and *wave-function collapse* when perceptual event is manifested.

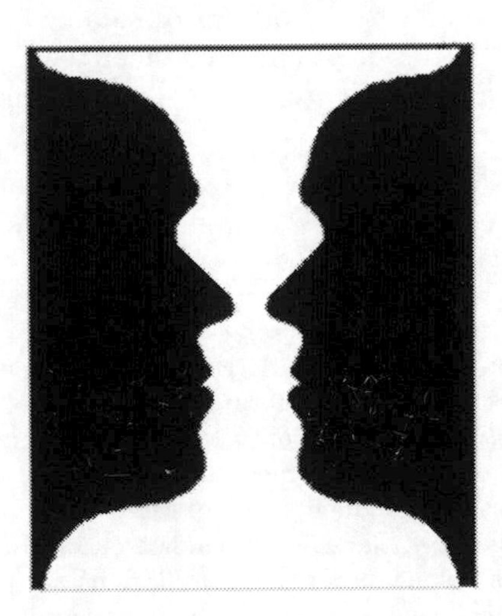

Fig. 17.3. Rubin face: ambiguous figure inducing two different representations, vase or faces. The subject is aware that different representations are possible, but he can perceive only one of the two possible percepts at any one time. The bi–stable perception is described as: $|\psi\rangle = a_{\mathrm{vase}}|1\rangle + b_{\mathrm{fase}}|2\rangle$. In manifest or actual state of consciousness the subject will select or the vase or the faces [Con08].

---

be shown that this shaping overrides and is at a higher level than feature recognition processes that pull together the face and the vase images – one can think of the lower levels putting together distinct regions of the picture (each region of which makes sense in isolation), but when the brain tries to make sense of it as a whole, contradictions ensue, and patterns must be discarded.

On the other hand, *consciousness*, which has traditionally been based on Descartes' statement "Cogito ergo sum," refers to our capacity of asking questions about the so-called reality (see [Pop08]). It concerns the field of 'potential information' instead of the so-called 'actual information'.[9] In order to describe its contents, a measure of potential information was suggested in terms of the capacity of actualities in the possible contents. This definition includes a variety of relevant consequences, i.e., the appearance of a memory, the question of time down to relation of molecular *biophotons*.[10] Consciousness works for evolutionary purposes which is not a matter of fact but a process. The improvement and the optimization of consciousness is at the same time the most important healing power of life. In other words, the communication system of biological systems is based on cavity resonator waves and light guides. This example shows the identity of mitotic figures with the force pattern of cavity resonator waves within the cells. These forces regulate the migration of the molecules in such a way that no error takes place. At the same time such a pattern represents in the average one coherent biophoton. They are the origin of consciousness.

Popp distinguishes four different 'global' forms of consciousness [Pop08]:

1. Self confidence according to the reference point of Descartes: "I am," resulting from the transformation of the physical existence of the body into 'doubts' and back to the control, and so on. The end of this process is the adjustment of actual and potential information.
2. Identification (awareness) according to the transformation of the actual information of the physical existence of matter or radiation into 'doubts' and back to the control, and so on. This leads after adjusting actual and potential information to the confirmation of "You are" or "It is."
3. Prediction: Repeated transformations of the actual information of the physical existence of different objects into the possibility field of the potential information under adjustment of the potential information of the memory which provide boundary conditions in the control processes. The transformations are again finished as soon as there is an equality of actual and potential information. The result is a statement: "It will be."

[9] The simplest system where actual and potential information are mutually transformed are *cavity resonators*. The potential information (in bit) is identical to the $Q-$value of the resonator. The actual information reflects the distribution of the coded sequence on the emitted wave. The actual information cannot exceed there the potential one. It corresponds to the maximum number of switch–on/off processes of a wave that is reflected N times within the resonator.

[10] Biophotons (BPHs) are weak photons within or emitted from living organisms. BPH emission originates from a de-localized coherent electromagnetic field within living organisms and is regulated by the field. Based on experimental results concerning Poisson and sub-Poisson distributions of photocount statistics, the coherent properties of BPHs and their functions in cell communication are described in [Cha08]. Functions of BPH roles are discussed in some important processes, including DNA replication, transcription, protein synthesis, and cell signalling, and in the processes of oxidative phosphorylation and photosynthesis.

4. Memory and Inspiration. Repeated transformation and re-transformation of potential information of the past and the presence. After adjustment of these probability distributions the statement will follow: "It could have been," or "It could become."

17.2 The Mental Quantum Leap: *INTENTION ⇛ ACTION ⇛ MANIFESTATION*

The Mental Quantum Leap represents the Mental–to–Physical transition leap: $INTENTION \Rrightarrow ACTION \Rrightarrow MANIFESTATION$, which is formally defined by the Feynman path integral:

$$\langle Manifestation|Action|Intention\rangle := \oint\!\!\!\!\!\sum \mathcal{D}[\Phi]\, \mathrm{e}^{\mathrm{i}S[\Phi]}. \tag{17.10}$$

On the left-hand side of this quantum-field path-integral there is Dirac's transition amplitude: $\langle Manifestation|Action|Intention\rangle$, whose absolute square $|\langle Manifestation|Action|Intention\rangle|^2$ is the overall probability for the transition leap: $INTENTION \Rrightarrow ACTION \Rrightarrow MANIFESTATION$. In other words, the absolute square $P = |\langle Manifestation|Action|Intention\rangle|^2$ is the transition probability P of occurring the final state of *Physical Manifestation* given the initial state of *Mental Intention*, via the action-at-a distance within a quantum bio-electro-magnetic field.

On the right-hand side of (23.73) is the celebrated *Feynman path integral* of modern quantum field theory. In the words of Stephen Hawking, it is the 'wave function of the Universe'. Technically, it is the complex-valued integral over the bio-electro-magnetic quantum field Φ. The Dirac's symbol $\oint\!\!\!\!\!\sum$ denotes the Lebesgue integration (with the field measure $\mathcal{D}[\Phi]$) over the continuous field spectrum and summation Σ over the spectrum of discrete quantum particles.

Finally, the key term in the field path integral (23.73) is the physical-field action $S[\Phi]$, which is dimensionally: $Effort = Energy \times Time$. Therefore:

Given the strong Focus = Mental Intention, followed by the strong physical Effort, we shall get the desired Physical Manifestation.

Now, let's elaborate a little-bit on this creative quantum–leap process. Historically, the first quantum particle-field theory was developed in 1949 by John A. Wheeler and Richard P. Feynman as their famous *action–at–a–distance electrodynamics* (which is explained in 'plain English' in Feynman's Nobel lecture). Here, we give a *mental interpretation* to the relativistic action–at–a–distance field theory (occurring within the Einstein space-time manifold), based on two (or more) *thought–particles*, moving with light-like speed, each in in its own time. The key to our quantum leap is the *Mental Focus*, which we define as the *synchronization* of thoughts.

Formally (using Einstein's summation convention) the action–at–a–distance $S[\Phi] = S[x, y; t_i, t_j]$ reads:

$$
\begin{aligned}
S[x,y;t_i,t_j] &= \frac{1}{2}\int_{t_i}\int_{t_j} \delta(I_{ij}^2)\; \dot{x}^i(t_i)\,\dot{y}^j(t_j)\; dt_i dt_j + \frac{1}{2}\int_t g_{ij}\,\dot{x}^i(t)\dot{x}^j(t)\,dt \\
&\text{with} \qquad I_{ij}^2 = \left[x^i(t_i) - y^j(t_j)\right]^2, \qquad (17.11) \\
&Intention \leq t_i, t_j, t \leq Manifestation.
\end{aligned}
$$

Here $\dot{x}^i(t_i) = dx^i/dt_i$ and $\dot{y}^j(t_j) = dy^j/dt_j$ are the space-time velocities of two thought-particles $x^i(t_i)$ and $y^j(t_j)$, moving in their own times (t_i) and (t_j), respectively. The first term in (17.11) represents the *mental potential energy* of *Intention*, formalized as an interaction between the two elementary particles. It is a double integral over a delta function of the square of interval I^2 between the two thought–particles; thus, interaction occurs only when this interval, representing the cognitive distance between the two thoughts, vanishes; that is, when *mental synchronization*, or *mental focus*, occurs. Therefore, the subsequent physical action is efficient only after the mental focus. It is given by the second term in (17.11), which represents the *physical kinetic energy*:

$$
T = \frac{1}{2} g_{ij}\,\dot{x}^i \dot{x}^j,
$$

generated by the Riemannian *metric tensor* g_{ij} of the Einstein space-time manifold (see [II06c, II07e]).

18

The Model for a Future Tennis Champion

The tennis champion of the future is Roger Federer's hypothetical younger brother, who knows and can do everything that Roger knows and can do. In addition, he is both physically and mentally stronger and faster. Because of this addition, he will generally be much more efficient in the future tennis game, which will be much faster than today's tennis, due to improved racquet technology and players' improved psychophysical abilities.

18.1 Superb Tennis Weapons

The *superb tennis weapons* of the future champion will demonstrate their general efficiency in all game situations:

1. An aggressive 300+ km/h serve, targeting either the "T"–spot, or the weaker side of the opponent; the second serve is the same as the first 300+ km/h serve;
2. An aggressive, passing return, based on the 'mind/body–reading' *anticipation* and lightning–fast reaction (based on the stretch–reflex)
3. An aggressive, penetrating 240+ km/h forehand and single–handed 200+ km/h backhand (or, a two-handed 220+ km/h backhand), hit consistently from any body configuration and from any position on the court;
4. In addition, a serve–volley game can add-up to the general aggressive attitude.

A quick and accurate decision–making is crucial: both for the serve and for the aggressive return, as well as for the most appropriate shot selection in any other game situation. It is based on efficiently reading the ball and accurately predicting the opponent's next move. In two words, this is what we call the *mental speed*. It will be trained (for years) by means of blitz and lightning–chess.

Besides the superb tennis weapons, the most obvious characteristic of the future champion will be *strong concentration*, consistent throughout the whole

T.T. Ivancevic et al.: Paradigm Shift for Future Tennis, COSMOS 12, pp. 111–114.
springerlink.com

match/tournament. In two words, this is what we call the *mental strength*. It will be trained (for years) by means of *visualization* exercises.

In general, *high tennis efficiency* means: "When the opportunity comes – *finish* the point, or game, or set, or match!"

High ability of *learning on the spot* means adjusting/changing both technique and tactics in the most appropriate way according to the situation on the court. (For example, if playing against a baseline player – being able to come to the net and efficiently finish the point with the winning volley.)

The ten main characteristics of future tennis champion are:

- Natural talent;
- Commitment (to regular and consistent training/competition);
- Passion (desire to train, compete and win);
- Self–determination (responsibility for their performance);
- Self–belief (I can do it, self–confidence in their own capacity);
- Planned approach (clear goals; Federer himself admits that "scheduling has been a very important factor in my success – it helps me to heal from injuries and mentally get away");
- Quick adaptation (to the changing situation on the court);
- Mind–body balance;
- Competitive toughness (like Federer, Nadal, Henin); and
- Perseverance (persistence);

The three most prominent fitness characteristics of future tennis champion are the three 'A's:

1. *Acceleration* (fast sprinting up to 15 m, from any starting body configuration);
2. *Agility* (quick changing of directions in both sprinting and jumping); and
3. *Athleticism*, which has two components:
   (i) ability to throw the current (reconstructed) official javelin 60 m;[1] and
   (ii) ability to perform a somersault on a court.[2]

Optimal body height (posture) for the future male tennis champion will be around 1.85–1.87 m, (like Federer, Nadal, Roddick, Djokovic), while his optimal body–weight will be 80 kg (like Federer and Djokovic). Optimal age for a male champion will be 22–27 (+/-2) years old.

If we would like to make a hypothetical future champion out of the current players, then it would be a combination of Roger Federer with Andy Roddick's power–serve, Rafael Nadal's agility and Gael Monfils' power–forehand.

[1] This means doing several years of parallel javelin training during the teenage years; this would give a 300+ km/h serve.

[2] We do not suggest that the future tennis players will actually have to perform somersaults on the court. Rather, any gymnastics–trained kid that can perform a somersault on the street with a safe landing on his feet obviously have all the abilities commonly covered by the umbrella of *athleticism: superb jump take–off, superb coordination and superb balance.* Every elite tennis player can have this true athleticism if in his youth he had several years of parallel gymnastics training.

In addition, every champion is a perfectionist: "If you can't do it properly, don't do it at all."

18.2 Mind of a Champion

At the beginning we claimed that Roger Federer has the brain of the sports champion. What does it really mean? In this section we will answer to this question, using as an example Michael Schumacher,[3] the greatest driver in the car-racing history. How can someone win 91 Grand Prix races in a single lifetime? What is the secret behind the greatest F1 legend?

18.2.1 Nine Essential Characteristic of a Champion's Brain

Michael Schumacher was one of the brightest stars in sport history. He had the ultimate brain of a Champion. There are nine essential characteristics of the brain of a Champion in every complex, speed-characterized sport. These are as follows:

1. Will power The first mark of a World Champion in any sport is the indomitable will power. This ability must be extremely well developed as it was in the extremely rare case of seven-time World Champion, Michael Schumacher.
2. Focus/Concentration From the indomitable will power follows the sharpest focus. This is an essential ability of every Champion, and is a capability which allows one to fully concentrate on a single thing for several hours, as if nothing else exists. Schumi had the sharpest focus in the history of F1 racing.
3. Planning Schumacher's character made him an extremely thorough planner of every race. This was clear in his 19/94/95 seasons, while driving for Benetton. This was even more powerfully demonstrated in his "golden period," 20/00/04, with his unbeatable Ferrari team. They collectively planned and carefully executed every single detail before and during the race.
4. RiskTaking Certain amount of risk-taking is necessary for high achievements in any field. For example, gymnastic rules award 0.20 of a mark for high risk, which means that a perfectly executed exercise is not enough for a perfect mark. High risk is also required. Schumi was well-known for his high-level of risk-taking (which even got him disqualified from a Championship in 1997, after collision with his rival Jacques Villeneuve, who after that won the Championship).

[3] German car-racing driver Michael Schumacher, popularly called Schumi (or, Schuey), was a record seven-time Formula One World Champion (1994, 1995, 2000, 2001, 2002, 2003, 2004) with a record of 91 Grand Prix races won. He has been officially declared "statistically the greatest driver" in the history of Formula One (or, F1) racing. Besides winning drivers' Championships and race victories, Schumi also holds many other records in F1 racing, including: fastest laps, pole positions, points scored (1,369) and most races won in a single season.

5. Situation Awareness The modern buzzword, "situation awareness", has been defined (by M. Endsley) as "The perception of the elements in the environment within time and space, the comprehension of their meaning, and projection of their status in the near future." It is necessary for optimal decision-making. Situation awareness is one of the most important conditions for safety in everyday car driving; even more so, in F1 racing. This was one of Michael Schumacher's superior abilities as the greatest driver ever.

6. Decision-Making Decision-making is the "sine-qua-non" for winning in any complex game/situation. In the past, it was evident in soccer, when a single movement of Pele or Maradona decided the whole match and ultimately the world Championship. Today, similar decision-making capability can be seen in tennis in the most important winning shots from Roger Federer. The same ability was also one of Schumi's winning qualities, namely the ability to decide when and where to push the car to its maximum, as it is impossible for the car to be pushed to its maximum all the time.

7. Anticipation of Opponents' Moves One of most dominant characteristics of a master in any competitive field is their ability to read the mind of an opponent. Like Roger Federer in tennis, Schumacher also had this crucial ability to sense, predict, or anticipate when and were his rivals will attack. This enabled him to prepare and execute an effective counter-attack which helped him on more than one occasion.

8. Reflexes In all speed-dominated sports, the Champion must have lightning-fast reflexes. As F1 is probably the fastest sport, clearly Schumi was no exception to this rule. From his early youth in cart-driving, to the end of his Champion career, he had lightning-fast reflexes. When combined with his superior anticipation of his opponents' moves, this gave him the fastest and most effective complex reaction.

9. Complete SituationControl As a result of the above 8 characteristics, follows complete situation control. Schumi had the ability to control the whole race, from start to finish. This was most clearly demonstrated during his "golden period," 200004, when he won more races and Championships than any other driver in F1 history.

In summary, Michael Schumacher had the strongest will power, the most thorough planning, the sharpest focus/concentration, the highest risk-taking, the most complete situation awareness, the quickest and most appropriate decision making, the most accurate anticipation, and the fastest reflexes. As a result, he had the highest rate of complete situation-control (see Appendix for technical details), which enabled him to win 91 F1 Grand-Prix races and seven World Championships.

19

Tennis Evolution: An Intensive 12 Week Programme

Here we present an intensive *12–week programme* for quick development of the basic tennis skills.

19.1 Beginners Group

General recommendations:

Every lesson lasts 60 minutes.

Every lesson starts with a 5 min warm-up (skipping or various types of running).

Every lesson finishes with a 5-min cool-down (stretching).

Athletics on Thursdays includes sprints, long jumps, and discus throws. It is followed by javelin throws.

Gymnastics on Fridays includes: monkey bars, forward/backward rolls and assisted somersaults.

50 min 5Km running on Sundays means completion of 5km running under 50 minutes.

Interval swimming on Sundays includes 12.5 m sprints with strong breathing for recovery.

20-Min. Chess on Thursdays and Saturdays means 20 minutes of Blitz Chess (or Bullet Chess) with a clock, 30 sec per move.

19.2 Intermediate Group

General recommendations:

Every lesson lasts 60 minutes.

Every lesson starts with a 5 min warm-up (skipping or various types of running).

Every lesson finishes with a 5-min cool-down (stretching).

20-Min. Chess on Thursdays means 20 minutes of Blitz Chess (or Bullet Chess) with a clock, 15 sec per move.

T.T. Ivancevic et al.: Paradigm Shift for Future Tennis, COSMOS 12, pp. 115–121.
springerlink.com

Monday	Tuesday	Wednesday	Thursday	Friday	Saturday	Sunday
Day 1 Serve Drill, Forehand Drill, Backhand Drill	Day 2 50-Minute Athletics, Javelin Throw	Day 3 Serve Drill, Forehand Drill, Backhand Drill	Day 4 30-Minute Playing Points 20-Min. Chess	Day 5 50-Minute Gymnastics	Day 6 30-Minute Playing Points 20-Min. Chess	Day 7 50-Minute 5Km Running
Day 8 Serve Drill, Forehand Drill, Backhand Drill	Day 9 50-Minute Athletics, Javelin Throw	Day 10 Serve Drill, Forehand Drill, Backhand Drill	Day 11 30-Minute Playing Points 20-Min. Chess	Day 12 50-Minute Gymnastics	Day 13 30-Minute Playing Points 20-Min. Chess	Day 14 50-Minute Interval Swimming
Day 15 Serve Drill, Forehand Drill, Backhand Drill	Day 16 50-Minute Athletics, Javelin Throw	Day 17 Serve Drill, Forehand Drill, Backhand Drill	Day 18 30-Minute Playing Points 20-Min. Chess	Day 19 50-Minute Gymnastics	Day 20 30-Minute Playing Points 20-Min. Chess	Day 21 50-Minute 5Km Running
Day 22 Serve Drill, Forehand Drill, Backhand Drill	Day 23 50-Minute Athletics, Javelin Throw	Day 24 Serve Drill, Forehand Drill, Backhand Drill	Day 25 30-Minute Playing Points 20-Min. Chess	Day 26 50-Minute Gymnastics	Day 27 30-Minute Playing Points 20-Min. Chess	Day 28 50-Minute Interval Swimming
Day 29 Serve Drill, Forehand Drill, Backhand Drill	Day 30 50-Minute Athletics, Javelin Throw	Day 31 Serve Drill, Forehand Drill, Backhand Drill	Day 32 30-Minute Playing Points 20-Min. Chess	Day 33 50-Minute Gymnastics	Day 34 30-Minute Playing Points 20-Min. Chess	Day 35 50-Minute 5Km Running
Day 36 Serve Drill, Forehand Drill, Backhand Drill	Day 37 50-Minute Athletics, Javelin Throw	Day 38 Serve Drill, Forehand Drill, Backhand Drill	Day 39 30-Minute Playing Points 20-Min. Chess	Day 40 50-Minute Gymnastics	Day 41 30-Minute Playing Points 20-Min. Chess	Day 42 50-Minute Interval Swimming
Day 43 Serve Drill, Forehand Drill, Backhand Drill	Day 44 50-Minute Athletics, Javelin Throw	Day 45 Serve Drill, Forehand Drill, Backhand Drill	Day 46 30-Minute Playing Points 20-Min. Chess	Day 47 50-Minute Gymnastics	Day 48 30-Minute Playing Points 20-Min. Chess	Day 49 50-Minute 5Km Running
Day 50 Serve Drill, Forehand Drill, Backhand Drill	Day 51 50-Minute Athletics, Javelin Throw	Day 52 Serve Drill, Forehand Drill, Backhand Drill	Day 53 30-Minute Playing Points 20-Min. Chess	Day 54 50-Minute Gymnastics	Day 55 30-Minute Playing Points 20-Min. Chess	Day 56 50-Minute Interval Swimming
Day 57 Serve Drill, Forehand Drill, Backhand Drill	Day 58 50-Minute Athletics, Javelin Throw	Day 59 Serve Drill, Forehand Drill, Backhand Drill	Day 60 30-Minute Playing Points 20-Min. Chess	Day 61 50-Minute Gymnastics	Day 62 30-Minute Playing Points 20-Min. Chess	Day 63 50-Minute 5Km Running
Day 64 Serve Drill, Forehand Drill, Backhand Drill	Day 65 50-Minute Athletics, Javelin Throw	Day 66 Serve Drill, Forehand Drill, Backhand Drill	Day 67 30-Minute Playing Points 20-Min. Chess	Day 68 50-Minute Gymnastics	Day 69 30-Minute Playing Points 20-Min. Chess	Day 70 50-Minute Interval Swimming
Day 71 Serve Drill, Forehand Drill, Backhand Drill	Day 72 50-Minute Athletics, Javelin Throw	Day 73 Serve Drill, Forehand Drill, Backhand Drill	Day 74 30-Minute Playing Points 20-Min. Chess	Day 75 50-Minute Gymnastics	Day 76 30-Minute Playing Points 20-Min. Chess	Day 77 50-Minute 5Km Running
Day 78 Serve Drill, Forehand Drill, Backhand Drill	Day 79 50-Minute Athletics, Javelin Throw	Day 80 Serve Drill, Forehand Drill, Backhand Drill	Day 81 30-Minute Playing Points 20-Min. Chess	Day 82 50-Minute Gymnastics	Day 83 30-Minute Playing Points 20-Min. Chess	Day 84 50-Minute Interval Swimming

Fig. 19.1. 12 Week Programme – Beginners Group.

19.3 Advanced Group

General recommendations:

Every lesson lasts 60 minutes.

Every lesson starts with a 5 min warm-up (skipping or various types of running).

Every lesson finishes with a 5-min cool-down (stretching).

Monday	Tuesday	Wednesday	Thursday	Friday	Saturday	Sunday
Day 1 Serve Drill, Forehand Drill, Backhand Drill	**Day 2** Volley/Smash Drill, Playing Points	**Day 3** Serve Drill, Forehand Drill, Backhand Drill	**Day 4** 30-Minute Playing Doubles 20-Min. Chess	**Day 5** Serve Drill, Forehand Drill, Backhand Drill	**Day 6** 50-Minute Playing Singles	**Day 7** 50-Minute 8Km Running
Day 8 Serve Drill, Forehand Drill, Backhand Drill	**Day 9** Volley/Smash Drill, Playing Points	**Day 10** Serve Drill, Forehand Drill, Backhand Drill	**Day 11** 30-Minute Playing Doubles 20-Min. Chess	**Day 12** Serve Drill, Forehand Drill, Backhand Drill	**Day 13** 50-Minute Playing Singles	**Day 14** 50-Minute Interval Swimming
Day 15 Serve Drill, Forehand Drill, Backhand Drill	**Day 16** Volley/Smash Drill, Playing Points	**Day 17** Serve Drill, Forehand Drill, Backhand Drill	**Day 18** 30-Minute Playing Doubles 20-Min. Chess	**Day 19** Serve Drill, Forehand Drill, Backhand Drill	**Day 20** 50-Minute Playing Singles	**Day 21** 50-Minute 8Km Running
Day 22 Serve Drill, Forehand Drill, Backhand Drill	**Day 23** Volley/Smash Drill, Playing Points	**Day 24** Serve Drill, Forehand Drill, Backhand Drill	**Day 25** 30-Minute Playing Doubles 20-Min. Chess	**Day 26** Serve Drill, Forehand Drill, Backhand Drill	**Day 27** 50-Minute Playing Singles	**Day 28** 50-Minute Interval Swimming
Day 29 Serve Drill, Forehand Drill, Backhand Drill	**Day 30** Volley/Smash Drill, Playing Points	**Day 31** Serve Drill, Forehand Drill, Backhand Drill	**Day 32** 30-Minute Playing Doubles 20-Min. Chess	**Day 33** Serve Drill, Forehand Drill, Backhand Drill	**Day 34** 50-Minute Playing Singles	**Day 35** 50-Minute 8Km Running
Day 36 Serve Drill, Forehand Drill, Backhand Drill	**Day 37** Volley/Smash Drill, Playing Points	**Day 38** Serve Drill, Forehand Drill, Backhand Drill	**Day 39** 30-Minute Playing Doubles 20-Min. Chess	**Day 40** Serve Drill, Forehand Drill, Backhand Drill	**Day 41** 50-Minute Playing Singles	**Day 42** 50-Minute Interval Swimming
Day 43 Serve Drill, Forehand Drill, Backhand Drill	**Day 44** Volley/Smash Drill, Playing Points	**Day 45** Serve Drill, Forehand Drill, Backhand Drill	**Day 46** 30-Minute Playing Doubles 20-Min. Chess	**Day 47** Serve Drill, Forehand Drill, Backhand Drill	**Day 48** 50-Minute Playing Singles	**Day 49** 50-Minute 8Km Running
Day 50 Serve Drill, Forehand Drill, Backhand Drill	**Day 51** Volley/Smash Drill, Playing Points	**Day 52** Serve Drill, Forehand Drill, Backhand Drill	**Day 53** 30-Minute Playing Doubles 20-Min. Chess	**Day 54** Serve Drill, Forehand Drill, Backhand Drill	**Day 55** 50-Minute Playing Singles	**Day 56** 50-Minute Interval Swimming
Day 57 Serve Drill, Forehand Drill, Backhand Drill	**Day 58** Volley/Smash Drill, Playing Points	**Day 59** Serve Drill, Forehand Drill, Backhand Drill	**Day 60** 30-Minute Playing Doubles 20-Min. Chess	**Day 61** Serve Drill, Forehand Drill, Backhand Drill	**Day 62** 50-Minute Playing Singles	**Day 63** 50-Minute 8Km Running
Day 64 Serve Drill, Forehand Drill, Backhand Drill	**Day 65** Volley/Smash Drill, Playing Points	**Day 66** Serve Drill, Forehand Drill, Backhand Drill	**Day 67** 30-Minute Playing Doubles 20-Min. Chess	**Day 68** Serve Drill, Forehand Drill, Backhand Drill	**Day 69** 50-Minute Playing Singles	**Day 70** 50-Minute Interval Swimming
Day 71 Serve Drill, Forehand Drill, Backhand Drill	**Day 72** Volley/Smash Drill, Playing Points	**Day 73** Serve Drill, Forehand Drill, Backhand Drill	**Day 74** 30-Minute Playing Doubles 20-Min. Chess	**Day 75** Serve Drill, Forehand Drill, Backhand Drill	**Day 76** 50-Minute Playing Singles	**Day 77** 50-Minute 8Km Running
Day 78 Serve Drill, Forehand Drill, Backhand Drill	**Day 79** Volley/Smash Drill, Playing Points	**Day 80** Serve Drill, Forehand Drill, Backhand Drill	**Day 81** 30-Minute Playing Doubles 20-Min. Chess	**Day 82** Serve Drill, Forehand Drill, Backhand Drill	**Day 83** 50-Minute Playing Singles	**Day 84** 50-Minute Interval Swimming

Fig. 19.2. 12 Week Programme – Intermediate Group.

While performing serve, forehand and backhand drills, focus on a stretch-reflex based whipping movements (kinetic chains).

Once a week perform measuring the ball speed in serve, forehand and backhand using a police radar-gun.

20-Min. Chess on Thursdays means 20 minutes of Blitz Chess (or Bullet Chess) with a clock, 15 sec per move.

Monday	Tuesday	Wednesday	Thursday	Friday	Saturday	Sunday
Day 1 ________ Serve Drill, Forehand Drill, Backhand Drill	Day 2 ________ Volley/Smash Drill, Playing Points	Day 3 ________ Serve Drill, Forehand Drill, Backhand Drill	Day 4 ________ 30-Minute Playing Doubles 20-Min. Chess	Day 5 ________ Serve Drill, Forehand Drill, Backhand Drill	Day 6 ________ 50-Minute Playing Singles	Day 7 ________ 50-Minute 10Km Running
Day 8 ________ Serve Drill, Forehand Drill, Backhand Drill	Day 9 ________ Volley/Smash Drill, Playing Points	Day 10 ________ Serve Drill, Forehand Drill, Backhand Drill	Day 11 ________ 30-Minute Playing Doubles 20-Min. Chess	Day 12 ________ Serve Drill, Forehand Drill, Backhand Drill	Day 13 ________ 50-Minute Playing Singles	Day 14 ________ 50-Minute Interval Swimming
Day 15 ________ Serve Drill, Forehand Drill, Backhand Drill	Day 16 ________ Volley/Smash Drill, Playing Points	Day 17 ________ Serve Drill, Forehand Drill, Backhand Drill	Day 18 ________ 30-Minute Playing Doubles 20-Min. Chess	Day 19 ________ Serve Drill, Forehand Drill, Backhand Drill	Day 20 ________ 50-Minute Playing Singles	Day 21 ________ 50-Minute 10Km Running
Day 22 ________ Serve Drill, Forehand Drill, Backhand Drill	Day 23 ________ Volley/Smash Drill, Playing Points	Day 24 ________ Serve Drill, Forehand Drill, Backhand Drill	Day 25 ________ 30-Minute Playing Doubles 20-Min. Chess	Day 26 ________ Serve Drill, Forehand Drill, Backhand Drill	Day 27 ________ 50-Minute Playing Singles	Day 28 ________ 50-Minute Interval Swimming
Day 29 ________ Serve Drill, Forehand Drill, Backhand Drill	Day 30 ________ Volley/Smash Drill, Playing Points	Day 31 ________ Serve Drill, Forehand Drill, Backhand Drill	Day 32 ________ 30-Minute Playing Doubles 20-Min. Chess	Day 33 ________ Serve Drill, Forehand Drill, Backhand Drill	Day 34 ________ 50-Minute Playing Singles	Day 35 ________ 50-Minute 10Km Running
Day 36 ________ Serve Drill, Forehand Drill, Backhand Drill	Day 37 ________ Volley/Smash Drill, Playing Points	Day 38 ________ Serve Drill, Forehand Drill, Backhand Drill	Day 39 ________ 30-Minute Playing Doubles 20-Min. Chess	Day 40 ________ Serve Drill, Forehand Drill, Backhand Drill	Day 41 ________ 50-Minute Playing Singles	Day 42 ________ 50-Minute Interval Swimming
Day 43 ________ Serve Drill, Forehand Drill, Backhand Drill	Day 44 ________ Volley/Smash Drill, Playing Points	Day 45 ________ Serve Drill, Forehand Drill, Backhand Drill	Day 46 ________ 30-Minute Playing Doubles 20-Min. Chess	Day 47 ________ Serve Drill, Forehand Drill, Backhand Drill	Day 48 ________ 50-Minute Playing Singles	Day 49 ________ 50-Minute 10Km Running
Day 50 ________ Serve Drill, Forehand Drill, Backhand Drill	Day 51 ________ Volley/Smash Drill, Playing Points	Day 52 ________ Serve Drill, Forehand Drill, Backhand Drill	Day 53 ________ 30-Minute Playing Doubles 20-Min. Chess	Day 54 ________ Serve Drill, Forehand Drill, Backhand Drill	Day 55 ________ 50-Minute Playing Singles	Day 56 ________ 50-Minute Interval Swimming
Day 57 ________ Serve Drill, Forehand Drill, Backhand Drill	Day 58 ________ Volley/Smash Drill, Playing Points	Day 59 ________ Serve Drill, Forehand Drill, Backhand Drill	Day 60 ________ 30-Minute Playing Doubles 20-Min. Chess	Day 61 ________ Serve Drill, Forehand Drill, Backhand Drill	Day 62 ________ 50-Minute Playing Singles	Day 63 ________ 50-Minute 10Km Running
Day 64 ________ Serve Drill, Forehand Drill, Backhand Drill	Day 65 ________ Volley/Smash Drill, Playing Points	Day 66 ________ Serve Drill, Forehand Drill, Backhand Drill	Day 67 ________ 30-Minute Playing Doubles 20-Min. Chess	Day 68 ________ Serve Drill, Forehand Drill, Backhand Drill	Day 69 ________ 50-Minute Playing Singles	Day 70 ________ 50-Minute Interval Swimming
Day 71 ________ Serve Drill, Forehand Drill, Backhand Drill	Day 72 ________ Volley/Smash Drill, Playing Points	Day 73 ________ Serve Drill, Forehand Drill, Backhand Drill	Day 74 ________ 30-Minute Playing Doubles 20-Min. Chess	Day 75 ________ Serve Drill, Forehand Drill, Backhand Drill	Day 76 ________ 50-Minute Playing Singles	Day 77 ________ 50-Minute 10Km Running
Day 78 ________ Serve Drill, Forehand Drill, Backhand Drill	Day 79 ________ Volley/Smash Drill, Playing Points	Day 80 ________ Serve Drill, Forehand Drill, Backhand Drill	Day 81 ________ 30-Minute Playing Doubles 20-Min. Chess	Day 82 ________ Serve Drill, Forehand Drill, Backhand Drill	Day 83 ________ 50-Minute Playing Singles	Day 84 ________ 50-Minute Interval Swimming

Fig. 19.3. 12 Week Programme – Intermediate Group.

Part II

Modern Background for Tennis Science

20

Modern Tennis Physiology

20.1 Muscular System

20.1.1 Muscular Histology

Human skeletal and face muscles, accounting for more than 40% of the body weight in man, consist of bundles of elongated, cylindric cells called *muscle fibers*, 50 to 200 μ in diameter and often many centimeters long. Bundles of muscle fibers, each called *fasciculus*, are surrounded by a connective tissue covering, the *endomysium* (see, e.g., [Mou80, Mar98]).

A muscle consists of a number of fasciculi encased in a thick outer layer of connective tissue, the *perimysium*. At both ends of a muscle the connective tissue melds into a tendon by which the muscle is attached to the face or bony skeleton. In some muscles (*fusiform*), the muscle fibers run the whole length of muscle between the tendons, which form at opposite ends. In most muscles (*pennate*), one of the tendons penetrates through the center of the muscle; muscle fibers run at an angle to the axis of the whole muscle from the central tendon to the perimysium.

Like other cells, muscle cells are surrounded by a cell membrane, the *sarcolemma*. *Myofibrils*, the *contractile elements*, are numerous parallel, lengthwise threads 1 to 3 in diameter that fill most of the muscle fiber. The *cross striations*, seen in the skeletal and face muscles with electron microscope, are located in the myofibrils. Squeezed between the myofibrils and the sarcolemma is a small amount of cytoplasm, the *sarcoplasm*, in which are suspended multiple nuclei, numerous mitochondria, lysosomes, lipid droplets, glycogen granules, and other intracellular inclusions. The sarcoplasm contains glycogen, glycolytic enzymes, nucleotides, creatine phosphate, amino acids, and peptides.

Sarcoplasm also contains a well–developed endoplasmic reticulum, which in muscle is called *sarcoplasmic reticulum*. The sarcoplasmic reticulum forms an extensive hollow membranous system within the cytoplasm surrounding the myofibrils. Periodically, there are branching invaginations of the sarcolemma called *T tubules* or transverse tubules. The sarcoplasmic reticulum bulges out on either side of the T tubules to form large *lateral cisternae*. The T tubule and two sets of

T.T. Ivancevic et al.: Paradigm Shift for Future Tennis, COSMOS 12, pp. 121–182.
springerlink.com

lateral cisternae constitute a *triad*. The triads play an important role in muscle excitation-contraction coupling (by release of Ca^{++} ions).

Two types of muscle fibers are found in human skeletal and face muscles: *red* and *white muscle fibers*, being histochemically and functionally distinctive. Many muscles are mixed, containing both types of fibers, which can be distinguished by various histochemical stains. In addition to muscle cells and fibroblasts in the connective tissue, a whole muscle contains fat cells and histiocytes.

Each muscle fiber contains numerous contractile elements - *myofibrils* ($1-3\,\mu$ in diameter) which are biological machines that utilize chemical energy from metabolism of food in the form of *adenosine triphosphate*, ATP hydrolysis to produce mechanical work. An understanding of contractility and muscle function requires, thus, both histo–mechanical and bio–energetic insight.

Contractile machinery unit of the myofibril, *sarcomere* ($1.5-3.5\,\mu$ long; on electron microscope it is seen as bounded by two Z lines, with H zone in the middle of the A band) is constituted of a great number of longitudinal protein filaments of two kinds: thick, *myosin* filaments (about $120\,\dot{A}$ in diameter and about $1.8\,\mu$ long; they are located in the center of the sarcomere arranged in a hexagonal array about $450\,\dot{A}$ apart) and thin, *actin* filaments (about $80\,\dot{A}$ in diameter and about $1.0\,\mu$ long; they are anchored into the transverse filaments forming the Z line) (see Figure 20.1). Each myosin filament is surrounded by six actin filaments. Each myosin filament has two heads and two projections from opposite sides at about $143\,\dot{A}$ intervals along its length.

20.1.2 Classical Theories of Muscular Contraction

Huxley's Sliding Filament Theory

Essential for the contraction process are *cross bridges* (see Figure 20.1). They extend from myosin filaments to touch one of the adjacent actin filaments. Each thin filament receives cross bridges from the three adjacent thick filaments. During shortening the two sets of interdigitating filaments slide with respect to each other, cross and finally overlap each other. This process of muscle shortening involving progressive interdigitation of the two sets of protein filaments represents the *sliding filament mechanism*, discovered and mathematically formulated as a *microscopic theory of muscular contraction* in 1954–57 by A.F. Huxley [HN54, Hux57].

According to Huxley, the myosin heads and cross bridges are elastic elements with a mechanism for attaching themselves transiently to specific sites on the thin filaments. The following cyclic events take place during muscular contraction:

1. The cross bridges extend from myosin filaments and attach themselves to specific sites on actin filaments. The probability that attachment will occur is $f(x)$, where x is the instantaneous distance between the equilibrium position (0) and the maximum distance for attachment h along the myofibrillar axis.
2. The cross bridges detach with probability $g(x)$.

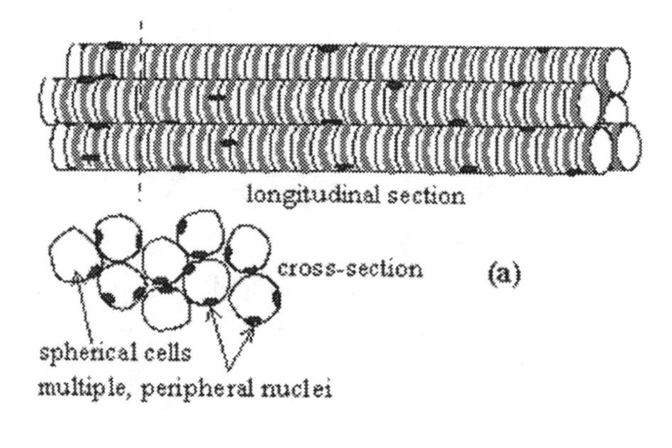

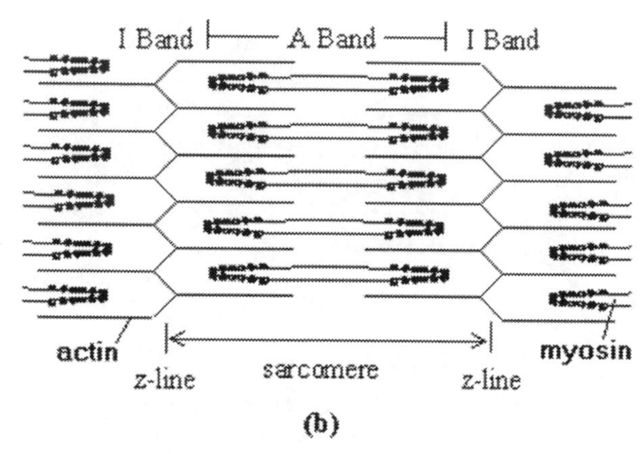

Fig. 20.1. Cellular structure of the voluntary (skeletal) human muscle: (a) Muscular fibers with their cross–sections; (b) Sarcomere with overlapping myofilaments.

If we let N equal the density of cross bridges and n the fraction of cross bridges that are attached, then nN equals the density of attached cross bridges. Huxley's rate equation for cross–bridge attachment–detachment, i.e. the *sliding filament model* of muscular contraction is now given by:

$$\dot{n} = f(x)[1 - n(x,t)] - g(x)n(x,t) = f(x) - [f(x) + g(x)]n(x,t). \qquad (20.1)$$

Huxley's model (20.1) leads to expressions for the force developed by the cross bridges. For an *isometric steady–state contraction* the *contraction tension* or *contraction force* is given by:

$$F_0 = 0.5\, N\, h^2 \frac{kf}{f+g}, \qquad (20.2)$$

where $k = k(x)$ is the stiffness of the cross–bridge spring. For *isotonic* steady states it recovers the classical *Hill's force–velocity* relation (20.3). The *static force* expression says that the force (or tension) generated in the muscle is the function of the interfilamentar overlap, and its maximum is about the middle of the shortening, where the acto–myosin overlap is maximal. This is the so–called *parabolic length–tension curve* of muscular contraction.

Hill's Force–Velocity Muscular Dynamics

The *dynamic force–velocity relation* of muscular contraction is firstly discovered in 1938, by A.V. Hill [Hil38], in his thermodynamic studies of muscular work, and put into the basis of *macroscopic muscle–load dynamics*. Hill's famous *hyperbolic force–velocity curve* has the equation:

$$(F + a)\,v = (F_0 + F)\,b, \tag{20.3}$$

and says that the muscle force is greatest in isometric conditions (without motion), while the velocity of shortening is maximal without external load; in other words, muscle is either 'strong' or 'fast', but no both. Constants a and b correspond respectively to the energy dissipated during the contraction and the velocity of the mechano–chemical processes.

Hill showed that energy change in muscle during contraction can be described by the following *thermodynamic relation*:

$$U = A + W + M, \tag{20.4}$$

where U is the total energy change associated with contraction, A is the *activation heat* (i.e., the heat production associated with the activation of the contractile elements), W is the mechanical work performed by the muscle by lifting a load, $\alpha \Delta x$ is the *shortening heat*, and M is the *maintenance heat* of contraction.

The activation heat begins and is almost completely liberated before any tension is developed, i.e. it is predominantly connected with the excitation–contraction coupling process, and corresponds in time to the *latency relaxation* of muscle. It is associated with the internal work required to transform the contractile elements from the resting to the active state. Part of the activation heat probably is associated with a change in the elastic properties of muscle, but about two thirds of it is associated with the release of Ca^{++} ions from the triads, its binding by troponin and the subsequent rearrangement of the thin filament proteins. The activation heat is greatest for the first twitch after a period rest and becomes smaller with succeeding twitches.

The maintenance heat begins at about the time tension begins and can be divided into two parts: the labile maintenance heat and the stable maintenance heat. For isometric contractions at shorter than rest length, both the labile and the stable heats diminish. For stretched muscle, the labile heat is approximately constant, whereas the stable heat diminishes with stretching and is roughly proportional to the degree of interfilamentar overlap. The stable heat has quite different values in functionally different muscles; it is law when the muscle maintains tension efficiently and vice versa.

The shortening heat is proportional mainly to the distance of shortening and does not depend greatly on the load, the speed of shortening, or the amount of work performed. Since mechanical work is $W = P\Delta x$, substituting this in the above thermodynamic relation (20.4) gives the *heat equation*:

$$U = A + (P + \alpha)\,\Delta x + M. \tag{20.5}$$

From the analogy of the term $(P + \alpha)$ in the heat equation (20.5) and the term $(P + a)$ in the force-velocity equation (20.3), Hill was able to show a rough equivalence between the coefficient of the shortening heat α and the force-velocity constant a. The shortening heat is greatest for the first twitch after a period of rest and is less for subsequent twitches.

Last, note should be made of *thermoelastic heat*. Generally speaking, resting muscle has rubberlike thermoelastic properties, whereas actively contracting muscle has springlike thermoelastic properties. During the development of tension the change in elastic properties is accompanied by an absorption of heat by the muscle. As tension falls during relaxation, an equivalent amount of heat is released by the muscle owing to its elastic properties. The various kinds of muscle heat must be corrected for the thermoelastic heat. However, for a complete cycle of contraction and relaxation, the net heat produced by thermoelastic mechanisms is zero.

In the same seminal paper [Hil38], Hill also proposed a three–element rheological model of the skeletal muscle–tendon complex (see Figure 20.2). In this model the length–tension property of muscle is represented by an active contractile element (CE) in parallel with a passive elastic element. Total isometric muscle force is assumed to be the sum of muscle force when it is inactive (passive) and when it is maximally excited (active). The muscle is in series with tendon, which is represented by a nonlinear spring. Pennation angle (α) is the angle between tendon and muscle fibers. Tendon slack length is the length of tendon at which force initially develops during tendon stretch. The model was

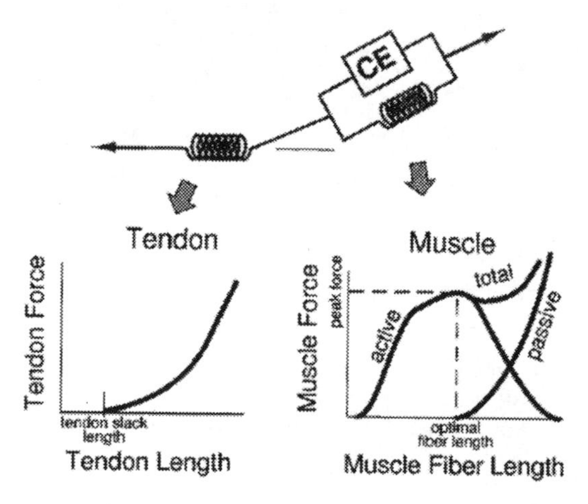

Fig. 20.2. Hill's model of the skeletal muscle–tendon complex.

scaled to represent each muscle by specifying the muscle's peak force, optimal fiber length, tendon slack length, and pennation angle based on data collected in anatomical experiments.

Hill's muscle–tendon model has been widely applied in biomechanical musculo–skeletal modelling.

Hatze's Myocybernetics

Dynamics of human skeletal and face muscles is in the most sophisticated form described in the series of papers of Hatze (see [Hat78]). His muscle–control model involves excitation dynamics of neuro–muscular inputs (motor units) and contraction dynamics based on Huxley's sliding–filament theory of muscle contraction. In brief, Hatze's *myocybernetics* can be divided into *excitation dynamics* and *contraction dynamics*. The excitation dynamics of a single muscle fibre stimulated by trains of normalized nerve impulses $\alpha(t)$ is represented by the system

$$\begin{aligned} \ddot{\beta} + c_4\dot{\beta} + c_5\beta &= c_6 V_N \alpha(t),\ \beta(0) = \dot{\beta}(0) = 0, \qquad (20.6)\\ \ddot{\gamma} + (c_1\dot{\gamma} + c_2\gamma)/\rho^*(\xi) &= c_3 V_T \beta(t),\ \gamma(0) = \dot{\gamma}(0) = 0,\\ \delta\dot{q} &= d_1\{d_2[1 - k^2(\xi)][h(\dot{xi}) - 1/(1-q_0)] - \delta q\}\, \delta q(t_s) = 0, \end{aligned}$$

where $\rho^*(\xi)$ is *normalized Ca density function*, $k(\xi)$ is *filamentary–overlap function*, $h(\dot{xi})$ is *velocity–dependence function*;
$c_1, \dots, c_6, d_1, d_2, V_N, V_T, q_0$ are defined constants; $V_T\beta(t)$ is *action potential* as appearing in the interior of the T–system of the fibre, while $\gamma(t)$ denotes the *free Ca–ion concentration* in the interfilamentary space; the variable δq expresses the *stretch potentiation* induced by an elongation of the tetanized fibre.

The variable ξ designates the *normalized length of the contractile element* of the fibre, and is defined by the contraction dynamics,

$$\begin{aligned} \dot{\xi} &= a_1[1/a_2 \operatorname{arcsin} h\, a_3 \ln(\frac{q^* k(\xi)}{b_2[f^{SE}/\bar{f} + b_1 k_1(\xi)]} - a_4)] - \frac{1}{2},\\ \xi(0) &= \xi_0, \qquad (20.7) \end{aligned}$$

where $a_1, \dots, a_4, b_1, b_2$ are defined constants, $f^{SE}/\bar{f}$ is the *normalized force across the series elastic element*, $b_1 k_1(\xi)$ is the *passive sarcomere tension*, and q^* is the *active state*.

Hodgkin–Huxley Theory of Neural Action Potential

The celebrated *Hodgkin–Huxley HH–neuron model* is described by the nonlinear coupled differential equations for the four variables, V for the membrane potential, and m, h and n for the gating variables of Na and K channels, and it is given by [HH52, Hod64]

$$
\begin{aligned}
C\dot{V} &= -g_{\text{Na}} m^3 h (V - V_{\text{Na}}) - g_{\text{K}} n^4 (V - V_{\text{K}}) - g_{\text{L}} (V - V_{\text{L}}) + I_j^{\text{ext}}, \\
\dot{m} &= -(a_m + b_m)\, m + a_m, \qquad \dot{h} = -(a_h + b_h)\, h + a_h, \qquad (20.8)\\
\dot{n} &= -(a_n + b_n)\, n + a_n, \qquad \text{where} \\
a_m &= 0.1\,(V + 40)/[1 - \mathrm{e}^{-(V+40)/10}], \qquad b_m = 4\,\mathrm{e}^{-(V+65)/18}, \\
a_n &= 0.01\,(V + 55)/[1 - \mathrm{e}^{-(V+55)/10}], \qquad b_n = 0.125\,\mathrm{e}^{-(V+65)/80}, \\
a_n &= 0.07\,\mathrm{e}^{-(V+65)/20}, \qquad b_n = 1/[1 + \mathrm{e}^{-(V+35)/10}].
\end{aligned}
$$

Here the reversal potentials of Na, K channels and leakage are $V_{\text{Na}} = 50$ mV, $V_{\text{K}} = -77$ mV and $V_{\text{L}} = -54.5$ mV; the maximum values of corresponding conductivities are $g_{\text{Na}} = 120$ mS/cm^2, $g_{\text{K}} = 36$ mS/cm^2 and $g_{\text{L}} = 0.3$ mS/cm^2; the capacity of the membrane is $C = 1$ μF/cm^2. The external, input current is given by

$$I_j^{\text{ext}} = g_{syn}(V_a - V_c) \sum_n \alpha(t - t_{in}), \qquad (20.9)$$

which is induced by the pre–synaptic spike–train input applied to the neuron i, given by

$$U_{\text{i}}(t) = V_{\text{a}} \sum_n \delta(t - t_{in}).$$

In equation (20.9), t_{in} is the nth firing time of the spike–train inputs, g_{syn} and V_c denote the conductance and the reversal potential, respectively, of the synapse, τ_{s} is the time constant relevant to the synapse conduction, and $\alpha(t)$ is the alpha function given by

$$\alpha(t) = (t/\tau_{\text{s}})\, \mathrm{e}^{-t/\tau_{\text{s}}} \Theta(t).$$

where $\Theta(t)$ is the Heaviside function. The HH model was originally proposed to account for the property of squid giant axons [HH52, Hod64] and it has been generalized with modifications of ion conductances. The HH–type models have been widely adopted for a study on activities of *transducer neurons* such as motor and thalamus relay neurons, which transform the amplitude–modulated input to spike–train outputs.

Muscular Action Potential

Hodgkin–Huxley theory of neural action potential was adapted by Noble [Nob62] as a model of muscular action potential. Noble model has the same form as the HH–neuron model (20.8), with changed the values of constants, so that the whole signal is about 10 times slower. Noble's model was later modified by Hatze's muscular excitation dynamics (20.6) and complemented by his contraction dynamics (20.7).

Now, to simplify Hatze's myiocybernetics, and yet to retain all the necessary excitation–contraction dynamics, as well as to establish the neuro–muscular inter–connection, we propose herein approach of recurrent diffusion physics. The EFS–response mapping $\mathcal{F}$ of a skeletal or face muscle, i.e., the response of the muscle system $\mathcal{M}$ to the efferent functional stimulation from the neural network system $\mathcal{N}$ – can be stated in the form of the *force generator* time behavior, $\mathcal{F} : \mathbb{R} \to Hom_t(\mathcal{N}, \mathcal{M})$, where: t denotes stimulation time, $\mathcal{N}$ and $\mathcal{M}$ correspond to the left $\mathbb{R}$–moduli of neural and muscular systems. The mapping $\mathcal{F}$ can be considered as an effect of a *fifth–order transmission cascade* ($\mathcal{F}_1 \mapsto \mathcal{F}_2 \mapsto \mathcal{F}_3 \mapsto \mathcal{F}_4 \mapsto \mathcal{F}_5$), where $\mathcal{F}_i\,(i = 1, \ldots, 5)$ represent *neural action potential, synaptic potential, muscular action potential, excitation–contraction coupling and muscle tension generating*, respectively (see [II06a, II06b]).

According to [Nob62, Hak93, Hak02], all transmission components of the system ($\mathcal{F}_1 \mapsto \mathcal{F}_2 \mapsto \mathcal{F}_3 \mapsto \mathcal{F}_4 \mapsto \mathcal{F}_5$), where $\mathcal{F}_i\,(i = 1, \ldots, 5)$ can be considered as being some kind of *diffusion processes*, forming the fifth–order transmission *flux* cascade.

Mapping $\mathcal{F}$ (for all included motor units in the particular muscle contraction) can be described by fifth order *recurrent, distributed parameter* diffusion system [II06a, II06b]

$$\begin{aligned} C_k \frac{\partial V_k}{\partial t} &= \frac{1}{R_k} \frac{\partial^2 V_{k-1}}{\partial z^2} - J_k(V_k), \qquad \text{with boundary condition at } z = 0, \\ V_k(0, t) &= V_0 \sin(2\pi f t) = S(t), \qquad (k = 1, \ldots, 5). \end{aligned}$$

The single element $\mathcal{F}_4$, $(k = 1, \ldots, 5)$ behavior is now given by

$$V_k(z, t) = V_0 \exp(-z_k/m) \sin(2\pi f(t - z_k/n)),$$
$$m = \frac{1}{R_k C_k f}, \qquad n = \frac{4\pi f}{R_k C_k}.$$

For muscle–mechanical purpose, the presented distributed map $\mathcal{F}$ can be first mathematically approximated with the corresponding lumped parameter $R_k C_k$ electric circuit (where the second circuit represents the *Eccles model of synaptic activation* (see [Ecc64, EIS67, Ecc86]) and the last one corresponds to the low-pass filter representing the contraction process itself), at $x = tendon$

$$\begin{aligned} \dot{z}_k &= \frac{1}{T_k}(b_k z_{k-1} - z_k), \qquad (k = 1, \ldots, 5), \\ z_k(0) &= 0, \qquad z_0 = S(t), \qquad z_5 = F(t), \end{aligned}$$

where $T_k = R_k C_k$ are time characteristics of the circuits in cascade, and b_k are corresponding input gains (conversion factors).

The single muscle behavior in the lumped approximation form is given by the recurrent sum of its transient and weighting terms (with time legs τ_k)

$$z_k(t) = b_k z_{k-1}(1 - \exp(-t/T_k)) + z_k \exp(-(t - \tau_k)/T_k).$$

The presented distributed mapping $\mathcal{F}$ can be further physically approximated with a second order forced–dumped linear oscillator in a Cauchy form

$$T\ddot{z} + 2aT\dot{z} + cz = bS, \qquad z(0) = \dot{z}(0) = 0,$$

where a (having dimension of force) corresponds to energy dissipated during the contraction, b (having dimension of velocity) is the phosphagenic energy transducing rate, while c corresponds to the second derivative of the stress–strain curve of the series viscoelastic element [Wil56] of the muscular actuator (assumed in exponential three–parameter form).

The complete efferent face and body neuro–muscular system $(\mathcal{N}, \mathcal{M})$ is now given by the set of equations

$$\dot{x}^i = -D^i_j x^i + T^i_j g^i(x^i) + S^i, \qquad (i,j = 1,\dots,n),$$
$$C_k \frac{\partial V_k}{\partial t} = \frac{1}{R_k}\frac{\partial^2 V_{k-1}}{\partial z^2} - J_k(V_k), \qquad (k = 1,\dots,5);$$

or, its discrete form

$$\dot{x}^i = (b - y^i)x^i, \qquad \dot{y}^i = -vy^i + g(x^i + T^i_j y^j), \tag{20.10}$$
$$\dot{z}_k = \frac{1}{T_k}(b_k z_{k-1} - z_k), \qquad (k = 1,\dots,5), \tag{20.11}$$
$$z_k(0) = 0, \qquad z_0 = S(t), \qquad z_5 = F(t). \tag{20.12}$$

Equations (20.10,20.12) constitute a $3n$–dimensional phase–space (for $n = 5$ or $k = i$) being a *hiper–cube* $\equiv$ *neuro–muscular control space*. The feedback control $\mathcal{F}^{-1}$ of the mapping $\mathcal{F}$ is performed by muscular *autogenetic motor servo*.

Houk's Autogenetic Motor Servo

It is now well–known (see [Hou79, HBB96]) that voluntary contraction force $\mathcal{F}$ of a skeletal or face muscle system $\mathcal{M}$ is reflexly excited (positive reflex feedback $+\mathcal{F}^{-1}$ by responses of its *spindle receptors* to stretch and is reflexly inhibited (negative reflex feedback $-\mathcal{F}^{-1}$ by responses of its *Golgi tendon organs* to contraction. Stretch and unloading reflexes are mediated by combined actions of several autogenetic neural pathways.

James Houk's term '*autogenetic*' means that the stimulus excites receptors located in the same face or body muscle that is the target of the reflex response. The most important of these muscle receptors are the primary and secondary endings in muscle–spindles, sensitive to length change – positive length feedback $+\mathcal{F}^{-1}$, and the Golgi tendon organs, sensitive to contractile force - negative force feedback $-\mathcal{F}^{-1}$.

The gain G of the length feedback $+\mathcal{F}^{-1}$ can be expressed as the *positional stiffness* (the ratio $G \approx S = dF/dx$ of the force $\mathcal{F}$–change to the length x–change) of the muscle system $\mathcal{M}$. The greater the stiffness S, the less will the muscle be disturbed by a change in load and the more reliable will be the performance of the muscle system $\mathcal{M}$ in executing controlled changes in length $+\mathcal{F}^{-1}$.

The autogenetic circuits $(+\mathcal{F}^{-1})$ and $(-\mathcal{F}^{-1})$ appear to function as *servoregulatory loops* that convey continuously graded amounts of excitation and inhibition to the large (*alpha*) skeletomotor neurons. Small (*gamma*) fusimotor neurons innervate the contractile poles of muscle spindles and function to modulate spindle–receptor discharge.

20.1.3 The Equivalent Muscular Actuator

A single skeletal muscle, (e.g., the triceps brachii muscle, see Figure 20.3), is attached at its *origin* to a large area of bone (the humerus in case of the triceps). At its other end, the *insertion*, it tapers into a glistening white *tendon* which, (in case of the triceps is attached to the ulna). As the triceps contracts, the insertion is pulled toward the origin and the arm is straightened or extended at the elbow. Thus the triceps is an *extensor*. Because skeletal muscle exerts force only when it contracts, a second muscle – a *flexor* – is needed to flex or bend the joint (e.g., the biceps brachii muscle is the flexor of the forearm). Together, they (the biceps and triceps) make up an antagonistic pair of muscles, which we will call forming the *equivalent muscular actuator*. Similar pairs, i.e., equivalent muscular actuators, working antagonistically across other joints, provide for almost all the movement of the skeleton. The equivalent muscular actuator has the role of 'driver' in biodynamics. It generates the equivalent muscular torque, which is the *primary cause* of human–like motion [II06a, II06b].

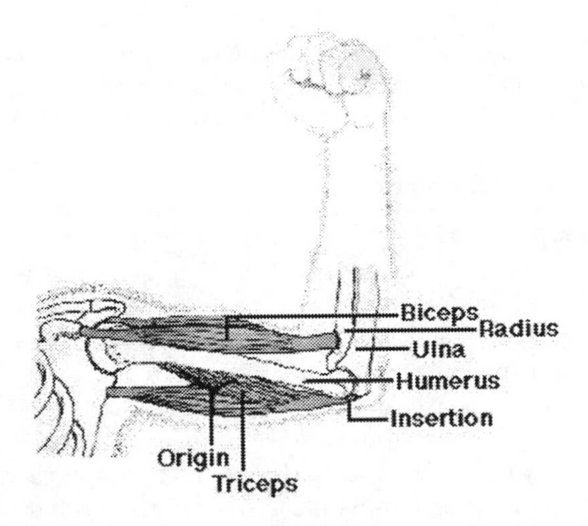

Fig. 20.3. An antagonistic pair of human skeletal muscles, one flexor and the other extensor (in the case of the forearm, biceps brachii and triceps brachii, respectively), forming the *equivalent muscular actuator* – the primary cause of the human–like motion.

20.1.4 Biochemistry of Muscular Contraction

The immediate energy source for contraction in human muscles is adenosine triphosphate ATP (see [Mou80, Mar98]). Muscle contains about 2 $\mu mole$ ATP/ gram wet weight. The myosin head is the only site of the major ATP hydrolysis in active muscle. At the concentrations of ATP, ADP (*adenosine diphosphate*), and P_i (*inorganic phosphate*) present in the sarcoplasm, ATP hydrolysis yields about 11.5 $kcal/mole$. About 0.3 $\mu mole$ of ATP/gram muscle is hydrolyzed by a single muscle twitch. The ATP hydrolysis overall scheme:

$$ATP \rightarrow ADP + P_i + \text{Energy} \tag{20.13}$$

represents actually the complex six–step chain–reaction $\{k_i\}$, $(i = 1, \ldots, 5)$, which can be summarized as follows:

1. Myosin reacts rapidly with ATP to form a complex; the myosin is from its resting (low–energy) form converted to an energy–rich form.
2. While complexed to the myosin, ATP is hydrolyzed to ADP and P_i. Reaction 2 is much more rapid than reaction 1. (This step is extremely temperature sensitive).
3. In reaction 3, while the ADP and P_i are still attached to the myosin, the latter is converted to a low-energy form. This step is slow, rate limiting in the sequence of reactions, and insensitive to temperature changes.
4. Reactions 4 and 5 are rapid.

Therefore, the reaction sequence proceeds as follows:

$$\begin{gathered} M + ATP \leftrightarrow M^* + ATP \leftrightarrow M^* + ADP + P \leftrightarrow \\ M + ADP + P, \leftrightarrow M + ADP + P_i \leftrightarrow M + ADP, \end{gathered} \tag{20.14}$$

where M is miosin in low–energy form, M^* is miosin in energy–rich form, and symbol $\leftrightarrow$ actually represents a pair of reversible reactions $\{k_i, k_{i-1}\}$, $(i = 1, \ldots, 5)$.

Muscles also contain about $20\,\mu mole$ CP/gram (*creatine phosphate*). Creatine phosphate can phosphorylate ADP to form ATP in a reversible reaction catalyzed by the enzyme *creatine kinase*.

Muscle contains large amounts of creatine kinase; it amounts to more than 25 percent of the soluble cytoplasmic protein. As soon as ATP is hydrolase, the ADP formed is very rapidly rephosphorylated by CP and the ATP is regenerated. Thus CP forms a reservoir of energy–rich phosphate bonds to quickly replenish the sarcoplasmic ATP.

Ultimately, ATP is produced by *glycolysis* and respiration [Mou80]. In glycolysis (the so called *Embden–Meyerhoff pathway*), glucose is degraded to pyruvate, or to lactic acid in the absence of O_2, yielding 2 moles $ATP/mole$ glucose metabolized. Intracellular glycogen granules provide a very readily available source of glucose. Muscles normally contain 9 to 16 gm/kg glycogen or, for a well–fed man of average height and weight, the total glycogen stores in muscle amount to $300 - 500\,gram$, with another $55 - 90\,gram$ in the liver. Glycogen breakdown in muscle begins immediately on stimulation, and the amount of muscle glycogen depleted is proportional to the mechanical work done. Glycogen is hydrolyzed by the enzyme phosphorylase to *glucose–1–phosphate*, which then enters the glycolytic pathway.

Red muscle fibers respond to a stimulus with a relatively slow twitch (maximum shortening velocity about 17 mm/sec) and therefore are also called *slow fibers*, whereas *white muscle fibers* react to a stimulus with a rapid twitch (maximum shortening velocity about $42 mm/sec$) and therefore are also called fast fibers. Red muscle has a more extensive blood supply than white muscle. Red muscle fibers are able to sustain activity for long periods of time whereas white muscle fibers characteristically produce short bursts of great tension followed by the rapid onset of fatigue.

Whole red and white muscles differ in *ATPase* activity, and, indeed, the purified contractile protein myosin extracted from red and white muscle differs in *ATPase* activity, a finding associated with different myosin light chains. White muscle and white muscle actomyosin show the greater *ATPase* activity. The innervation of red and white muscle differs, and, indeed, whether a given muscle is red or white results from trophic influences of the motor nerve.

Slow muscle fibers are generally thinner and possess many sarcosomes (mitochondria) containing large amounts of respiratory enzymes, as well as copious quantities of the O_2–carrying protein *myoglobin* in the sarcoplasm and man lipid droplets. The numerous sarcosomes and high level of myoglobin give slow fibers their red color. Fast (or white) muscle fibers, on the other hand, are generally of larger diameter and contain large amounts of phosphorylase and glycolytic enzymes and large deposits of glycogen. Slow muscles derive energy predominantly from respiration, whereas in fast muscle fibers, glycolysis and lactate production are more prominent.

20.2 Muscular Energy Systems and Their Flows

Human energy production spans the range of human movements from those requiring large bursts of energy over short periods of time (as in sprint running) – to those activities requiring small but sustained energy production (as in marathon running). Even within the same activity, the energy requirements change from one moment to the next (see any textbook on exercise physiology).

The Immediate Energy Source

Adenosine triphosphate (*ATP*) is the immediately usable form of chemical energy for muscular activity. It is stored in most cells, particularly muscle cells. Other forms of chemical energy, such as that available from the foods we eat, must be transferred into the *ATP* form before they can be utilized by the muscle cells.

The chemical structure of *ATP* is complicated, but for our purposes it can be simplified, saying that *ATP* consists of a large complex of molecules called adenosine and three simpler components called phosphate groups. The last two phosphate groups represent 'high energy bonds'. In other words, they store a high level of potential chemical energy. When the terminal phosphate bond is broken, energy is released, enabling the cell to perform work. The kind of work performed by the cell depends on the cell type. For example, mechanical work (contraction) is performed by muscle cells (skeletal, smooth and heart muscle), nerve conduction by nerve cells, secretion by secretory cells (e.g., endocrine cells), and so on. All 'biological' work performed by any cell requires the immediate energy derived from the breakdown of *ATP*.

The Principle of Coupled Reactions

Since energy is released when ATP is broken down, it is not too surprising that energy is required to rebuild or resynthesize ATP. The building blocks for ATP synthesis are the by-products of its breakdown, adenosine diphosphate (ADP) and inorganic phosphate (P_i). Therefore, there are two essential formulas, the direct: $ATP \to ADP + P_i + energy$, and the inverse one: $ADP + P_i \to ATP$. The energy for ATP resynthesis comes from three different series of chemical reactions that take place within the body. Two of the three depend upon the food we eat, whereas the other depends upon a chemical compound called phosphocreatine. The energy released from any one of these series reactions is coupled with the energy needs of the reaction that resynthesizes ATP. In other words, the separate reactions are functionally linked together in such a way that the energy released by the one is always used by the other. Biochemists refer to these functional links as coupled reactions, and it has been shown that such coupling is the fundamental principle involved in the metabolic production of ATP.

$ATP - PC$: The Phosphagen System

PC is an abbreviation for phosphocreatine, another one of those 'energy–rich' phosphate compounds closely related to ATP. For example, PC, like ATP, is stored in muscle cells, and when it is broken down (i.e., when its phosphate group is removed), a large amount of energy is released. The released energy, of course, is coupled to the energy requirement necessary for the resynthesis of ATP. In other words, as rapidly as ATP is broken down during muscular work, it is continuously reformed from ADP and P_i by the energy liberated during the breakdown of the stored PC: $ADP + P_i \to ATP$. For every mole of PC broken down, one mole of ATP is resynthesized.

The total muscular stores of both ATP and PC (collectively referred to as phosphagens) are very small – only about $0.3mole$ in females and $0.6mole$ in males. Thus, the amount of energy obtainable through this system is limited. In fact, if you were to run $100m$ as fast as you could, the phosphagen stores in the working muscles would probably be empty by the end of the sprint. However, the usefulness of the $ATP - PC$ system lies in the rapid availability of energy rather than in the quantity. For example, activities such as sprinting, jumping, swinging and other similar skills requiring only a few seconds to complete are all dependent upon the stored phosphagens for their primary energy source.

The Lactic Acid System

This system is also known as 'anaerobic glycolysis'. 'Glycolysis' refers to the breakdown of sugar; 'anaerobic' means without oxygen. In this system, the breakdown of sugar (a carbohydrate, one of the foodstuffs) supplies the necessary energy from which ATP is manufactured. When the sugar is only partially broken down, one of the end products is lactic acid (hence the name lactic acid system).

When lactic acid accumulates in the muscles and blood reaches very high levels, temporary muscular fatigue results. This is a very definite limitation, and is the main cause of the 'early' fatigue. Another limitation of the lactic acid system that relates to its anaerobic quality is that only a few moles of ATP can be resynthesized from the breakdown of sugar as compared to the yield possible when oxygen is present. For example, only three moles of ATP can be manufactured from the anaerobic breakdown of $180gr$ of glycogen (glycogen is the storage from glucose or sugar in the muscle). As we will soon see, the aerobic breakdown of $180gr$ of glycogen results in enough energy to resynthesize $39moles$ of ATP!

The lactic acid system, like the $ATP - PC$ system, is extremely important to us, primarily because it too provides for a rapid supply of ATP energy. For example, exercises that are performed at maximum rates for between $1min$ and $3min$, such as sprinting $400m$ and $800m$, depend heavily upon the lactic acid system for ATP energy. Also, in some performances, such as running $1500m$ or a mile, the lactic acid system is used predominantly for the 'kick' at the end of the race.

The Aerobic (O_2) System

In the presence of oxygen, the complete breakdown of $180gr$ of glycogen to carbon dioxide (CO_2) and water (H_2O) yields enough energy to manufacture $39moles$ of ATP. This series of reactions, like the anaerobic series, takes place within the muscle cell, but is confined to specialized subcellular compartments called mitochondria. Mitochondria are slipper–shaped cell bodies often referred to as the 'powerhouse' of the cell because they are the seat of the aerobic manufacture of ATP energy. Muscle cells are rich with mitochondria.

The aerobic breakdown of carbohydrates, fats, and even proteins provides energy for ATP resynthesis: $ADP + P_i \rightarrow ATP$. Since abundant ATP can be manufactured without yielding fatiguing by–product, the aerobic system is most suited for endurance activities. The carbon dioxide that is produced diffuses freely from the muscle cell into the blood and is carried to the lung, where it is exhaled. The water that is formed is useful within the cell itself, since the largest constituent of the cell is, in fact, water.

Another feature of the aerobic system that should be noticed is that concerned with the type of foodstuff required for breakdown. Not only glycogen but fats and proteins as well can be aerobically broken down to carbon dioxide and water, with energy released for ATP synthesis. For example, the breakdown of $256gr$ of fat will yield $130moles$ of ATP. During exercise, both glycogen and fats, but not protein, are important sources of ATP–yielding energy.

The amount of oxygen one needs to consume from the environment in order to synthesize one mole of ATP is approximately $3.5l$ if glycogen is the food fuel and about $4.0l$ with fat. At rest, most of us consume between $0.2l$ and $0.3l$ of oxygen per minute. In other words, a mole of ATP is aerobically manufactured every $12min$ to $20min$ under normal resting conditions. During maximal exercise, a mole of ATP can be aerobically supplied to the working muscles every minute by most of us. For the highly trained endurance athlete, more than $1.5moles$ of ATP can be aerobically synthesized and supplied to the muscles every minute during maximal effort.

In summary, then, the aerobic system is capable of utilizing both fats and glycogen for resynthesizing large amounts of ATP without simultaneously generating fatiguing by–products. Therefore, the aerobic system is particularly suited for manufacturing ATP during prolonged, endurance–type activities. For example, during marathon running, approximately $150moles$ of ATP are required over $2.5h$ the race takes. Such a large, sustained output of ATP energy is possible only because early fatigue can be avoided and large amounts of food (glycogen and fats) and oxygen are readily available.

The Energy Continuum Concept

Therefore, ATP is immediate form of muscular energy, and is supplied in three ways:

1. by the stored phosphagens, ATP and PC;
2. by the lactic acid system (anaerobic glycolysis); and
3. by the oxygen, or aerobic, system. And ability of each system to supply the major portion of the ATP required in any given activity is related to the specific kind of activity performed.

For example, in short–term, high–intensity types of activities, such as the $100m$ dash, most of the ATP is supplied by the readily available phosphagen system. In contrast, longer–term, lower intensity types of activities, such as the $42.2km-$marathon, are supported almost entirely by oxygen, or aerobic, system.

In the middle range of activities are those tasks that rely heavily on the lactic acid system for ATP production – e.g., the $400m$ and $800m$ dashes. Also in the middle are those activities that require a blend of both anaerobic and aerobic metabolism – e.g., the $1500m$ and mile runs. In these latter activities, the anaerobic systems supply the major portion of ATP during the sprint at both the start and finish of the race, with the aerobic system predominating during the middle or steady–state period of the run.

Just from these examples it can be seen that what is at work in physical activities is an energy continuum, a continuum relating the way in which ATP is made available and the type of physical activity performed.

20.3 Musculo–Skeletal Dynamics

Passive Joint Dynamics

All biological systems are *dissipative structures*, emphasizing *irreversible processes* inefficient energetically, but highly efficient in terms of information and control. In case of biomechanics, one has the passive damping contribution to the joint torques, $T_i(t,\, q^i,\, p_i)$, which has the basic stabilizing effect to the complex human movement. This effect can be described by (q,p)-quadratic form of the *Rayleigh – Van der Pol's dissipation function*,

$$R = \frac{1}{2}\sum_{i=1}^{9} p_i^2\,[a_i\,+\,b_i(q^i)^2], \tag{20.15}$$

where a_i and b_i denote dissipation parameters. Its partial derivatives $\partial R/\partial p$ give rise to viscous forces in the joints which are linear in p_i and quadratic in q^i. It is based on the unforced Van der Pol's oscillator

$$\ddot{x} - \left(a + b\,x^2\right)\,\dot{x} + x = 0,$$

where the damping force $F^{dmp}(\dot{x}) = -\partial R/\partial\dot{x}$ is given by the Rayleigh's dissipation function $R = \frac{1}{2}\left(a\,+\,b\,x^2\right)\,\dot{x}^2$ – with the velocity term $\dot{x}$ replaced by our momentum term p^2.

Using (20.15) one obtains the *dissipative joint Hamiltonian biodynamics equations*:

$$\begin{aligned} \dot{q}^i &= \frac{\partial H(q,p)}{\partial p_i} + \frac{\partial R(q,p)}{\partial p_i}, \\ \dot{p}_i &= -\frac{\partial H(q,p)}{\partial q^i} + \frac{\partial R(q,p)}{\partial q^i}, \qquad (i = 1,\ldots,N), \end{aligned} \tag{20.16}$$

which reduces to the gradient system in case $H = 0$ (as well as to the conservative system in case $R = 0$).

Active Muscular Dynamics

Muscular dynamics describes the internal *excitation* and *contraction* dynamics of *equivalent muscular actuators*, anatomically represented by resulting action of *antagonistic muscle-pairs* for each uniaxial joint. We attempt herein to describe the equivalent muscular dynamics in the simplest possible way (for example, Hatze used 51 nonlinear differential equations to derive his, arguably most elaborate, myocybernetic model), and yet to include the main excitation and contraction relations.

The active muscular contribution to the joint torques, $T_i(t,\, q^i,\, p_i)$, should describe the internal *excitation* and *contraction* dynamics of *equivalent muscular actuators*, anatomically represented by resulting action of *antagonistic muscle-pairs* per each active degree-of-freedom.

(a) *Excitation dynamics* can be described by impulse *force-time* relation

$$\begin{aligned} F_i^{imp} &= F_i^0(1\,-\,e^{-t/\tau_i}), \qquad &&\text{if stimul } > 0 \\ F_i^{imp} &= F_i^0 e^{-t/\tau_i}, &&\text{if stimul } = 0, \end{aligned}$$

where F_i^0 denote the maximal isometric muscular torques applied at i-th joint, while τ_i denote the time characteristics of particular muscular actuators. This is a rotational-joint form of the solution of the Wilkie's *muscular active-state element equation*,

$$\dot{x} + \beta x = \beta S A, \quad x(0) = 0, \quad 0 < S < 1,$$

where $x = x(t)$ represents the active state of the muscle, β denotes the element gain, A corresponds to the maximum tension the element can develop, and $S = S(r)$ is the 'desired' active state as a function of motor unit stimulus rate r.

(b) *Contraction dynamics* has classically been described by the Hill's *hyperbolic force-velocity* relation, which we propose here in the rotational (q,p)-form:

$$T_i^{Hill} = \frac{(T_i^0 b_i - a_i p_i)}{(p_i - b_i)},$$

where a_i (having dimension of torque) and b_i (having dimension of momentum) denote the *rotational Hill's parameters*, corresponding to the energy dissipated during the contraction and the phosphagenic energy conversion rate, respectively.

Therefore, one can describe the excitation/contraction dynamics for the ith equivalent muscle-joint actuator, i.e., antagonistic muscle pair, by the simple impulse-hyperbolic product–relation:

$$T_i(t,q,p) = T_i^{imp} \times T_i^{Hill}, \qquad (i = 1, \ldots, N). \tag{20.17}$$

Using (20.17) one can obtain the general formulation for the *forced dissipative Hamiltonian musculo-skeletal dynamics*,

$$\dot{q}^i = \frac{\partial H(q,p)}{\partial p_i} + \frac{\partial R(q,p)}{\partial p_i}, \qquad (i = 1, \ldots, N), \tag{20.18}$$

$$\dot{p}_i = T_i(t,q,p) - \frac{\partial H(q,p)}{\partial q^i} + \frac{\partial R(q,p)}{\partial q^i}.$$

20.3.1 Biodynamical Simulations

In this section we will present the essential tools for biomechanical simulations, using the free computer simulation package *Scilab*.[1]

Solving Equations of Motion in *Scilab*

Here we present the basic syntax for numerical solution of ordinary differential equations (ODEs for short) in *Scilab*.

[1] *Scilab* is a free simulation (for non-commercial use) package, developed by Consortium Scilab (DIGITEO, INRIA, ENPC). It is similar to $Matlab^{TM}$. Current version is 5.1.1. See http://www.scilab.org/.

```
The standard function for solving explicit ODE-systems
is "ode", defined by:  dy/dt=f(t,y).
It can be called using several formats.

The basic (simplest) calling sequence is:

  y = ode(y0,t0,t,f);

where parameters in the brackets have the following
meaning:
y0 = real vector or matrix of initial conditions;
t0 = real scalar denoting initial time;
t = real time-axis vector for the solution;
f = external (function or character string or list);
y = matrix of solution vectors:
    y=[y(t(1)),y(t(2)),...];

More general format for the calling sequence:

[y,w,iw] =
ode([type],y0,t0,t [,rtol [,atol]],f [,jac] [,w,iw]);

where square brackets [...] denote options:
"rtol" and "atol" are real constants or real
vectors of the same size as y;
jac = external jacobian;
w, iw = real vectors.

The "type" is one of the following strings of
characters:
[adams, stiff, rk, rkf, fix, discrete, roots].
It automatically selects between non-stiff "adams"
predictor-corrector method and "stiff" Backward
Differentiation Formula (BDF) method. It uses
non-stiff method initially and dynamically monitors
data in order to decide which method to use.
"rk": Runge-Kutta method of order 4 (RK4);
"rkf": adaptive Runge-Kutta-Fehlberg method of order
4 and 5 (RKF45);
this is the fastest algorithm for non-stiff and
mildly stiff problems.
"fix": this is "rkf" with simple interface
(only "rtol" and "atol" parameters); this is
the simplest method to try.
"discrete": performs discrete-time simulation, e.g.:
```

```
    y = ode("discrete",y0,k0,kvect,f);
"roots": ODE-solver with root-finding capabilities.

The most general format for the calling sequence:

[y,rd,w,iw] =
ode("root",y0,t0,t[,rtol [,atol]],f[,jac],ng,g[,w,iw]);

where:
root = ODE solver with root-finding capabilities;
ng = integer;
g = external (function or character string or list);
k0 = integer (initial time). kvect : integer vector;

In all cases, the input argument "f" defines the right
hand side (RHS) of the first order ODE-system:
    dy/dt = f(t,y).
It is an external i.e. a function with specified
syntax, or the name of a Fortran subroutine or a
C-function (string) with specified calling sequence
or a list. If f is a Scilab function, its syntax must
be:  ydot = f(t,y);
where t is a real time scalar,
y is a real state vector, and
ydot is a real vector equal dy/dt.

The "f" argument can also be a list with the following
structure:
    lst = list(real f,u1,u2,...un),
where real f is a Scilab function with syntax:
    ydot = f(t,y,u1,u2,...,un);
This syntax allows to use parameters as the arguments
of real f. The function f can return a pxq matrix
instead of a vector.

Optional input parameters can be given to control
the error of the solution: "rtol" and "atol" are
thresholds for relative and absolute estimated errors.
The estimated error on y(i) is:
    rtol(i)*abs(y(i)) + atol(i)
and integration is carried out as far as this error
is small for all components of the state.
If rtol and/or atol is a constant,
rtol(i) and/or atol(i) are set to this constant value.
Default values for rtol and atol are:
```

```
    rtol=1.d-5 and atol=1.d-7
for most solvers and
rtol=1.d-3 and atol=1.d-4 for "rfk" and "fix".
For stiff problems, it is better to give the Jacobian
of the RHS function as the optional argument "jac".
It is an external. If "jac" is a function, the syntax
should be:
    J = jac(t,y),
where t is a real time scalar and y a real state-vector.
The resulting matrix J must evaluate to df/dx, i.e.
J(k,i) = dfk/dxi,  with fk = k-th component of f.
```

Basic Euler's Rigid–Body Dynamics

Euler's homogenous rigid body equations (without external torques) read:

$$\dot{x}_1 = I_1 x_2 x_3; \qquad \dot{x}_2 = -I_2 x_3 x_1; \qquad \dot{x}_3 = -I_3 x_1 x_2,$$

where $(x_1, x_2, x_3) = (\rho, phi, theta) =$ Euler angles (roll, pitch and yau), while I_1, I_2, I_3 are local inertia moments (see The Newton-Euler Biomechanics Law for biomechanical details). Numerical solution of this ODE-system for the initial conditions:

$$IC: \;\; x_1(0) = 0, \;\; x_2(0) = 1, \;\; x_3(0) = 1$$

and inertia moments:

$$I_1 = 1, \;\; I_2 = 1, \;\; I_3 = 0.51,$$

is given in Figure 20.4.

The Scilab-code used for this simulation reads:

```
// Euler's rigid body equations
// Consider the following system:
// dx1 = x2*x3; dx2 = -x3*x1; dx3 = -0.51*x1*x2
// with IC:  [x1(0); x2(0); x3(0)] = [0; 1; 1].

// ODE system defined in matrix form
function xdot=EulerBody(t,x);
  xdot=[I1*x(2)*x(3); -I2*x(3)*x(1); -I3*x(1)*x(2)];
endfunction);

// initial Euler angles:
x0 = [0,1,1]';

// inertia moments:
I1 = 1;  I2 = 1;  I3 = 0.51;
```

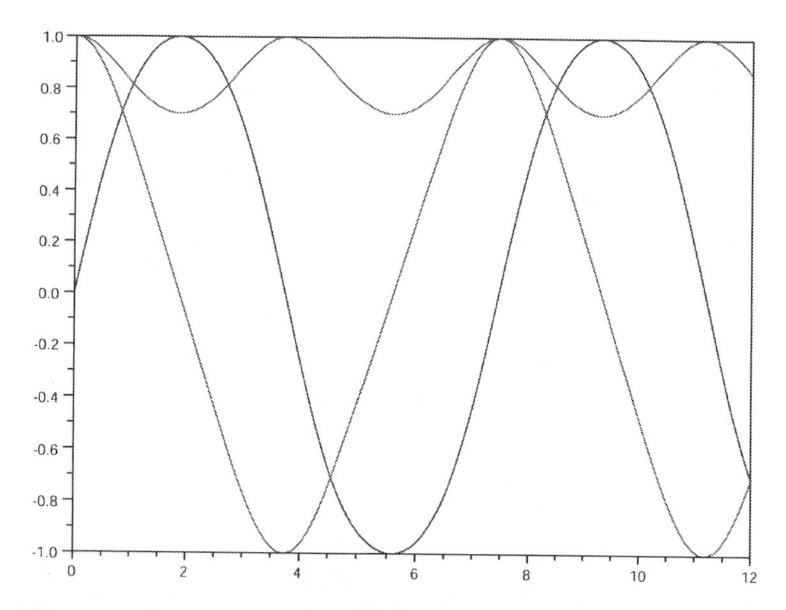

Fig. 20.4. Scilab simulation of the basic Euler rigid body equations.

```
// time axis:
t0 = 0;  t = linspace(0,12,1200);

// numerical solution and plot:
scf(1); clf; // open and clear new figure window
sol = ode(x0,t0,t,EulerBody); // numerical solution
plot(t',sol');  // basic plot
```

20.4 Stretch Reflex and Motor Servo

The myotatic stretch–reflex, also known as the monosynaptic reflex,[2] is the simplest reflex known: it involves only two neurons (one sensory, or afferent and one motor, or efferent), with one synapse between them. Therefore, it depends only on the monosynaptic connection between primary afferent fibers from *muscle spindles* and motor neurons innervating the same muscle.

The muscle spindle (see Figure 20.5) contains specialized elements that sense muscle length and velocity of length change. The spindle is innervated by large myelinated afferent fibers known as type Ia afferent fibers. The cell bodies of these neurons are clustered near the spinal cord in the dorsal root ganglia. They are

[2] It is called a monosynaptic reflex because it depends only on the simple connection between primary afferent fibers from muscle spindles and motor neurons innervating the same muscle. In other spinal reflexes such as those produced by cutaneous stimuli, one or more interneurons may be interposed between the primary afferent fibers and the motor neurons.

an example of a bipolar cell: one branch of the cell's axon goes out to the muscle and the other runs into the spinal cord. In the spinal cord the Ia afferent fibers make monosynaptic excitatory connections to alpha motor neurons innervating the same muscle from which they arise and motor neurons innervating synergistic muscles. They also inhibits motor neurons controlling antagonistic muscles through an inhibitory interneuron. Half the neurons in the brain are inhibitory. They release neurotransmitters that hyperpolarize the membrane potential of the postsynaptic cell, thus reducing the likelihood of firing. The muscle spindle is sensitive to stretch so that when the muscle is stretched the Ia afferent fibers increase the firing rate. This leads to the contraction of the same muscle and its synergists and relaxation of the antagonist muscle. The reflex therefore tends to counteract the stretch, enhancing the spring like properties of the muscles.

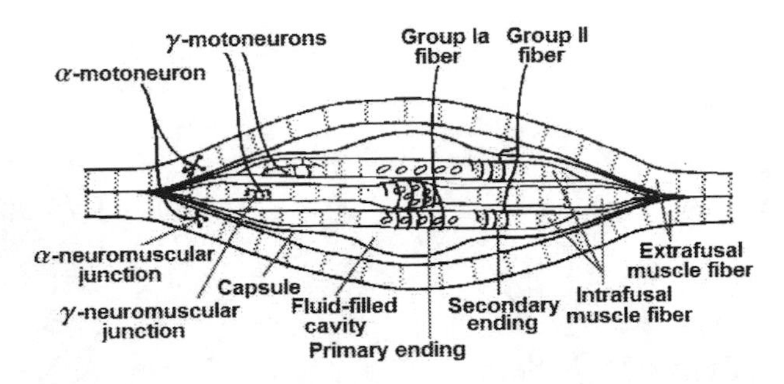

Fig. 20.5. Schematic of a muscular length–sensor, the muscle spindle.

A physician tapping the reflex hammer against the patellar tendon traditionally elicits the special case of myotatic stretch–reflex, called the knee jerk reflex. The tap of the reflex hammer causes stimulation of the muscle spindles afferents and Golgi tendon organs, by altering the stretch and length of the muscle. The afferent neuron synapses in the anterior horn of the spinal cord with an ipsilateral efferent neuron. The efferent neuron leads to a neuromuscular junction with the respective effector muscle the quadriceps femoris. The stretch–reflex also plays a central role in the maintenance of balance.

The intensity of the stretch–reflex is modulated by excitatory and inhibitory supraspinal input. Damage to supraspinal input, affecting descending input, precipitates abnormally high gain of the stretch–reflex. The effect is an increase in muscle stiffness, also referred as hypertonus.

About three/four decades ago, James Houk [Hou67, Hou78, Hou79] pointed out that stretch and unloading reflexes were mediated by combined actions of several autogenetic neural pathways. In this context, "autogenetic" (or, autogenic) means that the stimulus excites receptors located in the same muscle that is the target of the reflex response. The most important of these muscle receptors are the primary and secondary endings in muscle spindles, sensitive to length change, and the Golgi tendon organs, sensitive to contractile force. The autogenetic circuits appear to function as servo-regulatory loops that convey

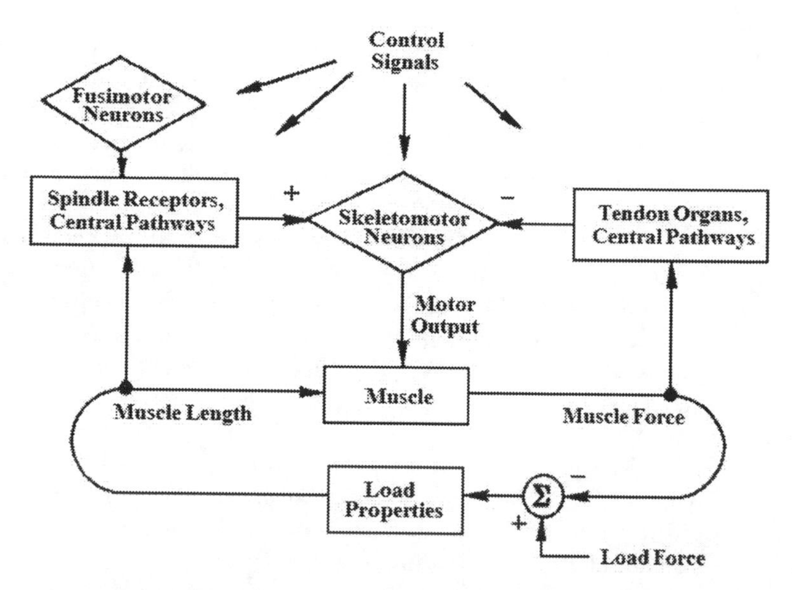

Fig. 20.6. Houk's autogenetic reflexes.

continuously graded amounts of excitation and inhibition to the large (alpha) skeletomotor neurons. Small (gamma) fusimotor neurons innervate the contractile poles of muscle spindles and function to modulate spindle–receptor discharge. Houk's term "motor servo" has been used to refer to this entire control system, summarized by the block diagram in Figure 20.6.

Prior to a study by Matthews, it was widely assumed that secondary endings belong to the mixed population of "flexor reflex afferents," so called because their activation provokes the flexor reflex pattern – excitation of flexor motoneurons and inhibition of extensor motoneurons. Matthews' results indicated that some category of muscle stretch receptor other than the primary ending provides important excitation to extensor muscles, and he argued forcefully that it must be the secondary ending.

The primary and secondary muscle spindle afferent fibers both arise from a specialized structure within the muscle, the *muscle spindle*, a fusiform structure 4–7 mm long and 80–200 μ in diameter. The spindles are located deep within the muscle mass, scattered widely through the muscle body, and attached to the tendon, the endomysium or the perimysium, so as to be in parallel with the extrafusal or regular muscle fibers. Although spindles are scattered widely in muscles, they are not found throughout. Muscle spindle (see Figure 20.5) contains two types of intrafusal muscle fibers (intrafusal means inside the fusiform spindle): the nuclear bag fibers and the nuclear chain fibers. The nuclear bag fibers are thicker and longer than the nuclear chain fibers, and they receive their name from the accumulation of their nuclei in the expanded bag-like equatorial region-the nuclear bag. The nuclear chain fibers have no equatorial bulge; rather their nuclei are lined up in the equatorial region-the nuclear chain. A typical spindle contains two nuclear bag fibers and 4-5 nuclear chain fibers.

The pathways from primary and secondary endings are treated commonly by Houk in Figure 20.6, since both receptors are sensitive to muscle length and both provoke reflex excitation. However, primary endings show an additional sensitivity to the dynamic phase of length change, called dynamic responsiveness, and they also show a much–enhanced sensitivity to small changes in muscle length.

The motor servo comprises three closed circuits (Figure 20.6), two neural feedback pathways, and one circuit representing the mechanical interaction between a muscle and its load. One of the feedback pathways, that from spindle receptors, conveys information concerning muscle length, and it follows that this loop will act to keep muscle length constant. The other feedback pathway, that from tendon organs, conveys information concerning muscle force, and it acts to keep force constant.

In general, it is physically impossible to maintain both muscle length and force constant when external loads vary; in this situation the action of the two feedback loops will oppose each other. For example, an increased load force will lengthen the muscle and cause muscular force to increase as the muscle is stretched out on its length-tension curve. The increased length will lead to excitation of motoneurons, whereas the increased force will lead to inhibition. It follows that the net regulatory action conveyed by skeletomotor output will depend on some relationship between force change and length change and on the strength of the feedback from muscle spindles and tendon organs. A simple mathematical derivation demonstrates that the change in skeletomotor output, the error signal of the motor servo, Should be proportional to the difference between a regulated stiffness and the actual stiffness provided by the mechanical properties of the muscle, where stiffness has the units of force change divided by length change. The regulated stiffness is determined by the ratio of the gain of length to force feedback.

It follows that the combination of spindle receptor and tendon organ feedback will tend to maintain the stiffness of the neuromuscular apparatus at some regulated level. If this level is high, due to a high gain of length feedback and a low gain of force feedback, one could simply forget about force feedback and treat muscle length as the regulated variable of the system. However, if the regulated level of stiffness is intermediate in value, i.e. not appreciably different from the average stiffness arising from muscle mechanical properties in the absence of reflex actions, one would conclude that stiffness, or its inverse, compliance, is the regulated property of the motor servo.

In this way, the autogenetic reflex motor servo provides the local, reflex feedback loops for individual muscular contractions. A voluntary contraction force F of human skeletal muscle is reflexly excited (positive feedback $+F^{-1}$) by the responses of its *spindle receptors* to stretch and is reflexly inhibited (negative feedback $-F^{-1}$) by the responses of its *Golgi tendon organs* to contraction. Stretch and unloading reflexes are mediated by combined actions of several autogenetic neural pathways, forming the *motor servo*.

In other words, branches of the afferent fibers also synapse with with interneurons that inhibit motor neurons controlling the antagonistic muscles – *reciprocal inhibition*. Consequently, the stretch stimulus causes the antagonists to relax so that they cannot resists the shortening of the stretched muscle caused by the main reflex arc. Similarly, firing of the Golgi tendon receptors causes inhibition of the muscle contracting too strong and simultaneous *reciprocal activation* of its antagonist.

20.5 Human Brain

Recall that *human brain* is the most complicated mechanism in the Universe. Roughly, it has its own physics and physiology. A closer look at the brain points to a rather *recursively hierarchical structure* and a very elaborate organization (see [II06b] for details). An average brain weighs about 1.3 kg, and it is made of: $\sim$ 77% water, $\sim$ 10% protein, $\sim$ 10% fat, $\sim$ 1% carbohydrates, $\sim$ 0.01% DNA/RNA. The largest part of the human brain, the cerebrum, is found on the top and is divided down the middle into left and right cerebral hemispheres, and front and back into *frontal lobe*, *parietal lobe*, *temporal lobe*, and *occipital lobe*. Further down, and at the back lies a rather smaller, spherical portion of the brain, the *cerebellum*, and deep inside lie a number of complicated structures like the *thalamus*, *hypothalamus*, *hippocampus*, etc.

Both the cerebrum and the cerebellum have comparatively thin outer surface layers of grey matter and larger inner regions of white matter. The grey regions constitute what is known as the *cerebral cortex* and the *cerebellar cortex*. It is in the *grey matter* where various kinds of *computational tasks* seem to be performed, while the *white matter* consists of long nerve fibers (axons) carrying signals from one part of the brain to another. However, despite of its amazing computational abilities, brain is not a computer, at least not a 'Von Neumann computer' [Neu58], but rather a huge, hierarchical, neural network. It is the cerebral cortex that is central to the higher brain functions, speech, thought, complex movement patterns, etc. On the other hand, the cerebellum seems to be more of an 'automaton'. It has to do more with precise coordination and control of the body, and with skills that have become our 'second nature'. Cerebellum actions seem almost to take place by themselves, without thinking about them. They are very similar to the unconscious reflex actions, e.g., reaction to pinching, which may not be mediated by the brain, but by the upper part of the spinal column.

Various regions of the cerebral cortex are associated with very specific functions. The *visual cortex*, a region in the occipital lobe at the back of the brain, is responsible for the reception and interpretation of vision. The *auditory cortex*, in the temporal lobe, deals mainly with analysis of sound, while the *olfactory cortex*, in the frontal lobe, deals with smell. The *somatosensory cortex*, just *behind* the division between frontal and parietal lobes, has to do with the sensations of touch. There is a very specific mapping between the various parts of the surface of the body and the regions of the somatosensory cortex. In addition, just in *front* of the division between the frontal and parietal lobes, in the frontal lobe, there is the *motor cortex*. The *motor cortex* activates the movement of different parts of the body and, again here, there is a very specific mapping between the various muscles of the body and the regions of the motor cortex. All the above mentioned regions of the cerebral cortex are referred to as *primary*, since they are the one most directly concerned with the input and output of the brain. Near to these primary regions are the *secondary* sensory regions of the cerebral cortex, where information is processed, while in the *secondary motor* regions, conceived plans of motion get translated into specific directions for actual muscle movement by the primary motor cortex. The most abstract and sophisticated activity of the brain is carried out in the remaining regions of the cerebral cortex, the *association cortex*.

The basic building blocks of the brain are the nerve cells or neurons. Among about 200 types of different basic types of human cells, the neuron is one of the most specialized, exotic and remarkably versatile cell. The neuron is highly unusual in three respects: its *variation in shape*, its *electrochemical function*, and its *connectivity*, i.e., its ability to link up with other neurons in networks. Let us start with a few elements of neuron microanatomy (see [II06b]). There is a central starlike bulb, called the *soma*, which contains the nucleus of the cell. A long nerve fibre, known as the *axon*, stretches out from one end of the soma. Its length, in humans, can reach up to few *cm*. The function of an axon is to transmit the neuron's output signal, in which it acts like a *coaxial cable*. The axon has the ability of multiple bifurcation, branching out into many smaller nerve fibers, and the very end of which there is always a *synaptic knob*. At the other end of the soma and often springing off in all directions from it, are the tree–like *dendrites*, along which input data are carried into the soma. The whole nerve cell, as basic unit, has a cell membrane surrounding soma, axon, synoptic knobs, and dendrites. Signals pass from one neuron to another at junctions known as *synapses*, where a synaptic knob of one neuron is attached to another neuron's soma or dendrites. There is very narrow gap, of a few *nm*, between the synaptic knob and the soma/dendrite to where the *synaptic cleft* is attached. The signal from one neuron to another has to propagate across this gap.

A nerve fibre is a cylindrical tube containing a mixed solution of NaCl and KCl, mainly the second, so there are Na^+, K^+, and Cl^- ions within the tube. Outside the tube the same type of ions are present but with more Na^+ than K^+. In the *resting* state there is an excess of Cl^- over Na^+ and K^+ inside the tube, giving it a negative charge, while it has positive charge outside. A nerve signal is a region of *charge reversal* traveling along the fibre. At its head, *sodium gates* open to allow the sodium to flow inwards and at its tail *potassium gates* open to allow potassium to flow outwards. Then, metabolic pumps act to restore order and establish the *resting state*, preparing the nerve fibre for another signal. There is no major material (ion) transport that produces the signal, just in and out local movements of ions, across the cell membranes, i.e., a *small* and *local* depolarization of the cell. Eventually, the nerve signal reaches the attached synaptic knob, at the very end of the nerve fibre, and triggers it to emit chemical substances, known as *neurotransmitters.* It is these substances that travel across the synaptic cleft to another neuron's soma or dendrite. It should be stressed that the signal here is not electrical, but a chemical one. What really is happening is that when the nerve signal reaches the synaptic knob, the local depolarization cause little bags immersed in the *vesicular grid,* the *vesicles* containing molecules of the neurotransmitter chemical (e.g., acetylcholine) to release their contents from the neuron into the synaptic cleft, the phenomenon of *exocytosis.* These molecules then diffuse across the cleft to interact with *receptor proteins* on receiving neurons. On receiving a neurotransmitter molecule, the receptor protein opens a gate that causes a local depolarization of the receiver neuron.

It depends on the nature of the synaptic knob and of the specific synaptic junction, if the next neuron would be encouraged to *fire,* i.e., to start a new signal along its own axon, or it would be discouraged to do so. In the former case we are talking about *excitatory synapses,* while in the latter case about *inhibitory synapses.* At any given moment, one has to add up the effect of all excitatory synapses and subtract the effect of all the inhibitory ones. If the net effect corresponds to a positive electrical potential difference between the inside and the outside of the neuron under consideration, *and* if it is bigger than a critical value, then the neuron *fires,* otherwise it stays mute.

The basic dynamical process of *neural communication* can be summarized in the following three steps: [II06b]

1. The neural axon is an *all or none* state. In the *all state* a signal, called a *spike* or *action potential* (AP), propagates indicating that the summation performed in the soma produced an amplitude of the order of tens of mV. In the *none state* there is no signal traveling in the axon, only the resting potential (~ -70mV). It is essential to notice that the presence of a traveling signal in the axon, *blocks* the possibility of transmission of a second signal.
2. The nerve signal, upon arriving at the ending of the axon, triggers the emission of neurotransmitters in the synaptic cleft, which in turn cause the receptors to open up and allow the penetration of ionic current into the *post synaptic* neuron. The *efficacy* of the synapse is a parameter specified by the amount of penetrating current per presynaptic spike.
3. The post synaptic potential (PSP) diffuses toward the soma, where all inputs in a short period, from all the presynaptic neurons connected to the postsynaptic are summed up. The amplitude of individual PSP's is about 1mV, thus quite a number of inputs is required to reach the 'firing' threshold, of tens of mV. Otherwise the postsynaptic neuron remains in the *resting* or *none* state.

The cycle–time of a neuron, i.e., the time from the emission of a spike in the presynaptic neuron to the emission of a spike in the postsynaptic neuron is of the order of $1-2$ ms. There is also some recovery time for the neuron, after it fired, of about $1-2$ ms, independently of how large the amplitude of the depolarizing potential would be. This period is called the absolute refractory period of the neuron. Clearly, it sets an upper bound on the spike frequency of $500-1000$/sec.

In the types of neurons that we will be interested in, the spike frequency is considerably lower than the above upper bound, typically in the range of 100/sec, or even smaller in some areas, at about 50/sec. It should be noticed that this rather exotic neural communication mechanism works very efficiently and it is employed universally, both by vertebrates and invertebrates. The vertebrates have gone even further in perfection, by protecting their nerve fibers by an insulating coating of myelin, a white fatty substance, which incidentally gives the white matter of the brain, discussed above, its color. Because of this insulation, the nerve signals may travel undisturbed at about 120 m/sec (see [II06b]).

A very important anatomical fact is that each neuron receives some 10^4 synaptic inputs from the axons of other neurons, usually one input per presynaptic neuron, and that each branching neural axon forms about the same number ($\sim 10^4$) of synaptic contacts on other, postsynaptic neurons. A closer look at our cortex then would expose a mosaic-type structure of assemblies of a few thousand densely connected neurons. These assemblies are taken to be the basic cortical processing *modules*, and their size is about $1(\mathrm{mm})^2$. The neural connectivity gets much sparser as we move to larger scales and with much less feedback, allowing thus for autonomous local collective, parallel processing and more serial and integrative processing of local collective outcomes. Taking into account that there are about 10^{11} nerve cells in the brain (about 7×10^{10} in the cerebrum and 3×10^{10} in the cerebellum), we are talking about 10^{15} synapses.

A very useful set of phenomenological rules has been put forward by Hebb [Heb49], the *Hebb rules*, concerning the underlying mechanism of brain plasticity. According to Hebb, a synapse between neuron 1 and neuron 2 would be strengthened whenever the firing of neuron 1 is followed by the firing of neuron 2, and weakened whenever it is not. It seems that *brain plasticity* is a *fundamental property* of the activity of the brain.

Many mathematical models have been proposed to try to simulate *learning process*, based upon the close resemblance of the dynamics of neural communication to computers and implementing, one way or another, the essence of the Hebb rules. These models are known as *neural networks*. They are closely related to *adaptive Kalman filtering* (see [II06c]).

Let us try to construct a *neural network* model for a set of N interconnected neurons (see e.g., [Kos92, II07b]). The activity of the neurons is usually parameterized by N functions $\sigma_i(t)$, $(i = 1, 2, \dots, N)$, and the synaptic strength, representing the synaptic efficacy, by $N \times N$ functions $j_{ik}(t)$. The total stimulus of the network on a given neuron (i) is assumed to be given simply by the sum of the stimuli coming from each neuron

$$S_i(t) = j_{ik}(t)\sigma_k(t), \qquad \text{(summation over } k\text{)}$$

where we have identified the individual stimuli with the product of the synaptic strength j_{ik} with the activity σ_k of the neuron producing the individual stimulus. The dynamic equations for the neuron are supposed to be, in the simplest case

$$\dot{\sigma}_i = F(\sigma_i, S_i), \tag{20.19}$$

with F a nonlinear function of its arguments. The dynamic equations controlling the time evolution of the synaptic strengths $j_{ik}(t)$ are much more involved and only partially understood, and usually it is assumed that the j–dynamics is such that it produces the synaptic couplings. The simplest version of a neural network model is the *Hopfield model* [Hop82]. In this model the neuron activities are conveniently and conventionally taken to be *switch*–like, namely ± 1, and the time t is also an integer–valued quantity. This $all(+1)$ or $none(-1)$ neural activity σ_i is based on the neurophysiology. The choice ± 1 is more natural the usual 'binary' one ($b_i = 1$ or 0), from a physicist's point of view corresponding to a two–state system, like the fundamental elements of the ferromagnet, i.e., the electrons with their spins up $(+)$ or $(-)$.

The increase of time t by one unit corresponds to one step for the dynamics of the neuron activities obtainable by applying (for all i) the rule

$$\sigma_i(t + \frac{i+1}{N}) = \mathrm{sign}(S_i(t + i/N)), \tag{20.20}$$

which provides a rather explicit form for (20.19). If, as suggested by the Hebb rules, the j matrix is *symmetric* ($j_{ik} = j_{ki}$), the Hopfield dynamics [Hop82] corresponds to a sequential algorithm for looking for the minimum of the Hamiltonian

$$H = -S_i(t)\sigma_i(t) = -j_{ik}\sigma_i(t)\sigma_k(t).$$

The Hopfield model, at this stage, is very similar to the dynamics of a statistical mechanics *Ising–type*, or, more generally a *spin–glass*, model (see [II07b]. This *mapping* of the Hopfield model to a spin–glass model is highly advantageous because we have now a justification for using the statistical mechanics language of phase transitions, like critical points or attractors, etc, to describe neural dynamics and thus brain dynamics. This simplified Hopfield model has many *attractors*, corresponding to many different *equilibrium* or *ordered* states, endemic in spin–glass models, and an unavoidable prerequisite for successful storage, in the brain, of many different patterns of activities. In the neural network framework, it is believed that an internal representation (i.e., a pattern of neural activities) is associated with each object or category that we are capable of recognizing and remembering. According to neurophysiology, it is also believed that an object is memorized by suitably changing the synaptic strengths. The so–called *associative memory* is generated in this scheme as follows [II06b]: An external stimulus, suitably involved, produces synaptic strengths such that a specific learned pattern $\sigma_i(0) = P_i$ is 'printed' in such a way that the neuron activities $\sigma_i(t) \sim P_i$ (II *learning*), meaning that the σ_i will remain for all times close to P_i, corresponding to a stable attractor point (III *coded brain*). Furthermore, if a *replication signal* is applied, pushing the neurons to σ_i values *partially* different from P_i, the neurons should evolve toward the P_i. In other words, the memory is able to retrieve the information on the whole object, from the knowledge of a part of it, or even in the presence of wrong information (IV *recall process*). Clearly, if the external stimulus is very different from any preexisting $\sigma_i = P_i$ pattern, it may either create a new pattern, i.e., create a new attractor point, or it may reach a chaotic, random behavior (I *uncoded brain*).

Despite the remarkable progress that has been made during the last few years in understanding brain function using the neural network paradigm, it is fair to say that neural networks are rather artificial and a very long way from providing a realistic model of brain function. It seems likely that the mechanisms controlling the changes in synaptic connections are much more complicated and involved than the ones considered in NN, as utilizing cytoskeletal restructuring of the sub–synaptic regions. *Brain plasticity* seems to play an essential, central role in the workings of the brain! Furthermore, the 'binding problem, i.e., how to *bind* together all the neurons firing to different features of the same object or category, especially when more than one object is perceived during a *single* conscious perceptual moment, seems to remain unanswered. In this way, we have come a long way since the times of the 'grandmother neuron', where a *single* brain location was invoked for self observation and control, identified with the pineal glands by Descartes [II06b].

It has been long suggested that different groups of neurons, responding to a common object/category, fire *synchronously*, implying *temporal correlations* [SG95]. If true, such correlated firing of neurons may help us in resolving the binding problem [Cri94]. Actually, brain waves recorded from the scalp, i.e., the EEGs, suggest the existence of some sort of *rhythms*, e.g., the '$\alpha-$rhythms' of a frequency of 10 Hz. More recently, oscillations were clearly observed in the visual cortex. Rapid oscillations, above EEG frequencies in the range of 35 to 75 Hz, called the '$\gamma-$oscillations' or the '40 Hz oscillations', have been detected in the cat's visual cortex [SG95]. Furthermore, it has been shown that these oscillatory responses can become *synchronized* in a stimulus–dependent manner. Studies of auditory–evoked responses in humans have shown inhibition of the 40 Hz coherence with *loss of consciousness* due to the induction of general anesthesia. These striking results have prompted Crick and Koch to suggest that this *synchronized firing* on, or near, the beat of a '$\gamma-$oscillation' (in the 35–75 Hz range) might be the *neural correlate* of *visual awareness* [Cri94]. Such a behavior would be, of course, a very special case of a much more general framework where coherent firing of *widely–distributed, non–local* groups of neurons, in the 'beats' of $x-$oscillation (of specific frequency ranges), *bind* them together in a mental representation, expressing the *oneness* of *consciousness* or *unitary sense of self*. While this is a bold suggestion [Cri94], it is should be stressed that in a physicist's language it corresponds to a phenomenological explanation, not providing the underlying physical mechanism, based on neuron dynamics, that triggers the synchronized neuron firing. On the other hand, the *Crick–Koch binding hypothesis* [CK90, Cri94] is very suggestive (see Figure 20.7) and in compliance with the *central biodynamic adjunction* [II06b]

$$\mathit{coordination} = \mathit{sensory} \dashv \mathit{motor} : \mathit{brain} \leftrightarrows \mathit{body}.$$

On the other hand, E.M. Izhikevich, Editor–in–Chief of the new Encyclopedia of Computational Neuroscience, considers brain as a *weakly–connected neural*

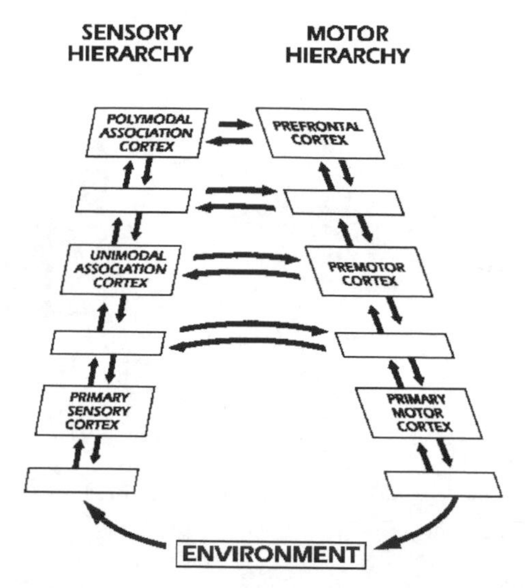

Fig. 20.7. Fiber connections between cortical regions participating in the *perception–action cycle*, reflecting again our sensory–motor adjunction. Empty rhomboids stand for intermediate areas or subareas of the labelled regions. Notice that there are connections between the two hierarchies at several levels, not just at the top level.

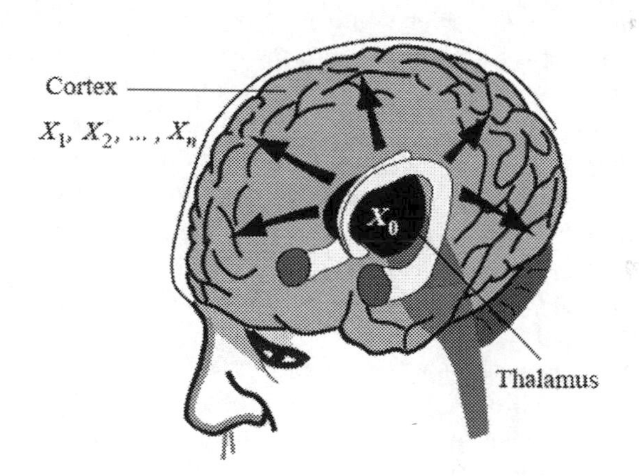

Fig. 20.8. A 1–to–many relation: *Thalamus* $\Rightarrow$ *Cortex* in the human brain (with permission from E. Izhikevich).

network [Izh99b], consisting of n quasi–periodic cortical oscillators $X_1, ..., X_n$ forced by the thalamic input X_0 (see Figure 20.8)

20.5.1 Basics of Brain Physiology

The nervous system consists basically of two types of cells: neurons and glia. *Neurons* (also called nerve cells, see Figure 20.9) are the primary cells, morphologic and functional units of the nervous system. They are found in the brain,

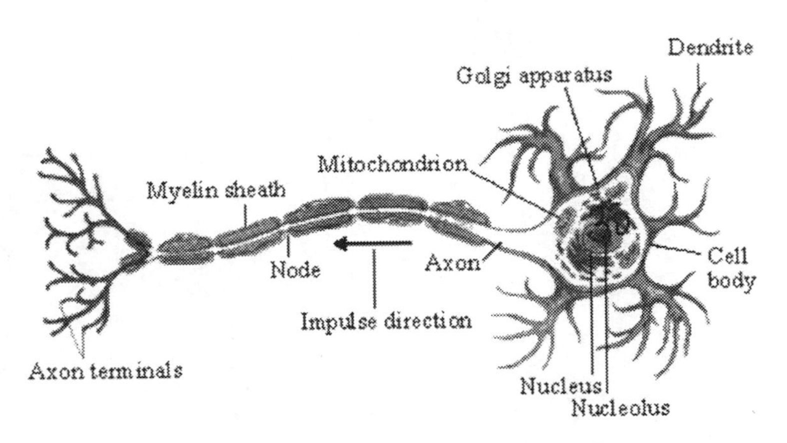

Fig. 20.9. A typical neuron, containing all of the usual cell organelles. However, it is highly specialized for the conductance of nerve impulse.

the spinal cord and in the peripheral nerves and ganglia. Neurons consist of four major parts, including the *dendrites* (shorter projections), which are responsible for receiving stimuli; the *axon* (longer projection), which sends the nerve impulse away from the cell; the *cell body*, which is the site of metabolic activity in the cell; and the *axon terminals*, which connect neurons to other neurons, or neurons to other body structures. Each neuron can have several hundred axon terminals that attach to another neuron multiple times, attach to multiple neurons, or both. Some types of neurons, such as Purkinje cells, have over 1000 dendrites. The body of a neuron, from which the axon and dendrites project, is called the soma and holds the nucleus of the cell. The nucleus typically occupies most of the volume of the soma and is much larger in diameter than the axon and dendrites, which typically are only about a micrometer thick or less. Neurons join to one another and to other cells through synapses.

A defining feature of neurons is their ability to become 'electrically excited', that is, to undergo an action potential–and to convey this excitation rapidly along their axons as an impulse. The narrow cross section of axons and dendrites lessens the metabolic expense of conducting action potentials, although fatter axons convey the impulses more rapidly, generally speaking.

Many neurons have insulating sheaths of myelin around their axons, which enable their action potentials to travel faster than in unmyelinated axons of the same diameter. Formed by glial cells, the myelin sheathing normally runs along the axon in sections about $1\,mm$ long, punctuated by unsheathed nodes of Ranvier. Neurons and glia make up the two chief cell types of the nervous system.

An action potential that arrives at its terminus in one neuron may provoke an action potential in another through release of neurotransmitter molecules across the synaptic gap.

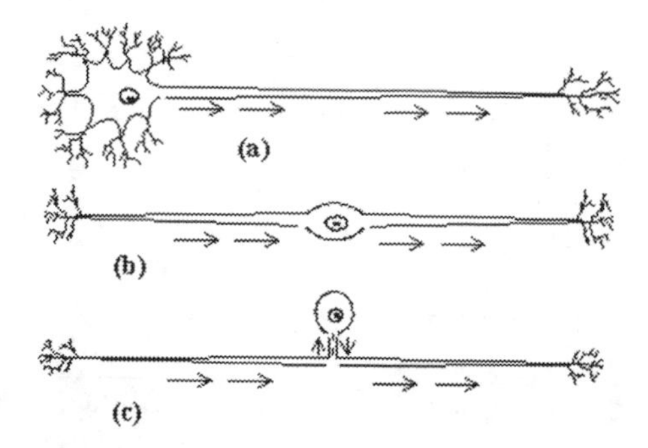

Fig. 20.10. Three structural classes of human neurons: (a) multipolar, (b) bipolar, and (c) unipolar.

There are three structural classes of neurons in the human body (see Figure 20.10):

1. The *multipolar neurons*, the majority of neurons in the body, in particular in the central nervous system.
2. The *bipolar neurons*, sensory neurons found in the special senses.
3. The *unipolar neurons*, sensory neurons located in dorsal root ganglia.

Neuronal Circuits

Figure 20.11 depicts a general model of a convergent circuit, showing two neurons converging on one neuron. This allows one neuron or *neuronal pool* to receive input from multiple sources. For example, the neuronal pool in the brain that regulates rhythm of breathing receives input from other areas of the brain, baroreceptors, chemoreceptors, and stretch receptors in the lungs.

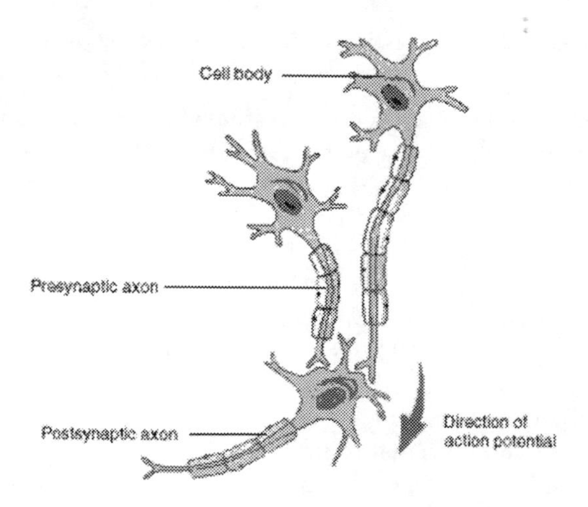

Fig. 20.11. A convergent neural circuit: nerve impulses arriving at the same neuron.

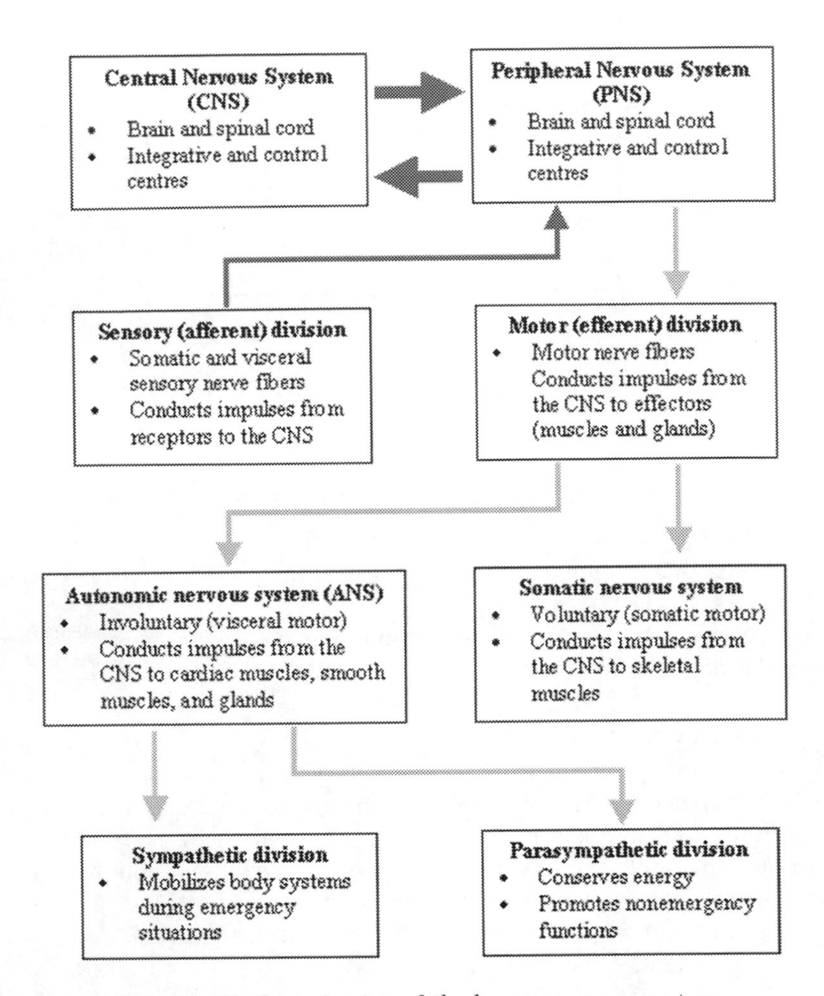

Fig. 20.12. Organization of the human nervous system.

Glia are specialized cells of the nervous system whose main function is to 'glue' neurons together. Specialized glia called *Schwann cells* secrete myelin sheaths around particularly long axons. Glia of the various types greatly outnumber the actual neurons.

The human nervous system consists of the central and peripheral parts (see Figure 20.12). The *central nervous system* (CNS) refers to the core nervous system, which consists of the *brain* and *spinal cord* (as well as *spinal nerves*). The *peripheral nervous system* (PNS) consists of the nerves and neurons that reside or extend outside the central nervous system–to serve the limbs and organs, for example. The peripheral nervous system is further divided into the somato–motoric nervous system and the autonomic nervous system (see Figure 20.13).

The CNS is further divided into two parts: the *brain* and the *spinal cord*. The average adult human brain weighs 1.3 to 1.4 kg (approximately 3 pounds). The

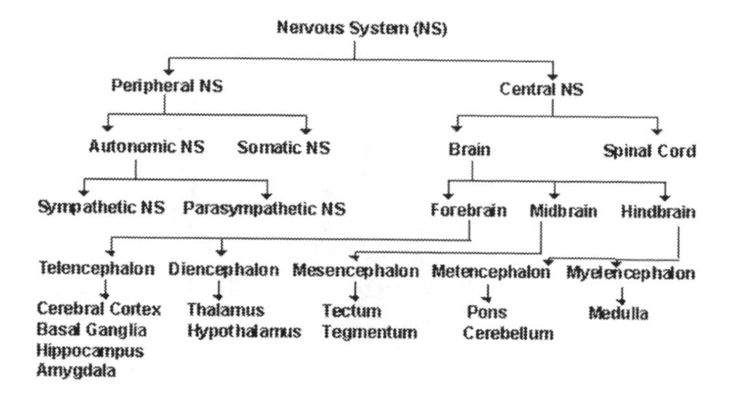

Fig. 20.13. Basic divisions of the human nervous system.

brain contains about 100 billion nerve cells (neurons) and trillions of 'support cells' called glia. Further divisions of the human brain are depicted in Figure 20.14. The spinal cord is about 43 cm long in adult women and 45 cm long in adult men and weighs about 35-40 grams. The vertebral column, the collection of bones (back bone) that houses the spinal cord, is about 70 cm long. Therefore, the spinal cord is much shorter than the vertebral column.

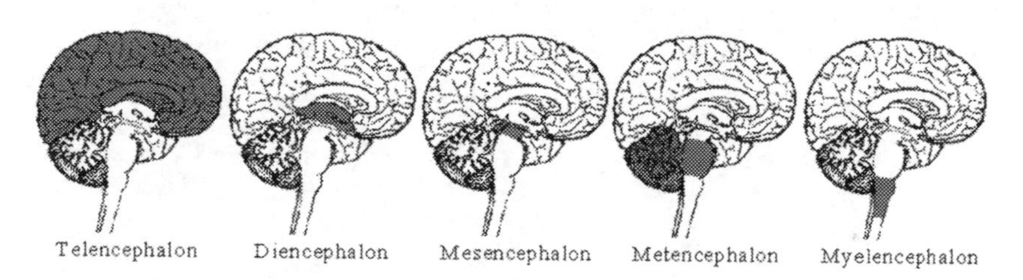

Fig. 20.14. Basic divisions of the human Brain.

The PNS is further divided into two major parts: the *somatic nervous system* and the *autonomic nervous system*.

The somatic nervous system consists of *peripheral nerve fibers* that send sensory information to the central nervous system and *motor nerve fibers* that project to skeletal muscle.

The autonomic nervous system (ANS) controls smooth muscles of the viscera (internal organs) and glands. In most situations, we are unaware of the workings of the ANS because it functions in an involuntary, reflexive manner. For example, we do not notice when blood vessels change size or when our heart beats faster. The ANS is most important in two situations:

1. In *emergencies* that cause stress and require us to 'fight' or take 'flight', and
2. In *non–emergencies* that allow us to 'rest' and 'digest'.

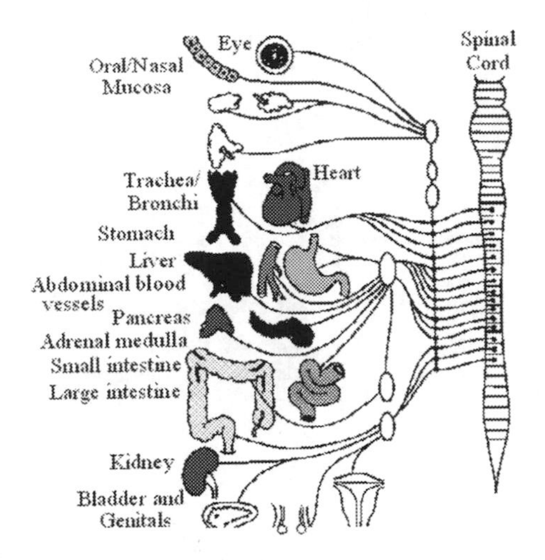

Fig. 20.15. Basic functions of the sympathetic nervous system.

The ANS is divided into three parts:

1. The *sympathetic nervous system* (see Figure 20.15),
2. The *parasympathetic nervous system* (see Figure 20.16), and
3. The *enteric nervous system*, which is a meshwork of nerve fibers that innervate the viscera (gastrointestinal tract, pancreas, gall bladder).

In the PNS, neurons can be functionally divided in 3 ways:

1. • *sensory* (*afferent*) neurons – carry information into the CNS from sense organs, and

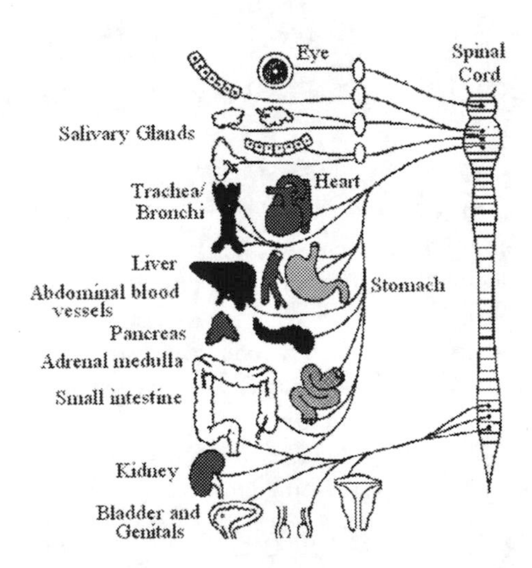

Fig. 20.16. Basic functions of the parasympathetic nervous system.

- *motor* (*efferent*) neurons – carry information away from the CNS (for muscle control).

2. - *cranial* neurons – connect the brain with the periphery, and
   - *spinal* neurons – connect the spinal cord with the periphery.
3. - *somatic* neurons – connect the skin or muscle with the central nervous system, and
   - *visceral* neurons – connect the internal organs with the central nervous system.

Some differences between the PNS and the CNS are:

1. - In the CNS, collections of neurons are called *nuclei.*
   - In the PNS, collections of neurons are called *ganglia.*
2. - In the CNS, collections of axons are called *tracts.*
   - In the PNS, collections of axons are called *nerves.*

Basic Brain Partitions and Their Functions

Cerebral Cortex. The word 'cortex' comes from the Latin word for 'bark' (of a tree). This is because the cortex is a sheet of tissue that makes up the outer layer of the brain. The thickness of the cerebral cortex varies from 2 to 6 mm. The right and left sides of the cerebral cortex are connected by a thick band of nerve fibers called the 'corpus callosum'. In higher mammals such as humans, the cerebral cortex looks like it has many bumps and grooves. A bump or bulge on the cortex is called a gyrus (the plural of the word gyrus is 'gyri') and a groove is called a sulcus (the plural of the word sulcus is 'sulci'). Lower mammals like rats and mice have very few gyri and sulci. The main cortical functions are: *thought, voluntary movement, language, reasoning,* and *perception.*

Cerebellum. The word 'cerebellum' comes from the Latin word for 'little brain'. The cerebellum is located behind the brain stem. In some ways, the cerebellum is similar to the cerebral cortex: the cerebellum is divided into hemispheres and has a cortex that surrounds these hemispheres. Its main functions are: *movement, balance,* and *posture.*

Brain Stem. The brain stem is a general term for the area of the brain between the thalamus and spinal cord. Structures within the brain stem include the medulla, pons, tectum, reticular formation and tegmentum. Some of these areas are responsible for the most basic functions of life such as breathing, heart rate and blood pressure. Its main functions are: *breathing, heart rate,* and *blood pressure.*

Hypothalamus. The *hypothalamus* is composed of several different areas and is located at the base of the brain. Although it is the size of only a pea (about 1/300 of the total brain weight), the hypothalamus is responsible for some very important functions. One important function of the hypothalamus is the control of body temperature. The hypothalamus acts like a 'thermostat' by sensing changes in body temperature and then sending signals to adjust the temperature. For example, if we are too hot, the hypothalamus detects this and then sends a signal to expand the capillaries in your skin. This causes blood to be cooled faster. The hypothalamus also controls the pituitary. Its main functions are: *body temperature*, *emotions*, *hunger*, *thirst*, *sexual instinct*, and *circadian rhythms*. The *hypothalamus* is 'the boss' of the ANS.

Thalamus. The thalamus receives sensory information and relays this information to the cerebral cortex. The cerebral cortex also sends information to the thalamus which then transmits this information to other areas of the brain and spinal cord. Its main functions are: *sensory processing* and *movement*.

Limbic System. The limbic system (or the limbic areas) is a group of structures that includes the amygdala, the hippocampus, mammillary bodies and cingulate gyrus. These areas are important for controlling the emotional response to a given situation. The hippocampus is also important for memory. Its main function is *emotions*.

Hippocampus. The hippocampus is one part of the limbic system that is important for memory and learning. Its main functions are: *learning* and *memory*.

Basal Ganglia. The basal ganglia are a group of structures, including the globus pallidus, caudate nucleus, subthalamic nucleus, putamen and substantia nigra, that are important in coordinating movement. Its main function is *movement*.

Midbrain. The midbrain includes structures such as the superior and inferior colliculi and red nucleus. There are several other areas also in the midbrain. Its main functions are: *vision*, *audition*, *eye movement* (see Figure 20.17), and *body movement*.

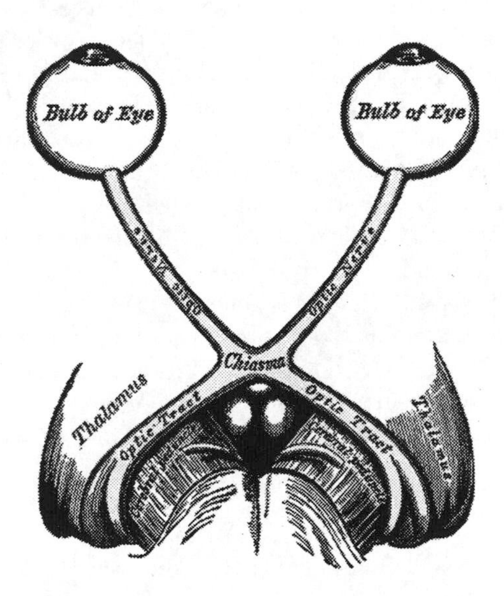

Fig. 20.17. Optical chiasma: the point of cross–over for optical nerves. By means of it, information presented to either left or right visual half–field is projected to the contralateral occipital areas in the visual cortex. For more details, see e.g. [II06b].

Nerves

A *nerve* is an enclosed, cable–like bundle of *nerve fibers* or *axons*, which includes the glia that ensheathe the axons in myelin (see [Mar98, II06b]).

Nerves are part of the peripheral nervous system. *Afferent nerves* convey sensory signals to the brain and spinal cord, for example from skin or organs, while *efferent* nerves conduct stimulatory signals from the *motor neurons* of the brain and spinal cord to the muscles and glands.

These signals, sometimes called nerve impulses, are also known as action potentials: Rapidly traveling electrical waves, which begin typically in the cell body of a neuron and propagate rapidly down the axon to its tip or terminus'.

Nerves may contain fibers that all serve the same purpose; for example *motor nerves*, the axons of which all terminate on muscle fibers and stimulate contraction. Or they be *mixed nerves.*

An *axon*, or 'nerve fibre', is a long slender projection of a nerve cell or neuron, which conducts electrical impulses away from the neuron's cell body or soma. Axons are in effect the primary transmission lines of the nervous system, and as bundles they help make up nerves. The axons of many neurons are sheathed in myelin.

On the other hand, a *dendrite* is a slender, typically branched projection of a nerve cell or neuron, which conducts the electrical stimulation received from other cells through synapses to the body or soma of the cell from which it projects.

Many dendrites convey this stimulation passively, meaning without action potentials and without activation of voltage–gated ion channels. In such dendrites the voltage change that results from stimulation at a synapse may extend both towards and away from the soma. In other dendrites, though an action potential may not arise, nevertheless voltage–gated channels help to propagate excitatory synaptic stimulation. This propagation is efficient only toward the soma due to an uneven distribution of channels along such dendrites.

The structure and branching of a neuron's dendrites strongly influences how it integrates the input from many others, particularly those that input only weakly (more at synapse). This integration is in aspects 'temporal'–involving the summation of stimuli that arrive in rapid succession–as well as 'spatial' – entailing the aggregation of excitatory and inhibitory inputs from separate branches or 'arbors'.

Spinal nerves take their origins from the spinal cord. They control the functions of the rest of the body. In humans, there are 31 pairs of spinal nerves: 8 *cervical*, 12 *thoracic*, 5 *lumbar*, 5 *sacral* and 1 *coccygeal*.

Neural Action Potential

As the traveling signals of nerves and as the localized changes that contract muscle cells, *action potentials* are an essential feature of animal life. They set the pace of thought and action, constrain the sizes of evolving anatomies and enable centralized control and coordination of organs and tissues (see [Mar98]).

Basic Features

When a biological cell or patch of membrane undergoes an action potential, the polarity of the transmembrane voltage swings rapidly from negative to positive and back. Within any one cell, consecutive action potentials typically are indistinguishable. Also between different cells the amplitudes of the voltage swings tend to be roughly the same. But the speed and simplicity of action potentials vary significantly between cells, in particular between different cell types.

Minimally, an action potential involves a *depolarization*, a *re–polarization* and finally a *hyperpolarization* (or 'undershoot'). In specialized muscle cells of the heart, such as the pacemaker cells, a 'plateau phase' of intermediate voltage may precede re–polarization.

Underlying Mechanism

The transmembrane voltage changes that take place during an action potential result from changes in the permeability of the membrane to specific ions, the internal and external concentrations of which are in imbalance. In the axon fibers of nerves, depolarization results from the inward rush of sodium ions, while re–polarization and hyperpolarization arise from an outward rush of potassium ions. Calcium ions make up most or all of the depolarizing currents at an axon's pre–synaptic terminus, in muscle cells and in some dendrites.

The imbalance of ions that makes possible not only action potentials but the resting cell potential arises through the work of pumps, in particular the sodium–potassium exchanger.

Changes in membrane permeability and the onset and cessation of ionic currents reflect the opening and closing of 'voltage–gated' ion channels, which provide portals through the membrane for ions. Residing in and spanning the membrane, these enzymes sense and respond to changes in transmembrane potential.

Initiation

Action potentials are triggered by an initial depolarization to the point of *threshold.* This threshold potential varies but generally is about 15 millivolts above the resting potential of the cell. Typically action potential initiation occurs at a synapse, but may occur anywhere along the axon. In his discovery of 'animal electricity', L. Galvani elicited an action potential through contact of his scalpel with the motor nerve of a frog he was dissecting, causing one of its legs to kick as in life.

Wave Propagation

In the fine fibers of simple (or 'unmyelinated') axons, action potentials propagate as waves, which travel at speeds up to 120 meters per second.

The propagation speed of these 'impulses' is faster in fatter fibers than in thin ones, other things being equal. In their *Nobel Prize* winning work uncovering the wave nature and ionic mechanism of action potentials, *Alan L. Hodgkin* and

Andrew F. Huxley performed their celebrated experiments on the 'giant fibre' of Atlantic squid [HH52]. Responsible for initiating flight, this axon is fat enough to be seen without a microscope (100 to 1000 times larger than is typical). This is assumed to reflect an adaptation for speed. Indeed, the velocity of nerve impulses in these fibers is among the fastest in nature.

Saltatory Propagation

Many neurons have insulating sheaths of myelin surrounding their axons, which enable action potentials to travel faster than in unmyelinated axons of the same diameter. The myelin sheathing normally runs along the axon in sections about 1 mm long, punctuated by unsheathed 'nodes of Ranvier'.

Because the salty cytoplasm of the axon is electrically conductive, and because the myelin inhibits charge leakage through the membrane, depolarization at one node is sufficient to elevate the voltage at a neighboring node to the threshold for action potential initiation. Thus in myelinated axons, action potentials do not propagate as waves, but recur at successive nodes and in effect hop along the axon. This mode of propagation is known as *saltatory conduction*. Saltatory conduction is faster than smooth conduction. Some typical action potential velocities are as follows:

Fiber	Diameter	AP Velocity
Unmyelinated	0.2–1.0 micron	0.2–2 m/sec
Myelinated	2–20 microns	12–120 m/sec

The disease called *multiple sclerosis* (MS) is due to a breakdown of myelin sheathing, and degrades muscle control by destroying axons' ability to conduct action potentials.

Detailed Features

Depolarization and re–polarization together are complete in about two milliseconds, while undershoots can last hundreds of milliseconds, depending on the cell. In neurons, the exact length of the roughly two–millisecond delay in re–polarization can have a strong effect on the amount of neurotransmitter released at a synapse. The duration of the hyperpolarization determines a nerve's 'refractory period' (how long until it may conduct another action potential) and hence the frequency at which it will fire under continuous stimulation. Both of these properties are subject to biological regulation, primarily (among the mechanisms discovered so far) acting on ion channels selective for potassium.

A cell capable of undergoing an action potential is said to be *excitable.*

Synapses

Synapses are specialized junctions through which cells of the nervous system signal to one another and to non-neuronal cells such as muscles or glands (see

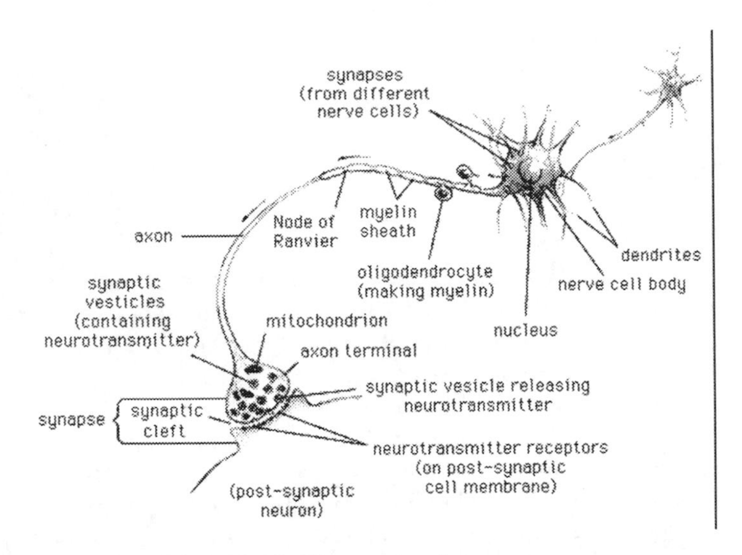

Fig. 20.18. Neuron forming a synapse.

Figure 20.18). Synapses define the circuits in which the neurons of the central nervous system interconnect. They are thus crucial to the biological computations that underlie perception and thought. They also provide the means through which the nervous system connects to and controls the other systems of the body (see [II06b]).

Anatomy and Structure. At a classical synapse, a mushroom-shaped bud projects from each of two cells and the caps of these buds press flat against one another (see Figure 20.19). At this interface, the membranes of the two cells flank each other across a slender gap, the narrowness of which enables signaling molecules known as neurotransmitters to pass rapidly from one cell to the other by diffusion. This gap is sometimes called the synaptic cleft.

Synapses are asymmetric both in structure and in how they operate. Only the so–called pre–synaptic neuron secretes the neurotransmittér, which binds to receptors facing into the synapse from the post–synaptic cell. The pre–synaptic nerve terminal generally buds from the tip of an axon, while the post–synaptic target surface typically appears on a dendrite, a cell body or another part of a cell.

Signalling across the Synapse

The release of neurotransmitter is triggered by the arrival of a nerve impulse (or action potential) and occurs through an unusually rapid process of cellular secretion: Within the pre–synaptic nerve terminal, vesicles containing neurotransmitter sit 'docked' and ready at the synaptic membrane. The arriving action potential produces an influx of calcium ions through voltage–dependent, calcium–selective ion channels, at which point the vesicles fuse with the membrane and release their contents to the outside. Receptors on the opposite side

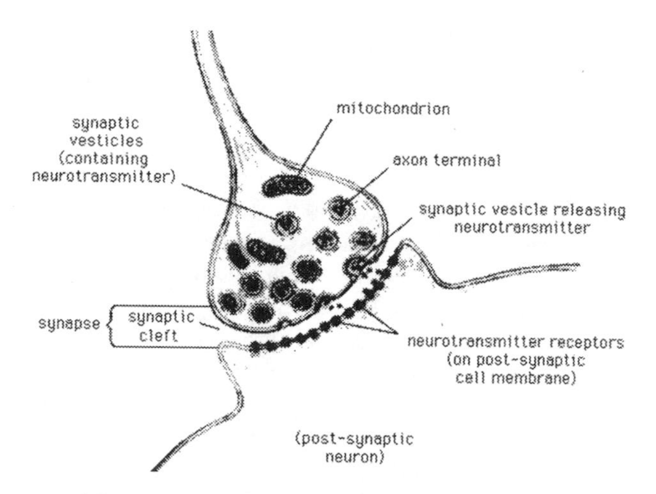

Fig. 20.19. Structure of a chemical synapse.

of the synaptic gap bind neurotransmitter molecules and respond by opening nearby ion channels in the post–synaptic cell membrane, causing ions to rush in or out and changing the local transmembrane potential of the cell. The result is *excitatory*, in the case of depolarizing currents, or *inhibitory* in the case of hyperpolarizing currents. Whether a synapse is excitatory or inhibitory depends on what type(s) of ion channel conduct the post–synaptic current, which in turn is a function of the type of receptors and neurotransmitter employed at the synapse.

Excitatory synapses in the brain show several forms of synaptic plasticity, including long–term potentiation (LTP) and long–term depression (LTD), which are initiated by increases in intracellular Ca^{2+} that are generated through NMDA (N-methyl-D-aspartate) receptors or voltage–sensitive Ca^{2+} channels. LTP depends on the coordinated regulation of an ensemble of enzymes, including Ca^{2+}/calmodulin–dependent protein kinase II, adenylyl cyclase 1 and 8, and calcineurin, all of which are stimulated by calmodulin, a Ca^{2+}–binding protein.

Synaptic Strength

The amount of current, or more strictly the change in transmembrane potential, depends on the 'strength' of the synapse, which is subject to biological regulation. One regulatory mechanism involves the simple coincidence of action potentials in the synaptically linked cells. Because the coincidence of sensory stimuli (the sound of a bell and the smell of meat, for example, in the experiments by *Nobel Laureate Ivan P. Pavlov*) can give rise to associative learning or conditioning, neuroscientists have hypothesized that synaptic strengthening through coincident activity in two neurons might underlie learning and memory. This is known as the *Hebbian theory* [Heb49]. It is related to *Pavlov's conditional–reflex learning*: it is learning that takes place when we come to associate two stimuli in the environment. One of these stimuli triggers a reflexive response. The second stimulus is originally neutral with respect to that response, but after it has been paired with the first stimulus, it comes to trigger the response in its own right.

Biophysics of Synaptic Transmission

Technically, synaptic transmission happens in transmitter–activated ion channels. Activation of a presynaptic neuron results in a release of neurotransmitters into the synaptic cleft. The transmitter molecules diffuse to the other side of the cleft and activate receptors that are located in the postsynaptic membrane. So–called ionotropic receptors have a direct influence on the state of an associated ion channel whereas metabotropic receptors control the state of the ion channel by means of a biochemical cascade of $g-$proteins and second messengers. In any case the activation of the receptor results in the opening of certain ion channels and, thus, in an excitatory or inhibitory postsynaptic current (EPSC or IPSC, respectively). The transmitter–activated ion channels can be described as an explicitly time–dependent conductivity $g_{syn}(t)$ that will open whenever a presynaptic spike arrives. The current that passes through these channels depends, as usual, on the difference of its reversal potential E_{syn} and the actual value of the membrane potential,

$$I_{syn}(t) = g_{syn}(t)(u - E_{syn}).$$

The parameter E_{syn} and the function $g_{syn}(t)$ can be used to characterize different types of synapse. Typically, a superposition of exponentials is used for $g_{syn}(t)$. For inhibitory synapses E_{syn} equals the reversal potential of potassium ions (about -75 mV), whereas for excitatory synapses $E_{syn} \approx 0$.

The effect of fast *inhibitory* neurons in the central nervous system of higher vertebrates is almost exclusively conveyed by a neuro–transmitter called $\gamma-$aminobutyric acid, or GABA for short. In addition to many different types of inhibitory interneurons, cerebellar Purkinje cells form a prominent example of projecting neurons that use GABA as their neuro–transmitter. These neurons synapse onto neurons in the deep cerebellar nuclei (DCN) and are particularly important for an understanding of cerebellar function.

The parameters that describe the conductivity of transmitter–activated ion channels at a certain synapse are chosen so as to mimic the time course and the amplitude of experimentally observed spontaneous postsynaptic currents. For example, the conductance $\bar{g}_{syn}(t)$ of inhibitory synapses in DCN neurons can be described by a simple exponential decay with a time constant of $\tau = 5$ ms and an amplitude of $\bar{g}_{syn} = 40$ pS,

$$g_{syn}(t) = \sum_f \bar{g}_{syn} \exp(-\frac{t - t^{(f)}}{\tau})\, \Theta(t - t^{(f)}),$$

where $t^{(f)}$ denotes the arrival time of a presynaptic action potential. The reversal potential is given by that of potassium ions, viz. $E_{syn} = -75$ mV (see [II06b]).

Clearly, more attention can be payed to account for the details of synaptic transmission. In cerebellar granule cells, for example, inhibitory synapses are also GABAergic, but their postsynaptic current is made up of two different components. There is a fast component, that decays with a time constant of about 5 ms, and there is a component that is ten times slower. The underlying postsynaptic conductance is thus of the form

$$g_{syn}(t) = \sum_f \left(\bar{g}_{fast} \exp(-\frac{t-t^{(f)}}{\tau_{fast}}) + \bar{g}_{slow} \exp(-\frac{t-t^{(f)}}{\tau_{slow}}) \right) \Theta(t-t^{(f)}).$$

Now, most of *excitatory* synapses in the vertebrate central nervous system rely on glutamate as their neurotransmitter. The postsynaptic receptors, however, can have very different pharmacological properties and often different types of glutamate receptors are present in a single synapse. These receptors can be classified by certain amino acids that may be selective agonists. Usually, NMDA (N–methyl–D–aspartate) and non–NMDA receptors are distinguished. The most prominent among the non–NMDA receptors are AMPA–receptors. Ion channels controlled by AMPA–receptors are characterized by a fast response to presynaptic spikes and a quickly decaying postsynaptic current. NMDA–receptor controlled channels are significantly slower and have additional interesting properties that are due to a voltage–dependent blocking by magnesium ions (see [GMK94]).

Excitatory synapses in cerebellar granule cells, for example, contain two different types of glutamate receptors, viz. AMPA– and NMDA–receptors. The time course of the postsynaptic conductivity caused by an activation of AMPA–receptors at time $t = t^{(f)}$ can be described as follows,

$$g_{AMPA}(t) = \bar{g}_{AMPA} \cdot N \cdot \left(\exp(-\frac{t-t^{(f)}}{\tau_{decay}}) - \exp(-\frac{t-t^{(f)}}{\tau_{rise}}) \right) \Theta(t-t^{(f)}),$$

with rise time $\tau_{rise} = 0.09$ ms, decay time $\tau_{decay} = 1.5$ ms, and maximum conductance $\bar{g}_{AMPA} = 720$ pS. The numerical constant $N = 1.273$ normalizes the maximum of the braced term to unity (see [GMK94]).

NMDA–receptor controlled channels exhibit a significantly richer repertoire of dynamic behavior because their state is not only controlled by the presence or absence of their agonist, but also by the membrane potential. The voltage dependence itself arises from the blocking of the channel by a common extracellular ion, Mg^{2+}. Unless Mg^{2+} is removed from the extracellular medium, the channels remain closed at the resting potential even in the presence of NMDA. If the membrane is depolarized beyond -50mV, then the Mg^{2+}–block is removed, the channel opens, and, in contrast to AMPA–controlled channels, stays open for 10–100 milliseconds. A simple ansatz that accounts for this additional voltage dependence of NMDA–controlled channels in cerebellar granule cells is

$$g_{NMDA}(t) = \bar{g}_{NMDA} \cdot N \cdot \left(\exp(-\frac{t-t^{(f)}}{\tau_{decay}}) - \exp(-\frac{t-t^{(f)}}{\tau_{rise}}) \right) g_\infty \Theta(t-t^{(f)}),$$

where $g_\infty = \left(1 + e^{\alpha u} \left[\mathrm{Mg}^{2+}\right]_o / \beta\right)$, $\tau_{rise} = 3$ ms, $\tau_{decay} = 40$ ms, $N = 1.358$, $\bar{g}_{NMDA} = 1.2$ nS, $\alpha = 0.062$ mV^{-1}, $\beta = 3.57$ mM, and the extracellular magnesium concentration $\left[\mathrm{Mg}^{2+}\right]_o = 1.2$ mM (see [GMK94]).

Finally, Though NMDA–controlled ion channels are permeable to sodium and potassium ions, their permeability to Ca^{2+} is even five or ten times larger. Calcium ions are known to play an important role in intracellular signaling and are probably also involved in long–term modifications of synaptic efficacy. Calcium influx through NMDA–controlled ion channels, however, is bound to the coincidence of presynaptic (NMDA release from presynaptic sites) and postsynaptic (removal of the Mg^{2+}–block) activity. Hence, NMDA–receptors operate as a kind of a molecular coincidence detectors as they are required for a biochemical implementation of Hebb's learning rule [Heb49].

Reflex Action: The Basis of CNS Activity

The basis of all CNS activity, as well as the simplest example of our sensory–motor adjunction, is the *reflex (sensory–motor) action*, *RA*. It occurs at all neural organizational levels. We are aware of some reflex acts, while others occur without our knowledge.

In particular, the spinal reflex action is defined as a composition of neural pathways, $RA = EN \circ CN \circ AN$, where EN is the efferent neuron, AN is the afferent neuron and $CN = CN_1, ..., CN_n$ is the chain of n connector neurons ($n = 0$ for the simplest, stretch, reflex, $n \geq 1$ for all other reflexes). In other words, we have the following *loop-diagram*:

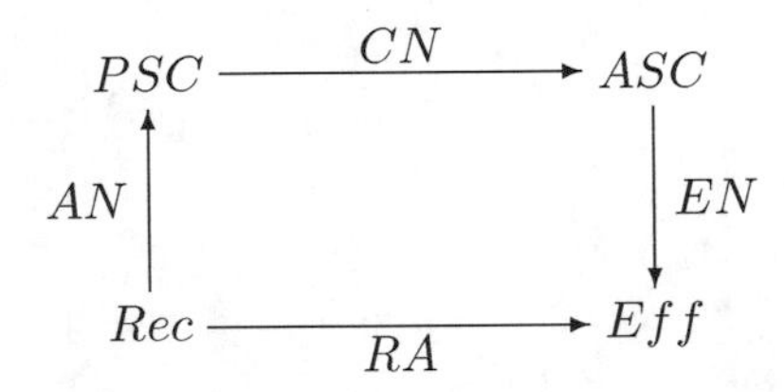

in which Rec is the receptor (for a complex–type receptor as eye, see Figure 20.17), Eff is the effector (e.g., muscle), PSC is the posterior (or, dorsal) horn of the spinal cord, and ASC is the anterior (or, ventral) horn of the spinal cord. In this way defined map $RA : Rec \to Eff$ is the simplest, *one–to–one* relation between one receptor neuron and one effector neuron (e.g., patellar reflex, see Figure 20.20).

Now, in the majority of human reflex arcs a chain CN of many connector neurons is found. There may be link–ups with various levels of the brain and spinal cord. Every receptor neuron is potentially linked in the CNS with a large number of effector organs all over the body, i.e., the map $RA : Rec \to Eff$ is *one–to–many*. Similarly, every effector neuron is potentially in communication with receptors all over the body, i.e., the map $RA : Rec \to Eff$ is *many–to–one*.

However, the most frequent form of the map $RA : Rec \to Eff$ is *many–to–many*. Other neurons synapsing with the effector neurons may give a complex

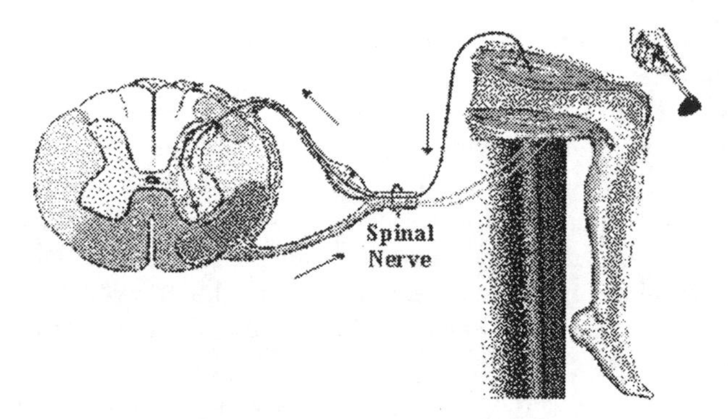

Fig. 20.20. Schematic of a simple knee–jerk reflex. Hammer strikes knee, generating sensory impulse to spinal cord. Primary neuron makes (monosynaptic, excitatory) synapse with anterior horn (motor) cell, whose axon travels via ventral root to quadriceps muscle, which contracts, raising foot. Hamstring (lower) muscle is simultaneously inhibited, via an internuncial neuron.

link–up with centers at higher and lower levels of the CNS. In this way, higher centers in the brain can modify reflex acts which occur through the spinal cord. These centers can send suppressing or facilitating impulses along their pathways to the cells in the spinal cord [II06b].

Through such 'functional' link–ups, neurons in different parts of the CNS, when active, can influence each other. This makes it possible for Pavlov's conditioned reflexes to be established. Such reflexes form the basis of all training, so that it becomes difficult to say where reflex (or involuntary) behavior ends and purely voluntary behavior begins.

In particular, the control of voluntary movements is extremely complex. Many different systems across numerous brain areas need to work together to ensure proper motor control. Our understanding of the nervous system decreases as we move up to higher CNS structures.

Bird's Look at the Brain

The *brain* is the supervisory center of the nervous system consisting of *grey matter* (superficial parts called *cortex* and deep brain nuclei) and *white matter* (deep parts except the brain nuclei). It controls and coordinates behavior, *homeostasis*[3] (i.e., negative feedback control of the body functions such as heartbeat,

[3] *Homeostasis* is the property of an open system to regulation its internal environment so as to maintain a stable state of structure and functions, by means of multiple dynamic equilibrium controlled by interrelated regulation mechanisms. The term was coined in 1932 by W. Cannon from two Greek words [*homeo–man*] and [*stasis–stationary*]. Homeostasis is one of the fundamental characteristics of living things. It is the maintenance of the internal environment within tolerable limits. All sorts of factors affect the suitability of our body fluids to sustain life; these include properties like temperature, salinity, acidity (carbon dioxide), and the concentrations

blood pressure, fluid balance, and body temperature) and mental functions (such as cognition, emotion, memory and learning) (see [Mar98, II06b]).

The *vertebrate brain* can be subdivided into: (i) *medulla oblongata* (or, *brain stem*); (ii) *myelencephalon*, divided into: *pons* and *cerebellum*; (iii) *mesencephalon* (or, *midbrain*); (iv) *diencephalon*; and (v) *telencephalon* (*cerebrum*).

Sometimes a gross division into three major parts is used: *hindbrain* (including medulla oblongata and myelencephalon), *midbrain* (mesencephalon) and *forebrain* (including diencephalon and telencephalon). The cerebrum and the cerebellum consist each of two *hemispheres*. The *corpus callosum* connects the two hemispheres of the cerebrum.

The cerebrum and the cerebellum consist each of two hemispheres. The corpus callosum connects the two hemispheres of the cerebrum. The cerebellum is a cauliflower–shaped section of the brain (see Figure 20.21). It is located in the hindbrain, at the bottom rear of the head, directly behind the pons. The cerebellum is a complex computer mostly dedicated to the intricacies of managing walking and balance. Damage to the cerebellum leaves the sufferer with a gait that appears drunken and is difficult to control.

The spinal cord is the extension of the central nervous system that is enclosed in and protected by the vertebral column. It consists of nerve cells and their connections (axons and dendrites), with both gray matter and white matter, with the former surrounded by the latter.

Cranial nerves are nerves which start directly from the brainstem instead of the spinal cord, and mainly control the functions of the anatomic structures of the head. In human anatomy, there are exactly *12 pairs* of them: (I) *olfactory nerve*, (II) *optic nerve*, (III) oculomotor nerve, (IV) Trochlear nerve, (V) Trigeminal nerve, (VI) Abducens nerve, (VII) Facial nerve, (VIII) Vestibulocochlear nerve (sometimes called the auditory nerve), (IX) Glossopharyngeal nerve, (X) Vagus nerve, (XI) Accessory nerve (sometimes called the spinal accessory nerve), and (XII) Hypoglossal nerve.

The optic nerve consists mainly of axons extending from the ganglionic cells of the eye's retina. The axons terminate in the lateral geniculate nucleus, pulvinar,

---

of nutrients and wastes (urea, glucose, various ion, oxygen). Since these properties affect the chemical reactions that keep bodies alive, there are built-in physiological mechanisms to maintain them at desirable levels. This control is achieved with various organs and glands in the body. For example [Mar98, II06b]: The hypothalamus monitors water content, carbon dioxide concentration, and blood temperature, sending nerve impulses to the pituitary gland and skin. The pituitary gland synthesizes ADH (anti–diuretic hormone) to control water content in the body. The muscles can shiver to produce heat if the body temperature is too low. Warm–blooded animals (homeotherms) have additional mechanisms of maintaining their internal temperature through homeostasis. The pancreas produces insulin to control blood–sugar concentration. The lungs take in oxygen and give out carbon dioxide. The kidneys remove urea and adjust ion and water concentrations. More realistic is dynamical homeostasis, or *homeokinesis*, which forms the basis of the *Anochin's theory of functional systems*

and superior colliculus, all of which belong to the primary visual center. From the lateral geniculate body and the pulvinar fibers pass to the visual cortex.

In particular, the *optic nerve* contains roughly one million nerve fibers. This number is low compared to the roughly 130 million receptors in the retina, and implies that substantial pre–processing takes place in the retina before the signals are sent to the brain through the optic nerve.

In most vertebrates the mesencephalon is the highest integration center in the brain, whereas in mammals this role has been adopted by the telencephalon. Therefore the cerebrum is the largest section of the mammalian brain and its surface has many deep fissures (sulci) and grooves (gyri), giving an excessively wrinkled appearance to the brain.

The *human brain* can be subdivided into several distinct regions:

The *cerebral hemispheres* form the largest part of the brain, occupying the anterior and middle cranial fossae in the skull and extending backwards over the tentorium cerebelli. They are made up of the cerebral cortex, the basal ganglia, tracts of synaptic connections, and the ventricles containing CSF.

The *diencephalon* includes the thalamus, hypothalamus, epithalamus and subthalamus, and forms the central core of the brain. It is surrounded by the cerebral hemispheres.

The *midbrain* is located at the junction of the middle and posterior cranial fossae.

The *pons* sits in the anterior part of the posterior cranial fossa; the fibres within the structure connect one cerebral hemisphere with its opposite cerebellar hemisphere.

The *medulla oblongata* is continuous with the spinal cord, and is responsible for automatic control of the respiratory and cardiovascular systems.

The *cerebellum* overlies the pons and medulla, extending beneath the tentorium cerebelli and occupying most of the posterior cranial fossa. It is mainly concerned with motor functions that regulate muscle tone, coordination, and posture.

Now, the two *cerebral hemispheres* (see Figure 20.21) can be further divided into *four lobes*:

The *frontal lobe* is concerned with higher intellectual functions, such as abstract thought and reason, speech (Broca's area in the left hemisphere only), olfaction, and emotion. Voluntary movement is controlled in the precentral gyrus (the primary motor area, see Figure 20.22).

The *parietal lobe* is dedicated to sensory awareness, particularly in the postcentral gyrus (the primary sensory area, see Figure 20.22). It is also associated with abstract reasoning, language interpretation and formation of a mental egocentric map of the surrounding area.

The *occipital lobe* is responsible for interpretation and processing of visual stimuli from the optic nerves, and association of these stimuli with other nervous inputs and memories.

The *temporal lobe* is concerned with emotional development and formation, and also contains the auditory area responsible for processing and discrimination

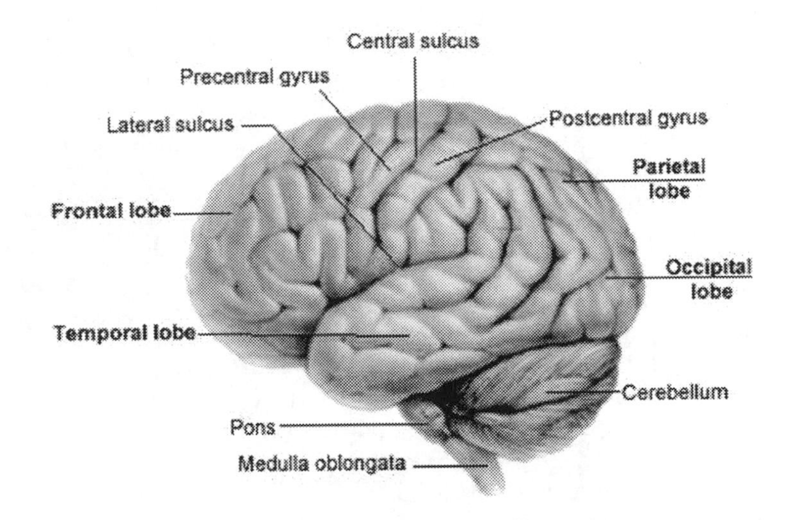

Fig. 20.21. The human cerebral hemispheres.

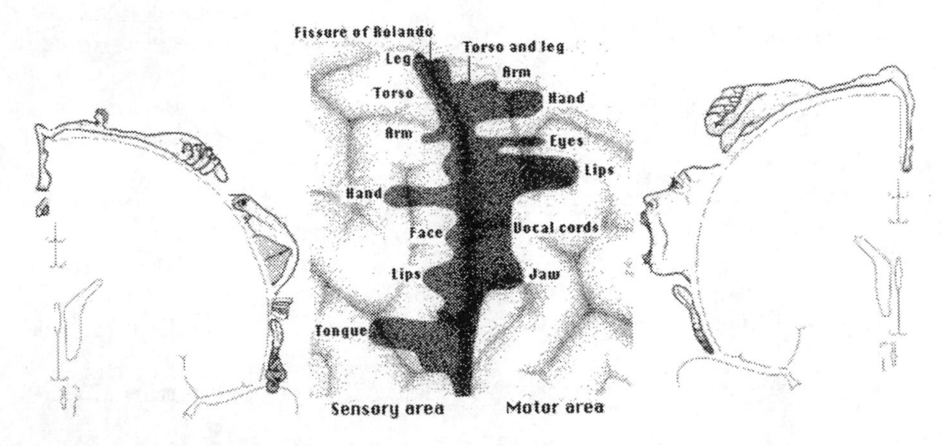

Fig. 20.22. Penfield's 'Homunculus', showing the primary somatosensory and motor areas of the human brain.

of sound. It is also the area thought to be responsible for the formation and processing of memories.

20.5.2 Neural Field Theory

The dynamical system from which the temporal evolution of neural activation fields is generated is constrained by the postulate that localized peaks of activation are stable objects, or, in mathematical terms, *fixed point attractors*. Such a field dynamics has the generic form (see [Ama77, II08c]):

$$\tau \dot{u}(x,t) = -u(x,t) + \text{resting level} + \text{input} + \text{interaction}, \tag{20.21}$$

where $u(x,t)$ is the activation field defined over the metric dimension x and time t. The first three terms define an input driven regime, in which attractor solutions have the form

$$u(x,t) = \text{resting level} + \text{input}.$$

The *rate of relaxation* is determined by the time scale parameter τ. The interaction stabilizes localized peaks of activation against decay by local excitatory interaction and against diffusion by global inhibitory interaction. In S. Amari's formulation, the conceptual model (20.21) is specified as a *continuous model for neural activity in cortical structures* [Ama77],

$$\tau\dot{u}(x,t) = -u(x,t) + h + S(x,t) + \int dx' w(x-x')\sigma(u(x',t)), \qquad (20.22)$$

where $h < 0$ is a constant resting level, $S(x,t)$ is spatially and temporally variable input function, $w(x)$ is an interaction kernel and $\sigma(u)$ is a sigmoidal nonlinear threshold function. The interaction term collects input from all those field sites x' at which activation is sufficiently large. The interaction kernel determines if inputs from those sites are positive, driving up activation (excitatory), or negative, driving down activation (inhibitory). Excitatory input from nearby location and inhibitory input from all field locations generically stabilizes localized peaks of activation. For this class of dynamics, detailed analytical results provide a framework for the inverse dynamics task facing the modeler, determining a dynamical system that has the appropriate attractor solutions.

20.5.3 Behavioral and Motivational Dynamics

The so-called *neural attractor dynamics* (NAD) is based on a discretization for single neurons of *Amari's neural field equation* (20.22). The so–called *discrete Amari equation* describes the temporal evolution of the activity of all single neurons considering positive and negative contributions from external input and internal neural interactions. Since only activated neurons can have an impact on other neurons, the neural attractor dynamics is nonlinear, and effects of bistability and hysteresis can be used for low-level memory and neural competition.

The NAD of behavioral robot activity describes the temporal rate of change of the dynamical variable u_i of neural activity for all behavioral neurons i. It is formulated as the following differential equation [IJP08]:

$$\tau\dot{u}_i = -u_i + h + s_i^{\text{beh}} + c_{\text{mot}} \cdot \sigma(m_i) + \alpha_{\text{selfexc},i}^{\text{beh}} + \alpha_{\text{exc},i}^{\text{beh}} - \alpha_{\text{inh},i}^{\text{beh}}, \qquad (20.23)$$

where the system parameters have the following meaning:

τ, the constant relaxation rate, i.e., the time scale on which the dynamics reacts to changes;

h, the constant negative resting level of neural activation;

$\sigma(.)$, a sigmoidal function, which maps the value of neural activity onto $[0,1]$, given by

$$\sigma(u) = \frac{1}{1 + \mathrm{e}^{-\beta u}},$$

where β (=100) parameterizes the slope of the resulting function;
s_i^{beh}, the adequate stimulus provided by sensory input of a certain duration;
u_i, activity of behavioral neuron i, i.e., activity of behavior i;
c_{mot}, a constant for weighting the motivational contribution, $c_{\mathrm{mot}} < |h|$;
$\alpha_{\mathrm{selfexc},i}^{\mathrm{beh}}$ excitatory contribution of neuron i's own activity u_i;
$\alpha_{\mathrm{exc},i}^{\mathrm{beh}}$, all excitatory contribution of active neurons connected to neuron i;
$\alpha_{\mathrm{inh},i}^{\mathrm{beh}}$, all inhibitory contribution of active neurons connected to neuron i
m_i, activity of motivational neuron i, i.e., motivation of behavior i is defined by the following NAD-equation, similar to (20.23):

$$\tau \dot{m}_i = -m_i + h + s_i^{\mathrm{mot}} + \alpha_{\mathrm{selfexc},i}^{\mathrm{mot}} + \alpha_{\mathrm{exc},i}^{\mathrm{mot}} - \alpha_{\mathrm{inh},i}^{\mathrm{mot}},$$

where
$\alpha_{\mathrm{selfexc},i}^{\mathrm{mot}}$, excitatory contribution of neuron i's own motivation m_i;
$\alpha_{\mathrm{exc},i}^{\mathrm{mot}}$, all excitatory contribution of motivation neurons connected to neuron i;
$\alpha_{\mathrm{inh},i}^{\mathrm{mot}}$, all inhibitory contribution of motivation neurons connected to neuron i.

In this framework, a nonlinear neural dynamical and control system generates the temporal evolution of behavioral variables, such that desired behaviors are fixed-point attractor solutions while un-desired behaviors are repellers.

This kind of *attractor & repeller dynamics* provides the basis for understanding *cognition*, both natural and artificial.

20.6 Neural Synchronization - REF

Recent studies have provided new evidence for the physiological relevance of oscillatory synchronization in human motor and cognitive functions. Synchronization in the beta frequency band seems to have a particularly important role in long–range communication. Electrophysiological recordings of basal ganglia–thalamocortical circuits in healthy monkeys, a monkey model of Parkinson's disease and patients with Parkinson's disease have provided new insights into the functional roles of oscillations and oscillatory synchronization in normal and disturbed motor behavior. Specifically, enhanced beta and reduced gamma oscillations are associated with the poverty and slowness of movement that is characteristic of Parkinson's disease. In addition, tremor seems to arise from abnormal synchronization of oscillations in several cortical and subcortical brain areas. Chronic high–frequency deep brain stimulation, which can be delivered through electrodes that have been implanted in specific basal ganglia target structures, greatly improves motor symptoms in patients with Parkinson's disease, probably through desynchronizing effects. Pathological changes in long–range synchronization are also evident in other movement disorders, as well as in neuropsychiatric diseases.

20.6.1 Oscillatory Phase Neurodynamics

In coupled oscillatory neuronal systems, under suitable conditions, the original dynamics can be reduced theoretically to a simpler phase dynamics. The state of the ith neuronal oscillatory system can be then characterized by a single phase variable φ_i representing the timing of the neuronal firings. The typical dynamics of *oscillator neural networks* are described by the *Kuramoto model*, consisting of N equally weighted, all–to–all, phase–coupled limit–cycle oscillators, where each oscillator has its own natural frequency ω_i drawn from a prescribed distribution function:

$$\dot{\varphi}_i = \omega_i + \frac{K}{N} J_{ij} \sin(\varphi_j - \varphi_i + \beta_{ij}). \tag{20.24}$$

Here, J_{ij} and β_{ij} are parameters representing the effect of the interaction, while $K \geq 0$ is the coupling strength. For simplicity, one can assume that all natural frequencies ω_i are equal to some fixed value ω_0. We can then eliminate ω_0 by applying the transformation $\varphi_i \to \varphi_i + \omega_0 t$. Using the complex representation $W_i = \exp(i\varphi_i)$ and $C_{ij} = J_{ij}\exp(i\beta_{ij})$ in (20.24), it is easily found that all neurons relax toward their stable equilibrium states, in which the relation $W_i = h_i/|h_i|$ $(h_i = C_{ij}W_j)$ is satisfied.

20.6.2 Kuramoto Synchronization Model

The microscopic individual level dynamics of the *Kuramoto model* (20.24) is easily visualized by imagining oscillators as points running around on the unit circle. Due to rotational symmetry, the average frequency $\Omega = \sum_{i=1}^{N} \omega_i/N$ can be set to 0 without loss of generality; this corresponds to observing dynamics in the co–rotating frame at frequency Ω.

The governing equation (20.24) for the ith oscillator phase angle φ_i can be simplified to

$$\dot{\varphi}_i = \omega_i + \frac{K}{N} \sum_{i=1}^{N} \sin(\varphi_j - \varphi_i), \quad 1 \leq i \leq N. \tag{20.25}$$

It is known that as K is increased from 0 above some critical value K_c, more and more oscillators start to get synchronized (or phase–locked) until all the oscillators get fully synchronized at another critical value of K_{tp}. In the choice of $\Omega = 0$, the fully synchronized state corresponds to an exact steady state of the 'detailed', fine–scale problem in the co–rotating frame.

Such synchronization dynamics can be conveniently summarized by considering the fraction of the synchronized (phase–locked) oscillators, and conventionally described by a *complex–valued order parameter*, $re^{i\psi} = \frac{1}{N} e^{i\varphi_j}$, where the radius r measures the phase coherence, and ψ is the average phase angle.

20.7 Cerebellar Movement Control

When someone compares learning a new skill to learning how to ride a bike they imply that once mastered, the task seems imbedded in our brain forever. Well, imbedded in the cerebellum to be exact. This brain structure is the commander of coordinated movement and possibly even some forms of cognitive learning. Damage to this area leads to motor or movement difficulties.

A part of a human brain that is devoted to the sensory-motor control of human movement, that is motor coordination and learning, as well as equilibrium and posture, is the cerebellum (which in Latin means "little brain"). It performs integration of sensory perception and motor output. Many neural pathways link the cerebellum with the motor cortex, which sends information to the muscles causing them to move, and the spino–cerebellar tract, which provides proprioception, or feedback on the position of the body in space. The cerebellum integrates these pathways, using the constant feedback on body position to fine–tune motor movements.

The human cerebellum has 7–14 million Purkinje cells. Each receives about 200,000 synapses, most onto dendritic splines. Granule cell axons form the *parallel fibers.* They make excitatory synapses onto Purkinje cell dendrites. Each parallel fibre synapses on about 200 Purkinje cells. They create a strip of excitation along the cerebellar folia.

Mossy fibers are one of two main sources of input to the cerebellar cortex (see Figure 20.23). A mossy fibre is an axon terminal that ends in a large, bulbous swelling. These mossy fibers enter the granule cell layer and synapse on the dendrites of granule cells; in fact the granule cells reach out with little 'claws' to grasp the terminals. The granule cells then send their axons up to the molecular layer, where they end in a T and run parallel to the surface. For this reason these axons are called *parallel fibers.* The parallel fibers synapse on the huge dendritic arrays of the Purkinje cells. However, the individual parallel fibers are not a strong drive to the Purkinje cells. The Purkinje cell dendrites fan out within a plane, like the splayed fingers of one hand. If one were to turn a Purkinje cell to the side, it would have almost no width at all. The parallel fibers run perpendicular to the Purkinje cells, so that they only make contact once as they pass through the dendrites.

Unless firing in bursts, parallel fibre EPSPs do not fire Purkinje cells. Parallel fibers provide excitation to all of the Purkinje cells they encounter. Thus, granule cell activity results in a strip of activated Purkinje cells.

Mossy fibers arise from the spinal cord and brainstem. They synapse onto granule cells and deep cerebellar nuclei. The Purkinje cell makes an inhibitory synapse (GABA) to the deep nuclei. Mossy fibre input goes to both cerebellar cortex and deep nuclei. When the Purkinje cell fires, it inhibits output from the deep nuclei.

The *climbing fibre* arises from the inferior olive. It makes about 300 excitatory synapses onto one Purkinje cell. This powerful input can fire the Purkinje cell.

The parallel fibre synapses are plastic—that is, they can be modified by experience. When parallel fibre activity and climbing fibre activity converge on the same Purkinje cell, the parallel fibre synapses become weaker (EPSPs are smaller). This is called long-term depression. Weakened parallel fibre synapses result in less Purkinje cell activity and less inhibition to the deep nuclei,

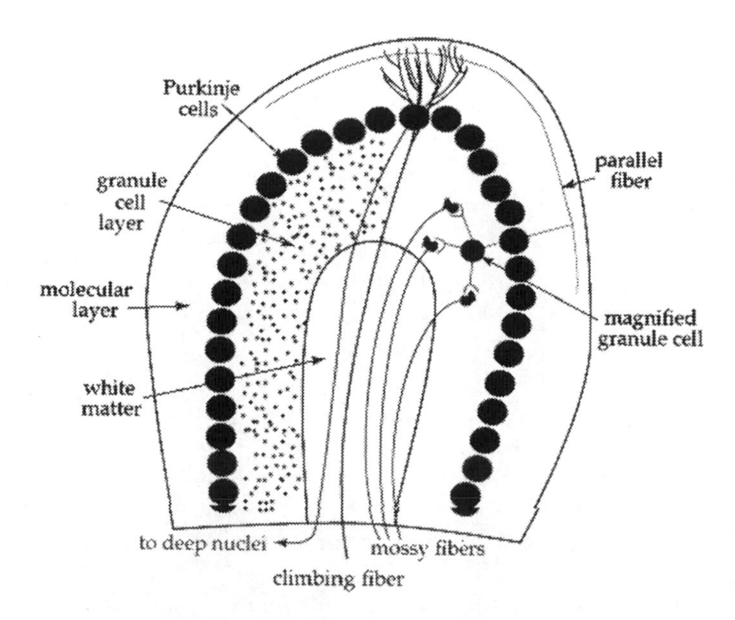

Fig. 20.23. Stereotypical ways throughout the cerebellum.

resulting in facilitated deep nuclei output. Consequently, the mossy fibre collaterals control the deep nuclei.

The *basket cell* is activated by parallel fibers afferents. It makes inhibitory synapses onto Purkinje cells. It provides lateral inhibition to Purkinje cells. Basket cells inhibit Purkinje cells lateral to the active beam.

Golgi cells receive input from parallel fibers, mossy fibers, and climbing fibers. They inhibit granule cells. Golgi cells provide feedback inhibition to granule cells as well as feedforward inhibition to granule cells. Golgi cells create a brief burst of granule cell activity.

Although each parallel fibre touches each Purkinje cell only once, the thousands of parallel fibers working together can drive the Purkinje cells to fire like mad.

The second main type of input to the folium is the *climbing fibre*. The climbing fibers go straight to the Purkinje cell layer and snake up the Purkinje dendrites, like ivy climbing a trellis. Each climbing fibre associates with only one Purkinje cell, but when the climbing fibre fires, it provokes a large response in the Purkinje cell.

The Purkinje cell compares and processes the varying inputs it gets, and finally sends its own axons out through the white matter and down to the *deep nuclei*. Although the inhibitory Purkinje cells are the main output of the cerebellar cortex, the output from the cerebellum as a whole comes from the deep nuclei. The three deep nuclei are responsible for sending excitatory output back to the thalamus, as well as to postural and vestibular centers.

There are a few other cell types in cerebellar cortex, which can all be lumped into the category of inhibitory interneuron. The *Golgi cell* is found among the granule cells. The *stellate* and *basket cells* live in the molecular layer. The basket cell (right) drops axon branches down into the Purkinje cell layer where the branches wrap around the cell bodies like baskets.

The cerebellum operates in 3's: there are 3 highways leading in and out of the cerebellum, there are 3 main inputs, and there are 3 main outputs from 3 deep nuclei. They are:

The 3 highways are the *peduncles*. There are 3 pairs:

1. The *inferior cerebellar peduncle* (restiform body) contains the dorsal spinocerebellar tract (DSCT) fibers. These fibers arise from cells in the ipsilateral Clarke's column in the spinal cord (C8–L3). This peduncle contains the cuneo–cerebellar tract (CCT) fibers. These fibers arise from the ipsilateral accessory cuneate nucleus. The largest component of the inferior cerebellar peduncle consists of the olivo–cerebellar tract (OCT) fibers. These fibers arise from the contralateral inferior olive. Finally, vestibulo–cerebellar tract (VCT) fibers arise from cells in both the vestibular ganglion and the vestibular nuclei and pass in the inferior cerebellar peduncle to reach the cerebellum.
2. The *middle cerebellar peduncle* (brachium pontis) contains the pontocerebellar tract (PCT) fibers. These fibers arise from the contralateral pontine grey.
3. The *superior cerebellar peduncle* (brachium conjunctivum) is the primary efferent (out of the cerebellum) peduncle of the cerebellum. It contains fibers that arise from several deep cerebellar nuclei. These fibers pass ipsilaterally for a while and then cross at the level of the inferior colliculus to form the decussation of the superior cerebellar peduncle. These fibers then continue ipsilaterally to terminate in the red nucleus ('ruber–duber') and the motor nuclei of the thalamus (VA, VL).

The 3 inputs are: *mossy fibers* from the *spinocerebellar* pathways, climbing fibers from the *inferior olive*, and more mossy fibers from the *pons*, which are carrying information from *cerebral cortex* (see Figure 20.24). The mossy fibers from the spinal cord have come up ipsilaterally, so they do not need to cross. The fibers coming down from cerebral cortex, however, do need to cross (as the cerebrum is concerned with the opposite side of the body, unlike the cerebellum). These fibers synapse in the pons (hence the huge block of fibers in the cerebral peduncles labelled 'cortico–pontine'), cross, and enter the cerebellum as mossy fibers.

The 3 deep nuclei are the *fastigial*, *interposed*, and *dentate nuclei*. The fastigial nucleus is primarily concerned with balance, and sends information mainly to vestibular and reticular nuclei. The dentate and interposed nuclei are concerned more with voluntary movement, and send axons mainly to thalamus and the red nucleus.

The main function of the cerebellum as a motor controller is depicted in Figure 22.9. A coordinated movement is easy to recognize, but we know little about how it is achieved. In search of the neural basis of coordination, a model of spinocerebellar interactions was recently presented in which the structure-functional organizing principle is a division of the cerebellum into discrete micro–complexes.

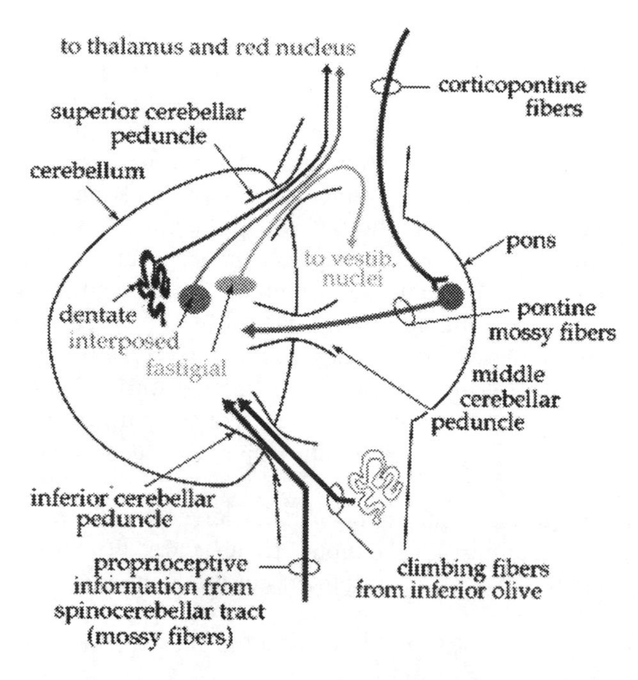

Fig. 20.24. Inputs and outputs of the cerebellum.

Each micro–complex is the recipient of a specific motor error signal - that is, a signal that conveys information about an inappropriate movement. These signals are encoded by spinal reflex circuits and conveyed to the cerebellar cortex through climbing fibre afferents. This organization reveals salient features of cerebellar information processing, but also highlights the importance of systems level analysis for a fuller understanding of the neural mechanisms that underlie behavior.

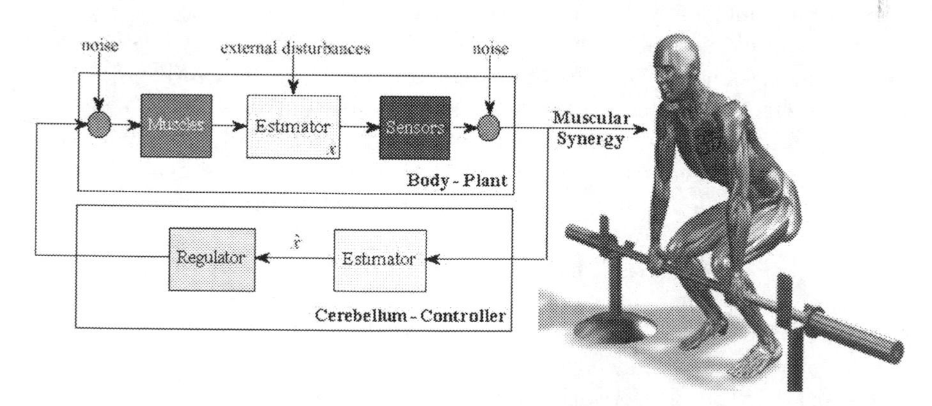

Fig. 20.25. The cerebellum as a motor controller.

The cerebellum is crucial for the coordination of movement. The authors presented a model of the cerebellar paravermis, a region concerned with the control of voluntary limb movements through its interconnections with the spinal cord. They particularly focused on the olivo-cerebellar climbing fibre system.

Climbing fibres are proposed to convey motor error signals (signals that convey information about inappropriate movements) related to elementary limb movements that result from the contraction of single muscles. The actual encoding of motor error signals is suggested to depend on sensorimotor transformations carried out by spinal modules that mediate nociceptive withdrawal reflexes.

The termination of the climbing fibre system in the cerebellar cortex subdivides the paravermis into distinct microzones. Functionally similar but spatially separate microzones converge onto a common group of cerebellar nuclear neurons. The processing units formed as a consequence are termed 'multizonal micro-complexes' (MZMCs), and are each related to a specific spinal reflex module.

The distributed nature of microzones that belong to a given MZMC is proposed to enable similar climbing fibre inputs to integrate with mossy fibre inputs that arise from different sources. Anatomical results consistent with this notion have been obtained.

Within an individual MZMC, the skin receptive fields of climbing fibres, mossy fibres and cerebellar cortical inhibitory interneurons appear to be similar. This indicates that the inhibitory receptive fields of Purkinje cells within a particular MZMC result from the activation of inhibitory interneurons by local granule cells.

On the other hand, the parallel fibre–mediated excitatory receptive fields of the Purkinje cells in the same MZMC differ from all of the other receptive fields, but are similar to those of mossy fibres in another MZMC. This indicates that the excitatory input to Purkinje cells in a given MZMC originates in non–local granule cells and is mediated over some distance by parallel fibres.

The output from individual MZMCs often involves two or three segments of the ipsilateral limb, indicative of control of multi–joint muscle synergies. The distal–most muscle in this synergy seems to have a roughly antagonistic action to the muscle associated with the climbing fibre input to the MZMC.

The cerebellar paravermis system could provide the control of both single– and multi–joint movements. Agonist-antagonist activity associated with single–joint movements might be controlled within a particular MZMC, whereas coordination across multiple joints might be governed by interactions between MZMCs, mediated by parallel fibres.

Two main theories address the function of the cerebellum, both dealing with motor coordination. One claims that the cerebellum functions as a regulator of the "timing of movements." This has emerged from studies of patients whose timed movements are disrupted.

The second, "Tensor Network Theory" provides a mathematical model of transformation of sensory (covariant) space-time coordinates into motor (contravariant) coordinates by cerebellar neuronal networks.

Studies of motor learning in the vestibulo–ocular reflex and eye-blink conditioning demonstrate that the timing and amplitude of learned movements are encoded by the cerebellum. Many synaptic plasticity mechanisms have been found throughout the cerebellum. The *Marr–Albus model* mostly attributes motor learning to a single plasticity mechanism: the long-term depression of parallel fiber synapses. The Tensor Network Theory of sensory–motor transformations by the cerebellum has also been experimentally supported.

20.8 Brain Topology versus Small–World Topology

Now, from a physiological point of view, a *neural–networks complexity* has been studied as an interplay between *dynamics and topology of brain networks* [LBM05]. This approach is of interest from several points of view: the brain and its structural features can be seen as a prototype of a physical system capable of highly complex and adaptable patterns in connectivity, selectively improved through evolution; architectural organization of brain cortex is one of the key features of how brain system evolves, adapts itself to the experience, and to possible injuries. Brain activity can be modeled as a dynamical process acting on a network; each vertex of the structure represents an elementary component, such as brain areas, groups of neurons or individual cells. Recall that a *complexity measure*, has been introduced (see [LBM05] and reference therein) with the purpose to get a sensible measure of two important features of the brain activity: segregation and integration. The former is a measure of the relative statistical independence of small subsets; the latter is the measure of statistical deviation from independence of large subsets. Complexity is based on the values of the *Shannon entropy* calculated over the dynamics of the different sized subgraphs of the whole network. It is sensitive both to the statistical properties of the dynamics and to the connectivity. It has been shown by means of genetic algorithms that the graphs showing high values of complexity are characterized by being both segregated and integrated; the complexity is low when the system is either completely independent (segregated), or completely dependent (integrated) [LBM05].

In general this approach has been developed within the framework of *small–world networks*, comprising World Wide Web structure etc. which have received much attention from researchers in various disciplines, since they were introduced by Watts and Strogatz [WS98] as models of real networks that lie somewhere between being random and being regular.

Watts and Strogatz introduced a simple model for tuning collections of coupled dynamical systems between the two extremes of random and regular networks. In this model, connections between nodes in a regular array are randomly rewired with a probability p, such that $p = 0$ means the network is regularly connected, while $p = 1$ results in a random connection of nodes. For a range of intermediate values of p between these two extremes, the network retains a property of regular networks (a large clustering coefficient) and also acquires a property of random networks (a short characteristic path length between nodes).

Many examples of such small worlds, both natural and human–made, have been discussed [Str01]. For example, a model of a *social network* is given in Figure 20.26, where (a) denotes people (dots) belonging to groups (ellipses), which in turn belong to groups of groups and so on. The largest group corresponds to the entire community. As we go down in this hierarchical organization, each group represents a set of people with increasing social affinity. In the example, there are $l = 3$ hierarchical levels, each representing a subdivision in $b = 3$ smaller groups, and the lowest groups are composed of $g = 11$ people, on average. This defines a *social hierarchy*. The distance between the highlighted individuals i and j in this hierarchy is 3. (b) Each hierarchy can be represented as a tree-like structure. Different hierarchies are correlated, in the sense that distances that are short along one of them are more likely to be short along the others as well. The figure shows an example with $H = 2$ hierarchies, where highlighted in the second hierarchy are those people belonging to group A in the first one. (c) Pairs of people at shorter social distances are more likely to be linked by social ties, which can represent either friendship or acquaintanceship ties (we do not distinguished them here because the ones that are relevant for the problem in question may depend on the social context). The figure shows, for a person in the network, the distribution of acquaintances at social distance $D = 1, 2$, and 3, where D is the minimum over the distances along all the hierarchies.

Not surprisingly, there has been much interest in the synchronization of dynamical systems connected in a *small–world geometry* [NML03]. Generically, such studies have shown that the presence of small–world connections make it easier for a network to synchronize, an effect generally attributed to the reduced path length between the linked systems. This has also been found to be true for the special case in which the dynamics of each oscillator is described by a *Kuramoto model* [HCK02a, HCK02b].

Small–world networks are characterized by two numbers: the average path length L and the clustering coefficient C. L, which measures efficiency of communication or passage time between nodes, is defined as being the average number of links in the shortest path between a pair of nodes in the network. C represents the degree of local order, and is defined as being the probability that two nodes connected to a common node are also connected to each other.

Many real networks are sparse in the sense that the number of links in the network is much less than $N(N-1)/2$, the number of all possible (bidirectional) links. On one hand, random sparse networks have short average path length (i.e., $L \sim \log N$), but they are poorly clustered (i.e., $C \ll 1$). On the other hand, regular sparse networks are typically highly clustered, but L is comparable to N. (All–to–all networks have $C = 1$ and $L = 1$, so they are most efficient, but most expensive in the sense that they have all $N(N-1)/2$ possible connections and so they are dense rather than sparse.) The small–world network models have advantages of both random and regular sparse networks: they have small L for fast communication between nodes, and they have large C, ensuring sufficient redundancy for high fault tolerance. The models of small–world networks are constructed from a regular lattice by adding a relatively small number of

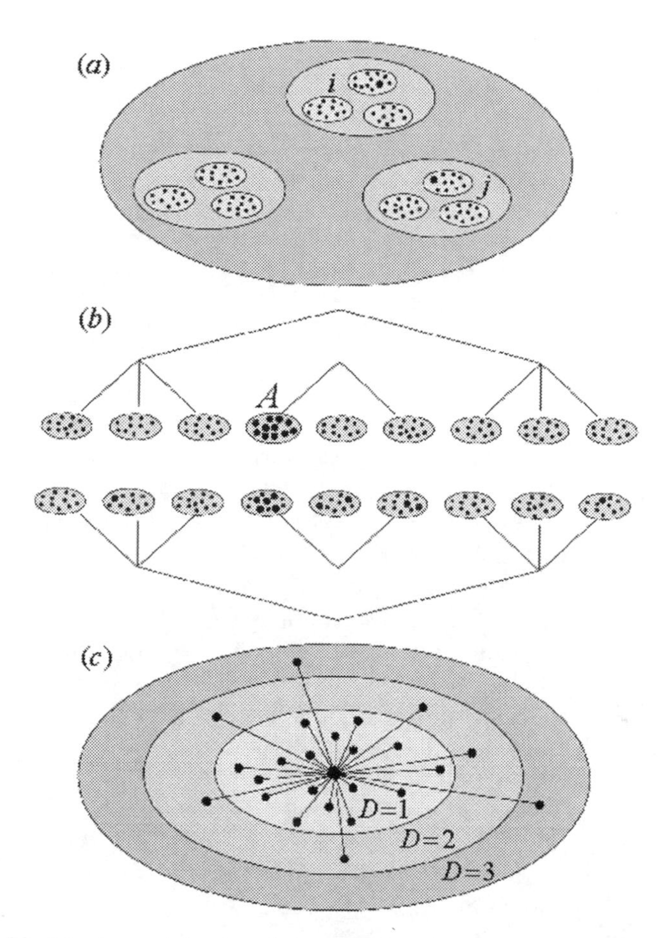

Fig. 20.26. Model of a *social network* (modified and adapted from [MNL03] – see text for explanation.

shortcuts at random, where a link between two nodes u and v is called a *shortcut* if the shortest path length between u and v in the absence of the link is more than two [Wat99]. The regularity of the underlying lattice ensures high clustering, while the shortcuts reduce the size of L.

Most work has focused on average properties of such models over different realizations of *random* shortcut configurations. However, a different point of view is necessary when a network is to be designed to optimize its performance with a restricted number of long–range connections. For example, a transportation network should be designed to have the smallest L possible, so as to maximize the ability of the network to transport people efficiently, while keeping a reasonable cost of building the network. The same can be said about communication networks for efficient exchange of information between nodes.

Most random choices of shortcuts result in a suboptimal configuration, since they do not have any special structures or organizations. On the contrary, many

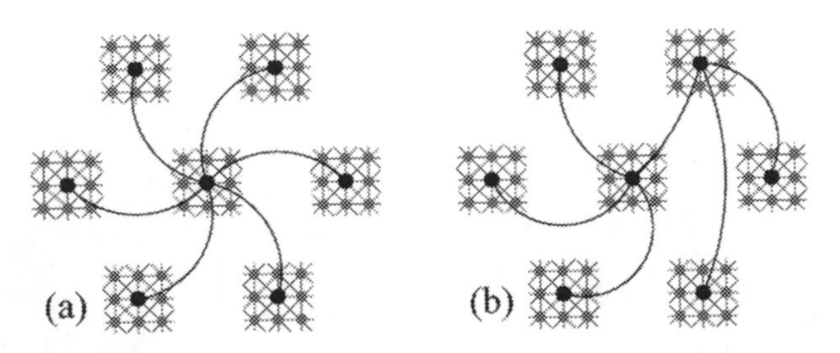

Fig. 20.27. Examples of shortcut configuration with (a) a single center and (b) two centers.

real networks have highly structured configurations of shortcuts. For example, in long–range transportation networks, the airline connections between major cities which can be regarded as shortcuts, are far from being random, but they are organized around hubs. Efficient travel involves ground transportation to a nearest airport, then flights through a hub to an airport closest to the destination, and ground transportation again at the end.

It has been shown in [NML02] that the average path length L of a small–world network with a fixed number of shortcuts attains its minimum value when there exists a 'center' node, from which all shortcuts are connected to uniformly distributed nodes in the network (see Figure 20.27(a)). If a small–world network has several 'centers' and its subnetwork of shortcuts is *connected*, then L is almost as small as the minimum value (see Figure 20.27(b)).

For example, a small–world geometry has been explored in a family of 1D networks of N integrate–and–fire neurons arranged on a ring [AT06]. To simplify the formalism, the geometry of a neuron is reduced to a point, and the time evolution of the membrane potential V_i of cell i, $i = 1, \dots, N$ during an inter-spike interval follows the equation

$$\dot{V}_i(t) = -[V_i(t) - V^{res}]/\tau_m + \frac{R_m}{\tau_m}[I^{syn}(t) + I^{inh}(t) + I_i(t)], \qquad (20.26)$$

where τ_m is the time constant of the neural–cell membrane, R_m its passive resistance, and V^{res} the resting potential. The total current entering cell i consists of a synaptic current $I_i^{syn}(t)$ due to the spiking of all other cells in the network that connect to cell i, an inhibitory current $I^{inh}(t)$, which depends on the network activity, and a small external current $I_i(t)$, which represents all other inputs into cell i.

21

Modern Tennis Physics: Nonlinear and Quantum Dynamics

21.1 Basics of Nonlinear Dynamics and Chaos Theory

Nonlinear dynamics depicts an *irregular and unpredictable* time evolution of both simple and complex deterministic dynamical systems, characterized by nonlinear coupling of its variables. Given an initial condition, the dynamic equation determines the dynamic process, i.e., every step in the evolution. However, the initial condition, when magnified, reveals a cluster of values within a certain error bound. For a regular dynamic system, processes issuing from the cluster are bundled together, and the bundle constitutes a predictable process with an error bound similar to that of the initial condition. In a unpredictable nonlinear dynamical system, processes issuing from the cluster diverge from each other exponentially, and after a while the error becomes so large that the dynamic equation losses its predictive power (see Figure 21.1).

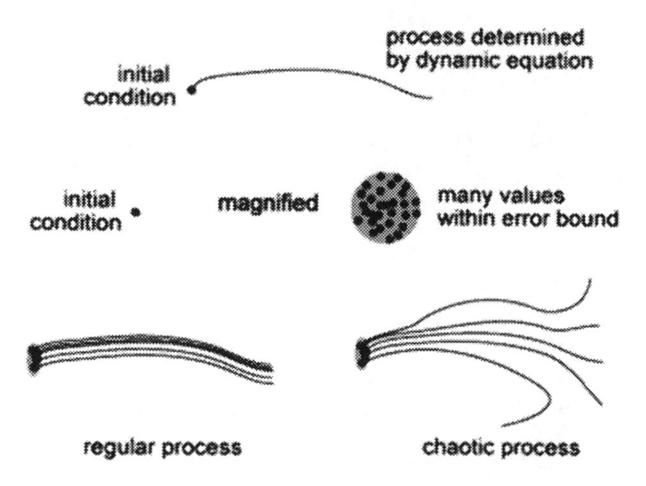

Fig. 21.1. Regular (or, linear) v.s. unpredictable nonlinear dynamical process.

T.T. Ivancevic et al.: Paradigm Shift for Future Tennis, COSMOS 12, pp. 183–225.
springerlink.com

For example, in a *pinball game*, any two trajectories that start out very close to each other separate exponentially with time, and in a finite (and in practice, a very small) number of bounces their separation $\delta x(t)$ attains the magnitude of L, the characteristic linear extent of the whole system. This property of sensitivity to initial conditions can be quantified as

$$|\delta x(t)| \approx e^{\lambda t} |\delta x(0)|,$$

where λ, the *mean rate of separation of trajectories* of the system, is called the *Lyapunov exponent*. For any finite accuracy $|\delta x(0)| = \delta x$ of the initial data, the dynamics is predictable only up to a finite *Lyapunov time*

$$T_{Lyap} \approx -\frac{1}{\lambda} \ln |\delta x/L|,$$

despite the deterministic and infallible simple laws that rule the pinball motion.

When a physicist says that a certain system "exhibits completely unpredictable nonlinear dynamics", they mean that the system obeys deterministic laws of evolution, but that the outcome is highly sensitive to small uncertainties in the specification of the initial state. If a deterministic system is locally unstable (positive Lyapunov exponent) and globally mixing (positive entropy), it is said to be *completely unpredictable* (see [Str94, II07a]).

In the remainder of this section we will outline the main historical pillars of nonlinear dynamics and chaos theory.

21.1.1 Poincaré's Qualitative Dynamics, Topology and Chaos

Chaos theory really started with *Henry Jules Poincaré*, the last mathematical universalist, the father of both *dynamical systems* and *topology* (which he considered to be the two sides of the same coin). Together with the four dynamics giants mentioned above, Poincaré has been considered as one of the great scientific geniuses of all time.[1]

[1] Recall that Henri Poincaré (April 29, 1854–July 17, 1912), was one of France's greatest mathematicians and theoretical physicists, and a philosopher of science. Poincaré is often described as the last 'universalist' (after Gauss), capable of understanding and contributing in virtually all parts of mathematics. As a mathematician and physicist, he made many original fundamental contributions to pure and applied mathematics, mathematical physics, and celestial mechanics. He was responsible for formulating the Poincaré conjecture, one of the most famous problems in mathematics. In his research on the three-body problem, Poincaré became the first person to discover a *deterministic chaotic system*. Besides, Poincaré introduced the modern principle of relativity and was the first to present the Lorentz transformations in their modern symmetrical form (Poincaré group). Poincaré discovered the remaining relativistic velocity transformations and recorded them in a letter to Lorentz in 1905. Thus he got perfect invariance of all of Maxwell's equations, the final step in the discovery of the theory of special relativity. As a mathematician and physicist, he made many original fundamental contributions to pure and applied mathematics,

Poincaré conjectured and proved a number of theorems. Two of them related to chaotic dynamics are:

1. The *Poincaré–Bendixson theorem* says: Let F be a dynamical system on the real plane defined by

$$(\dot{x}, \dot{y}) = (f(x, y), g(x, y)),$$

where f and g are continuous differentiable functions of x and y. Let S be a closed bounded subset of the 2D phase–space of F that does not contain a stationary point of F and let C be a trajectory of F that never leaves S. Then C is either a limit–cycle or C converges to a limit–cycle. The Poincaré–Bendixson theorem limits the types of long term behavior that can be exhibited by continuous planar dynamical systems. One important implication is that a 2D continuous dynamical system cannot give rise to a *strange attractor*. If a strange attractor C did exist in such a system, then it could be enclosed in a closed and bounded subset of the phase–space. By making this subset small enough, any nearby stationary points could be excluded. But then the Poincaré–Bendixson theorem says that C is not a strange attractor at all – it is either a limit–cycle or it converges to a limit–cycle. The Poincaré–Bendixson theorem says that chaotic behavior can only arise in continuous dynamical systems whose phase–space has 3 or more dimensions. However, this restriction does not apply to discrete dynamical systems, where chaotic behavior can arise in two or even one–dimensional.

2. The *Poincaré–Hopf index theorem* says: Let M be a compact differentiable manifold and v be a vector–field on M with isolated zeroes. If M has boundary, then we insist that v be pointing in the outward normal direction along the boundary. Then we have the formula

$$\sum_i index_v = \chi(M),$$

where the sum is over all the isolated zeroes of v and $\chi(M)$ is the *Euler characteristic* of M. systems.

mathematical physics, and celestial mechanics. He was responsible for formulating the Poincaré conjecture, one of the most famous problems in mathematics. In his research on the three-body problem, Poincaré became the first person to discover a *deterministic chaotic system*. Besides, Poincaré introduced the modern principle of relativity and was the first to present the Lorentz transformations in their modern symmetrical form (Poincaré group). Poincaré discovered the remaining relativistic velocity transformations and recorded them in a letter to Lorentz in 1905. Thus he got perfect invariance of all of Maxwell's equations, the final step in the discovery of the theory of special relativity.

Poincaré had the opposite philosophical views of Bertrand Russell and Gottlob Frege, who believed that mathematics were a branch of logic. Poincaré strongly disagreed, claiming that *intuition* was the *life of mathematics*. Poincaré gives an interesting point of view in his book 'Science and Hypothesis': "For a superficial observer, scientific truth is beyond the possibility of doubt; the logic of science is infallible, and if the scientists are sometimes mistaken, this is only from their mistaking its rule."

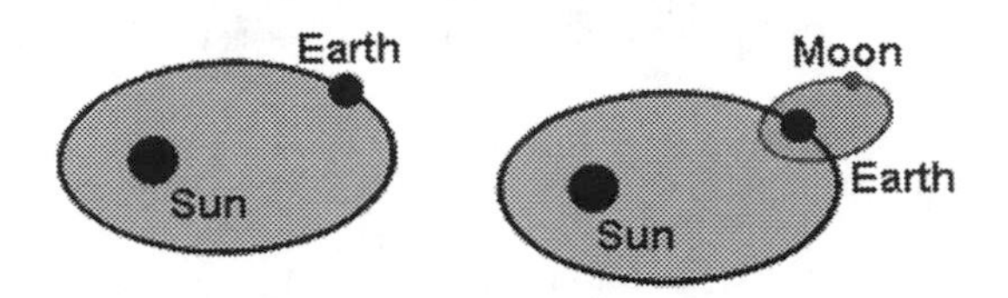

Fig. 21.2. The 2–body, problem solved by Newton (left), and the 3–body problem, first attacked by Poincaré, and still the point of active research (right).

In 1887, in honor of his 60th birthday, Oscar II, King of Sweden offered a prize to the person who could answer the question "Is the Solar system stable?" Poincaré won the prize with his famous work on the *3–body problem*. He considered the Sun, Earth and Moon orbiting in a plane under their mutual gravitational attractions (see Figure 21.2). Like the pendulum, this system has some unstable solutions. Introducing a *Poincaré section*, he saw that *homoclinic tangles* must occur. These would then give rise to *chaos* and *unpredictability*.

Recall that trying to predict the motion of the Moon has preoccupied astronomers since antiquity. Accurate understanding of its motion was important for determining the longitude of ships while traversing open seas. The *Rudolphine Tables* of *Johannes Kepler* had been a great improvement over previous tables, and Kepler was justly proud of his achievements. Bernoulli used Newton's work on mechanics to derive the elliptic orbits of Kepler and set an example of how equations of motion could be solved by integrating. But the motion of the Moon is not well approximated by an ellipse with the Earth at a focus; at least the effects of the Sun have to be taken into account if one wants to reproduce the data the classical Greeks already possessed. To do that one has to consider the motion of three bodies: the Moon, the Earth, and the Sun. When the planets are replaced by point particles of arbitrary masses, the problem to be solved is known as the 3–body problem. The 3–body problem was also a model to another concern in astronomy. In the Newtonian model of the Solar system it is possible for one of the planets to go from an elliptic orbit around the Sun to an orbit that escaped its domain or that plunged right into it. Knowing if any of the planets would do so became the problem of the stability of the Solar system. A planet would not meet this terrible end if Solar system consisted of two celestial bodies, but whether such fate could befall in the 3–body case remained unclear.

After many failed attempts to solve the 3–body problem, natural philosophers started to suspect that it was impossible to integrate. The usual technique for integrating problems was to find the conserved quantities, quantities that do not change with time and allow one to relate the momenta and positions different times. The first sign on the impossibility of integrating the 3–body problem came from a result of Burns that showed that there were no conserved quantities that were polynomial in the momenta and positions. Burns' result did not preclude the possibility of more complicated conserved quantities. This problem was settled by Poincaré and Sundman in two very different ways.

In an attempt to promote the journal *Acta Mathematica*, *Gustaf Mittag–Leffler* got the permission of the King Oscar II of Sweden and Norway to establish a mathematical competition. Several questions were posed (although the king would have preferred only one), and the prize of 2500 kroner would go to the best submission. One of the questions was formulated by the 'father of modern analysis', *Karl Weierstrass*:

> "Given a system of arbitrary mass points that attract each other according to Newton's laws, under the assumption that no two points ever collide, try to find a representation of the coordinates of each point as a series in a variable that is some known function of time and for all of whose values the series converges uniformly.
> This problem, whose solution would *considerably extend our understanding of the Solar system...*"

Poincaré's submission won the prize. He showed that *conserved quantities that were analytic in the momenta and positions could not exist.* To show that he introduced methods that were very geometrical in spirit: the importance of phase flow, the role of *periodic orbits* and their cross sections, the *homoclinic points* (see [CAM05]).[2]

Poincaré pointed out that the problem was not correctly posed, and proved that a complete solution to it could not be found. His work was so impressive that in 1888 the jury recognized its value by awarding him the prize. He found that the evolution of such a system is often chaotic in the sense that a small perturbation in the initial state, such as a slight change in one body's initial position, might lead to a radically different later state. If the slight change is not detectable by our measuring instruments, then we will not be able to predict which final state will occur. One of the judges, the distinguished Karl Weierstrass, said, "This work cannot indeed be considered as furnishing the complete solution of the question proposed, but that it is nevertheless of such importance that its publication will inaugurate a new era in the history of celestial mechanics." Weierstrass did not know how accurate he was. In Poincaré's paper, he described new mathematical ideas such as homoclinic points. The memoir was about to

[2] The interesting thing about Poincaré's work was that it did not solve the problem posed. He did not find a function that would give the coordinates as a function of time for all times. He did not show that it was impossible either, but rather that it could not be done with the Bernoulli technique of finding a conserved quantity and trying to integrate. Integration would seem unlikely from Poincaré's prize–winning memoir, but it was accomplished by the Finnish–born Swedish mathematician Sundman, who showed that to integrate the 3–body problem one had to confront the 2–body collisions. He did that by making them go away through a trick known as regularization of the collision manifold. The trick is not to expand the coordinates as a function of time t, but rather as a function of 3 t. To solve the problem for all times he used a conformal map into a strip. This allowed Sundman to obtain a series expansion for the coordinates valid for all times, solving the problem that was proposed by Weirstrass in the King Oscar II's competition. Though Sundman's work deserves better credit than it gets, it did not live up to Weirstrass's expectations, and the series solution did not 'considerably extend our understanding of the Solar system.' The work that followed from Poincaré did.

be published in Acta Mathematica when an error was found by the editor. This error in fact led to further discoveries by Poincaré, which are now considered to be the beginning of *chaos theory*. The memoir was published later in 1890. Poincaré's research into orbits about Lagrange points and low-energy transfers was not utilized for more than a century afterwards.

In 1889 Poincaré proved that for the restricted three body problem no integrals exist apart from the Jacobian. In 1890 Poincaré proved his famous recurrence theorem, namely that in any small region of phase–space trajectories exist which pass through the region infinitely often. Poincaré published 3 volumes of 'Les méthods nouvelle de la mécanique celeste' between 1892 and 1899. He discussed convergence and uniform convergence of the series solutions discussed by earlier mathematicians and proved them not to be uniformly convergent. The stability proofs of Lagrange and Laplace became inconclusive after this result.

Poincaré introduced further topological methods in 1912 for the theory of stability of orbits in the 3–body problem. It fact Poincaré essentially invented topology in his attempt to answer stability questions in the three body problem. He conjectured that there are infinitely many periodic solutions of the restricted problem, the conjecture being later proved by George *Birkhoff*. The stability of the orbits in the three body problem was also investigated by Levi–Civita, Birkhoff and others (see [II06c] for technical details).

To examine chaos, Poincaré used the idea of a section, today called the *Poincaré section*, which cuts across the orbits in phase–space. While the original dynamical system always *flows in continuous time*, on the Poincaré section we can observe *discrete–time steps*. More precisely, the original *phase–space flow* (see [II06c]) is replaced by an *iterated map*, which reduces the dimension of the phase–space by one (see Figure 21.3). Later, to show what a Poincaré section would look like, Hénon devised a simple 2D–map, which is today called the *Hénon map*: $x_{new} = 1 - ax^2 + by, \quad y_{new} = x,$ with parameters

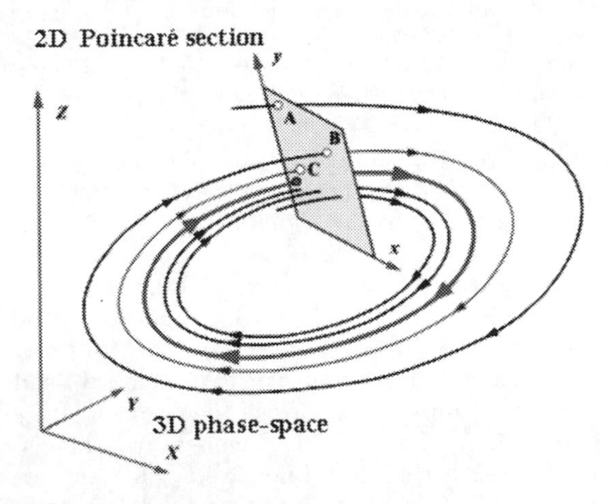

Fig. 21.3. The 2D Poincaré section, reducing the 3D phase–space, using the iterated map: $x_{new} = F(x, y), \quad y_{new} = G(x, y)$.

$a = 1.4$, $b = 0.3$. Given any starting point, this map generates a sequence of points settling onto a chaotic attractor.

As an inheritance of Poincaré work, the chaos of the Solar system has been recently used for the SOHO project,[3] to minimize the fuel consumption need for the space flights. Namely, in a rotating frame, a spacecraft can remain stationary at 5 *Lagrange's points* (see Figure 21.4).

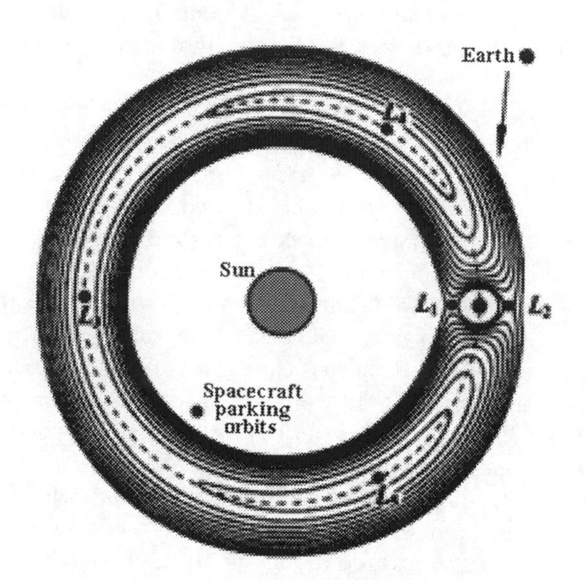

Fig. 21.4. The *Lagrange's points* ($L_1, ...L_5$) used for the space flights. Points L_1, L_2, L_3 on the Sun–Earth axis are unstable. The SOHO spacecraft used a *halo orbit* around L_1 to observe the Sun. The triangular points L_4 and L_5 are often stable. A Japanese rescue mission used a chaotic Earth–Moon trajectory.

Poincaré had two protegés in the development of chaos theory in the new world: *George D. Birkhoff* (see [Bir15, Bir27, Bir17] for the *Birkhoff curve shortening flow*), and *Stephen Smale*.

[3] The SOHO project is being carried out jointly by ESA (European Space Agency) and NASA (US National Aeronautics and Space Administration), as a cooperative effort between the two agencies in the framework of the Solar Terrestrial Science Program (STSP) comprising SOHO and CLUSTER, and the International Solar–Terrestrial Physics Program (ISTP), with Geotail (ISAS–Japan), Wind, and Polar. SOHO was launched on December 2, 1995. The SOHO spacecraft was built in Europe by an industry team led by Matra, and instruments were provided by European and American scientists. There are nine European Principal Investigators (PI's) and three American ones. Large engineering teams and more than 200 co–investigators from many institutions support the PI's in the development of the instruments and in the preparation of their operations and data analysis. NASA is responsible for the launch and mission operations. Large radio dishes around the world which form NASA's Deep Space Network are used to track the spacecraft beyond the Earth's orbit. Mission control is based at Goddard Space Flight Center in Maryland.

In some detail, the theorems of John von Neumann and George Birkhoff on the *ergodic hypothesis* (see footnote 6 above) were published in 1912 and 1913. This line of enquiry developed in two directions. One direction took an abstract approach and considered *dynamical systems as transformations of measurable spaces into themselves.* Could we classify these transformations in a meaningful way? This lead *Andrey N. Kolmogorov* to the introduction of the fundamental concept of *entropy* for dynamical systems. With entropy as a *dynamical invariant* it became possible to classify a set of abstract dynamical systems known as the *Bernoulli systems.*

The other line that developed from the ergodic hypothesis was in trying to find mechanical systems that are ergodic. *An ergodic system could not have stable orbits, as these would break ergodicity.* So, in 1898 *Jacques S. Hadamard* published a paper on *billiards*, where he showed that the *motion of balls on surfaces of constant negative curvature is everywhere unstable.* This dynamical system was to prove very useful and it was taken up by Birkhoff.

Marston Morse in 1923 showed that it was possible to enumerate the orbits of a ball on a surface of constant negative curvature.[4] He did this by introducing a symbolic code to each orbit and showed that the number of possible codes grew exponentially with the length of the code. With contributions by E. Artin, G. Hedlund, and *Heinz Hopf* it was eventually proven that the motion of a ball on a surface of constant negative curvature was ergodic. The importance of this result escaped most physicists, one exception being N.M. *Krylov*, who understood that a physical billiard was a dynamical system on a surface of negative curvature, but with the curvature concentrated along the lines of collision. Sinai, who was the first to show that a physical billiard can be ergodic, knew Krylov's work well.

On the other hand, the work of Lord Rayleigh also received vigorous development. It prompted many experiments and some theoretical development by B. *Van der Pol*, G. *Duffing*, and D. *Hayashi*. They found other systems in which the nonlinear oscillator played a role and classified the possible motions of these systems. This concreteness of experiments, and the possibility of analysis was too much of temptation for M. L. *Cartwright* and J.E. *Littlewood*, who set out to prove that many of the structures conjectured by the experimentalists and theoretical physicists did indeed follow from the equations of motion.

Also, G. Birkhoff had found a 'remarkable curve' in a 2D map; it appeared to be non–differentiable and it would be nice to see if a smooth flow could generate

---

[4] Recall from [II06c] that in differential topology, the techniques of *Morse theory* give a very direct way of analyzing the topology of a manifold by studying differentiable functions on that manifold. According to the basic insights of Marston Morse, a differentiable function on a manifold will, in a typical case, reflect the topology quite directly. Morse theory allows one to find the so–called CW–structures and handle decompositions on manifolds and to obtain substantial information about their homology. Before Morse, Arthur Cayley and James Clerk Maxwell developed some of the ideas of Morse theory in the context of topography. Morse originally applied his theory to geodesics (critical points of the energy functional on paths). These techniques were later used by *Raoul Bott* in his proof of the celebrated Bott periodicity theorem.

such a curve. The work of Cartwright and Littlewood lead to the work of N. Levinson, which in turn provided the basis for the horseshoe construction of Steve Smale.

In Russia, *Aleksandr M. Lyapunov* paralleled the methods of Poincaré and initiated the strong Russian dynamical systems school. A. *Andronov*[5] carried on with the study of nonlinear oscillators and in 1937 introduced together with *Lev S. Pontryagin*[6] the notion of *coarse systems*. They were formalizing the understanding garnered from the study of nonlinear oscillators, the understanding that many of the details on how these oscillators work do not affect the overall picture of the phase–space: there will still be limit cycles if one changes the dissipation or spring force function by a little bit. And changing the system a little bit has the great advantage of eliminating exceptional cases in the mathematical analysis. Coarse systems were the concept that caught Smale's attention and enticed him to study dynamical systems (see [CAM05]).

The path traversed from ergodicity to entropy is a little more confusing. The general character of entropy was understood by *Norbert Wiener*,[7] who seemed to have spoken to *Claude E. Shannon*.[8] In 1948 Shannon published his results on *information theory*, where he discusses the entropy of the shift transformation.

In Russia, *Andrey N. Kolmogorov* went far beyond and suggested a definition of the metric entropy of an area preserving transformation in order to classify Bernoulli shifts. The suggestion was taken by his student Ya.G. *Sinai* and the results published in 1959. In 1967 D.V. Anosov[9] and Sinai applied the notion of entropy to the study of dynamical systems. It was in the context of studying the entropy associated to a dynamical system that Sinai introduced *Markov partitions* (in 1968), which allow one *to relate dynamical systems and statistical mechanics*; this has been a very fruitful relationship. It adds measure notions to the topological framework laid down in Smale's dynamical systems paper. Markov partitions *divide the phase–space* of the dynamical system into nice little

[5] Recall that both the *Andronov–Hopf bifurcation* and a crater on the Moon are named after Aleksandr Andronov.

[6] The father of modern optimal control theory (see [II06c])

[7] The father of cybernetics

[8] The father of information theory

[9] Recall that the *Anosov map* on a manifold M is a certain type of mapping, from M to itself, with rather clearly marked local directions of 'expansion' and 'contraction'. More precisely:

- If a differentiable map f on M has a hyperbolic structure on the tangent bundle, then it is called an *Anosov map*. Examples include the *Bernoulli map*, and *Arnold cat map*.
- If the Anosov map is a diffeomorphism, then it is called an *Anosov diffeomorphism*. Anosov proved that Anosov diffeomorphisms are *structurally stable*.
- If a flow on a manifold splits the tangent bundle into three invariant subbundles, with one subbundle that is exponentially contracting, and one that is exponentially expanding, and a third, non–expanding, non–contracting 1D sub–bundle, then the flow is called an *Anosov flow*.

boxes that map into each other. Each box is labelled by a code and the dynamics on the phase–space maps the codes around, inducing a *symbolic dynamics*. From the number of boxes needed to cover all the space, Sinai was able to define the notion of entropy of a dynamical system. However, the relations with statistical mechanics became explicit in the work of *David Ruelle*.[10] Ruelle understood that the topology of the orbits could be specified by a symbolic code, and that one could associate an 'energy' to each orbit. The energies could be formally combined in a *partition function* (see [II06c]) to generate the invariant measure of the system.

21.1.2 Smale's Horseshoe: Chaos of Stretching and Folding

The first deliberate, coordinated attempt to understand how global system's behavior might differ from its local behavior, came from topologist Steve Smale from the University of California at Berkeley. A young physicist, making a small talk, asked what Smale was working on. The answer stunned him: "Oscillators." It was absurd. Oscillators (pendulums, springs, or electric circuits) where the sort of problem that a physicist finished off early in his training. They were easy. Why would a great mathematician be studying elementary physics? However, Smale was looking at nonlinear oscillators, chaotic oscillators – and seing things that physicists had learned no to see [Gle87].

Smale's 1966 Fields Medal honored a famous piece of work in high–dimensional topology, proving *Poincaré conjecture* for all dimensions greater than 4; he later generalized the ideas in a 107 page paper that established the *H–cobordism theorem* (this seminal result provides algebraic algebraic topological criteria for establishing that higher–dimensional manifolds are diffeomorphic).

After having made great strides in topology, Smale then turned to the study of nonlinear dynamical systems, where he made significant advances as well.[11] His first contribution is the famous *horseshoe map* [Sma67] that started–off

[10] David Ruelle is a mathematical physicist working on statistical physics and dynamical systems. Together with *Floris Takens*, he coined the term *strange attractor*, and founded a modern *theory of turbulence*. Namely, in a seminal paper [RT71] they argued that, as a function of an external parameter, the *route to chaos in a fluid flow* is a transition sequence leading from stationary (S) to single periodic (P), double periodic (QP_2), triple periodic (QP_3) and, possibly, quadruply periodic (QP_4) motions, before the flow becomes chaotic (C).

[11] In the fall of 1961 Steven Smale was invited to Kiev where he met V.I. Arnol'd, (one of the fathers of modern geometrical mechanics [II06c]), D.V. Anosov, Sinai, and Novikov. He lectured there, and spent a lot of time with Anosov. He suggested a series of conjectures, most of which Anosov proved within a year. It was Anosov who showed that there are dynamical systems for which all points (as opposed to a nonwandering set) admit the hyperbolic structure, and it was in honor of this result that Smale named them *Axiom–A systems*. In Kiev Smale found a receptive audience that had been thinking about these problems. Smale's result catalyzed their thoughts and initiated a chain of developments that persisted into the 1970's.

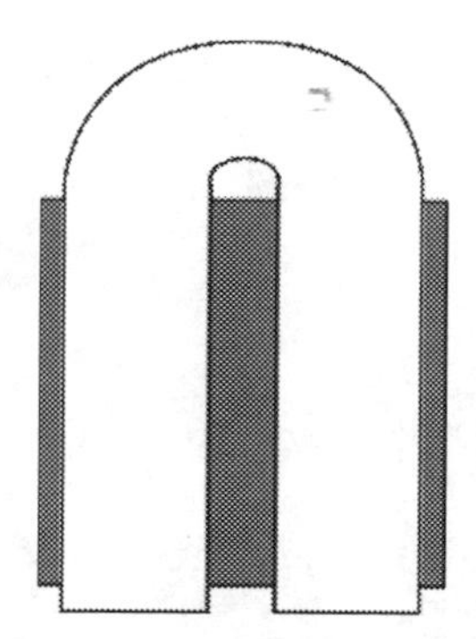

Fig. 21.5. The *Smale horseshoe map* consists of a sequence of operations on the unit square. First, stretch in the y–direction by more than a factor of two, then squeeze (compress) in the x–direction by more than a factor of two. Finally, fold the resulting rectangle and fit it back onto the square, overlapping at the top and bottom, and not quite reaching the ends to the left and right (and with a gap in the middle), as illustrated in the diagram. The shape of the stretched and folded map gives the horseshoe map its name. Note that it is vital to the construction process for the map to overlap and leave the middle and vertical edges of the initial unit square uncovered.

significant research in dynamical systems and chaos theory.[12] Smale also outlined a mathematical research program carried out by many others. Smale is also known for injecting *Morse theory* into mathematical economics, as well as recent explorations of various theories of computation. In 1998 he compiled a list of 18 problems in mathematics to be solved in the 21st century. This list was compiled in the spirit of Hilbert's famous list of problems produced in 1900. In fact, Smale's list includes some of the original Hilbert problems. Smale's problems include the Jacobian conjecture and the Riemann hypothesis, both of which are still unsolved.

The *Smale horseshoe map* (see Figure 21.5) *is any member of a class of chaotic maps of the square into itself.* This topological transformation provided a basis for understanding the chaotic properties of dynamical systems. Its basis are simple: A space is stretched in one direction, squeezed in another, and then folded. When the process is repeated, it produces something like a many–layered

[12] In his landmark 1967 Bulletin survey article entitled 'Differentiable dynamical systems' [Sma67], Smale presented his program for hyperbolic dynamical systems and stability, complete with a superb collection of problems. The major theorem of the paper was the Ω–Stability Theorem: the global foliation of invariant sets of the map into disjoint stable and unstable parts, whose proof was a tour de force in the new dynamical methods. Some other important ideas of this paper are the existence of a horseshoe and enumeration and ordering of all its orbits, as well as the use of zeta functions to study dynamical systems. The emphasis of the paper is on the global properties of the dynamical system, on how to understand the topology of the orbits. Smale's account takes us from a local differential equation (in the form of vector fields) to the global topological description in terms of horseshoes.

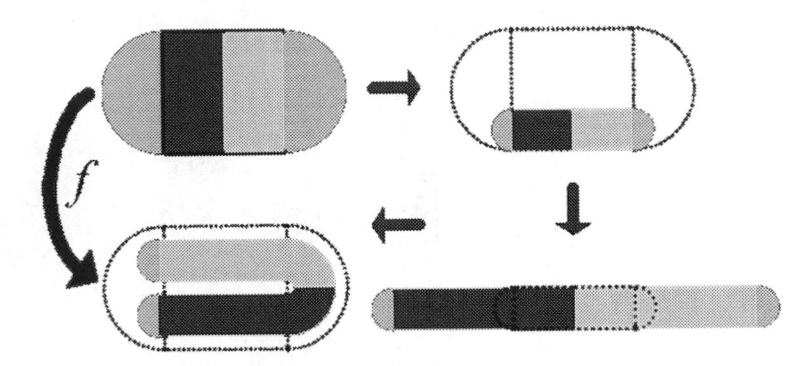

Fig. 21.6. The *Smale horseshoe map* f, defined by *stretching, folding* and *squeezing* of the system's phase–space.

pastry dough, in which a pair of points that end up close together may have begun far apart, while two initially nearby points can end completely far apart.[13]

The horseshoe map was introduced by Smale while studying the behavior of the orbits of the *relaxation Van der Pol oscillator*. The action of the map is defined geometrically by squishing the square, then stretching the result into a long strip, and finally folding the strip into the shape of a horseshoe.

Most points eventually leave the square under the action of the map f. They go to the side caps where they will, under iteration, converge to a *fixed–point* in one of the caps. The points that remain in the square under repeated iteration form a *fractal set* and are part of the *invariant set* of the map f (see Figure 21.6).

The *stretching, folding* and *squeezing* of the horseshoe map are the essential elements that must be present in any chaotic system. In the horseshoe map the squeezing and stretching are uniform. They compensate each other so that the area of the square does not change. The folding is done neatly, so that the orbits that remain forever in the square can be simply described.

Repeating this generates the horseshoe attractor. If one looks at a cross section of the final structure, it is seen to correspond to a *Cantor set*.

The Smale horseshoe map is the set of basic topological operations for constructing an attractor consist of stretching (which gives sensitivity to initial conditions) and folding (which gives the attraction). Since *trajectories in phase–space cannot cross*, the repeated stretching and folding operations result in an object of great topological complexity. For any horseshoe map we have:

- There is an infinite number of periodic orbits;
- Periodic orbits of arbitrarily long period exist;
- The number or periodic orbits grows exponentially with the period; and

[13] Originally, Smale had hoped to explain all dynamical systems in terms of *stretching* and *squeezing* – with no folding, at least no folding that would drastically undermine a system's stability. But *folding* turned out to be necessary, and folding allowed sharp changes in dynamical behavior [Gle87].

- Close to any point of the fractal invariant set there is a point of a periodic orbit.

More precisely, the horseshoe map f is a *diffeomorphism* defined from a region S of the plane into itself. The region S is a square capped by two semi–disks. The action of f is defined through the composition of three geometrically defined transformations. First the square is contracted along the vertical direction by a factor $a < 1/2$. The caps are contracted so as to remain semi-disks attached to the resulting rectangle. Contracting by a factor smaller than one half assures that there will be a gap between the branches of the horseshoe. Next the rectangle is stretched by a factor of $1/a$; the caps remain unchanged. Finally the resulting strip is folded into a horseshoe–shape and placed back into S.

The interesting part of the dynamics is the image of the square into itself. Once that part is defined, the map can be extended to a diffeomorphism by defining its action on the caps. The caps are made to contract and eventually map inside one of the caps (the left one in the figure). The extension of f to the caps adds a fixed–point to the *non–wandering set* of the map. To keep the class of horseshoe maps simple, the curved region of the horseshoe should not map back into the square.

The horseshoe map is one–to–one (1–1, or injection): any point in the domain has a unique image, even though not all points of the domain are the image of a point. The inverse of the horseshoe map, denoted by f^{-1}, cannot have as its domain the entire region S, instead it must be restricted to the image of S under f, that is, the domain of f^{-1} is $f(S)$.

By folding the contracted and stretched square in different ways, other types of horseshoe maps are possible (see Figure 21.7). The contracted square cannot overlap itself to assure that it remains 1–1. When the action on the square is extended to a diffeomorphism, the extension cannot always be done on the plane. For example, the map on the right needs to be extended to a diffeomorphism of the sphere by using a 'cap' that wraps around the equator.

The horseshoe map is an Axiom A diffeomorphism that serves as a model for the general behavior at a transverse *homoclinic point*, where the *stable and unstable manifolds* of a periodic point intersect.

The horseshoe map was designed by Smale to reproduce the chaotic dynamics of a *flow* in the neighborhood of a given periodic *orbit*. The neighborhood is chosen to be a small disk perpendicular to the orbit. As the system evolves,

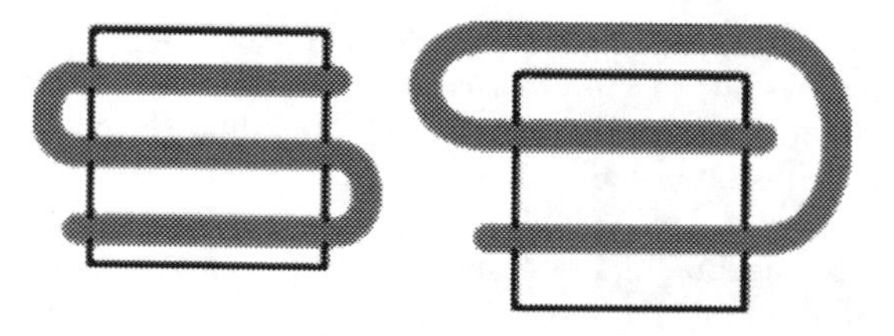

Fig. 21.7. Other types of horseshoe maps can be made by folding the contracted and stretched square in different ways.

points in this disk remain close to the given periodic orbit, tracing out orbits that eventually intersect the disk once again. Other orbits diverge.

The behavior of all the orbits in the disk can be determined by considering what happens to the disk. The intersection of the disk with the given periodic orbit comes back to itself every period of the orbit and so do points in its neighborhood. When this neighborhood returns, its shape is transformed. Among the points back inside the disk are some points that will leave the disk neighborhood and others that will continue to return. The set of points that never leaves the neighborhood of the given periodic orbit form a fractal.

A symbolic name can be given to all the orbits that remain in the neighborhood. The initial neighborhood disk can be divided into a small number of regions. Knowing the sequence in which the orbit visits these regions allows the orbit to be pinpointed exactly. The visitation sequence of the orbits provide the so–called *symbolic dynamics*[14].

It is possible to describe the behavior of all initial conditions of the horseshoe map. An initial point $u_0 = x, y$ gets mapped into the point $u_1 = f(u_0)$. Its iterate is the point $u_2 = f(u_1) = f^2(u_0)$, and repeated iteration generates the orbit $u_0, u_1, u_2, ...$ Under repeated iteration of the horseshoe map, most orbits end up at the fixed–point in the left cap. This is because the horseshoe maps the left cap into itself by an *affine transformation*, which has exactly one fixed–point. Any orbit that lands on the left cap never leaves it and converges to the fixed–point in the left cap under iteration. Points in the right cap get mapped into the left cap on the next iteration, and most points in the square get mapped into the caps. Under iteration, most points will be part of orbits that converge to the fixed–point in the left cap, but some points of the square never leave.

Under forward iterations of the horseshoe map, the original square gets mapped into a series of horizontal strips. The points in these horizontal strips come from vertical strips in the original square. Let S_0 be the original square, map it forward n times, and consider only the points that fall back into the square S_0, which is a set of horizontal stripes $H_n = f^n(S_0) \cap S_0$. The points in the horizontal stripes came from the vertical stripes $V_n = f^{-n}(H_n)$, which are the horizontal strips H_n mapped backwards n times. That is, a point in V_n will, under n iterations of the horseshoe map, end up in the set H_n of vertical strips (see Figure 21.8).

Now, if a point is to remain indefinitely in the square, then it must belong to an *invariant set* Λ that maps to itself. Whether this set is empty or not has to be

[14] Symbolic dynamics is the practice of modelling a dynamical system by a space consisting of infinite sequences of abstract symbols, each sequence corresponding to a state of the system, and a shift operator corresponding to the dynamics. Symbolic dynamics originated as a method to study general dynamical systems, now though, its techniques and ideas have found significant applications in data storage and transmission, linear algebra, the motions of the planets and many other areas. The distinct feature in symbolic dynamics is that time is measured in discrete intervals. So at each time interval the system is in a particular state. Each state is associated with a symbol and the evolution of the system is described by an infinite sequence of symbols (see text below).

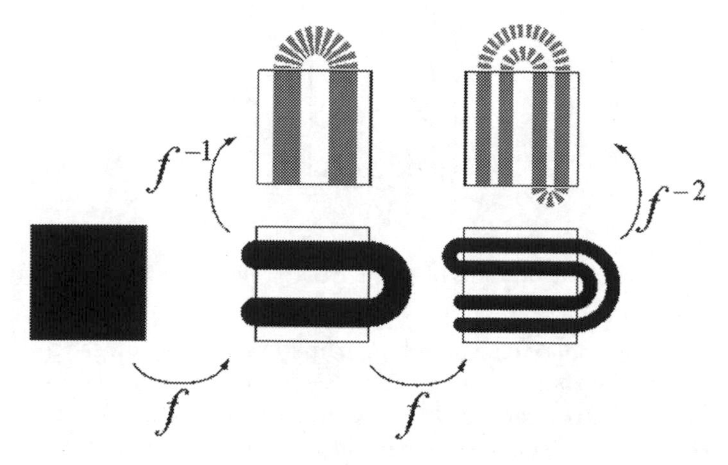

Fig. 21.8. Iterated horseshoe map: pre–images of the square region.

Fig. 21.9. Intersections that converge to the invariant set Λ.

determined. The vertical strips V_1 map into the horizontal strips H_1, but not all points of V_1 map back into V_1. Only the points in the intersection of V_1 and H_1 may belong to Λ, as can be checked by following points outside the intersection for one more iteration. The intersection of the horizontal and vertical stripes, $H_n \cap V_n$, are squares that converge in the limit $n \to \infty$ to the invariant set Λ (see Figure 21.9).

The structure of invariant set Λ can be better understood by introducing a system of labels for all the intersections, namely a *symbolic dynamics*. The intersection $H_n \cap V_n$ is contained in V_1. So any point that is in Λ under iteration must land in the left vertical strip A of V_1, or on the right vertical strip B. The lower horizontal strip of H_1 is the image of A and the upper horizontal strip is the image of B, so $H_1 = f(A) \cap f(B)$. The strips A and B can be used to label the four squares in the intersection of V_1 and H_1 (see Figure 21.10) as:

$$\Lambda_{A\bullet A} = f(A) \cap A, \qquad \Lambda_{A\bullet B} = f(A) \cap B,$$
$$\Lambda_{B\bullet A} = f(B) \cap A, \qquad \Lambda_{B\bullet B} = f(B) \cap B.$$

The set $\Lambda_{B\bullet A}$ consist of points from strip A that were in strip B in the previous iteration. A dot is used to separate the region the point of an orbit is in from the region the point came from.

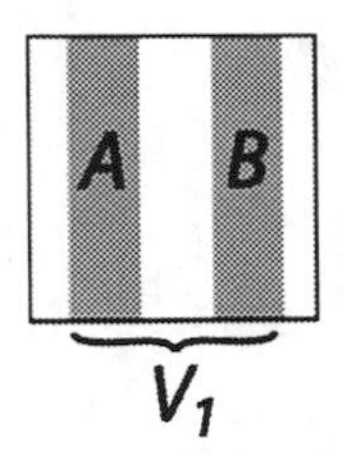

Fig. 21.10. The basic domains of the horseshoe map in symbolic dynamics.

This notation can be extended to higher iterates of the horseshoe map. The vertical strips can be named according to the sequence of visits to strip A or strip B. For example, the set $ABB \subset V_3$ consists of the points from A that will all land in B in one iteration and remain in B in the iteration after that:

$$ABB = \{x \in A | f(x) \in B \ \text{ and } \ f^2(x) \in B\}.$$

Working backwards from that trajectory determines a small region, the set ABB, within V_3.

The horizontal strips are named from their vertical strip pre–images. In this notation, the intersection of V_2 and H_2 consists of 16 squares, one of which is

$$\Lambda_{AB\bullet BB} = f^2(AB) \cap BB.$$

All the points in $\Lambda_{AB\bullet BB}$ are in B and will continue to be in B for at least one more iteration. Their previous trajectory before landing in BB was A followed by B.

Any one of the intersections $\Lambda_{P\bullet F}$ of a horizontal strip with a vertical strip, where P and F are sequences of As and Bs, is an affine transformation of a small region in V_1. If P has k symbols in it, and if $f^{-k}(\Lambda_{P\bullet F})$ and $\Lambda_{P\bullet F}$ intersect, then the region $\Lambda_{P\bullet F}$ will have a *fixed–point*. This happens when the sequence P is the same as F. For example, $\Lambda_{ABAB\bullet ABAB} \subset V_4 \cap H_4$ has at least one fixed–point. This point is also the same as the fixed–point in $\Lambda_{AB\bullet AB}$. By including more and more ABs in the P and F part of the label of intersection, the area of the intersection can be made as small as needed. It converges to a point that is part of a *periodic orbit of the horseshoe map.* The periodic orbit can be labelled by the simplest sequence of As and Bs that labels one of the regions the periodic orbit visits. For every sequence of As and Bs there is a periodic orbit.

The Smale horseshoe map is the same topological structure as the *homoclinic tangle*. To dynamically introduce homoclinic tangles, let us consider a classical engineering problem of *escape from a potential well.* Namely, if we have a motion, $x = x(t)$, of a damped particle in a well with potential energy $V = x^2/2 - x^3/3$ (see Figure 21.11) excited by a periodic driving force, $F\cos(wt)$ (with the period $T = 2\pi/w$), we are dealing with a nonlinear dynamical system given by [TS01]

$$\ddot{x} + a\dot{x} + x - x^2 = F\cos(wt). \tag{21.1}$$

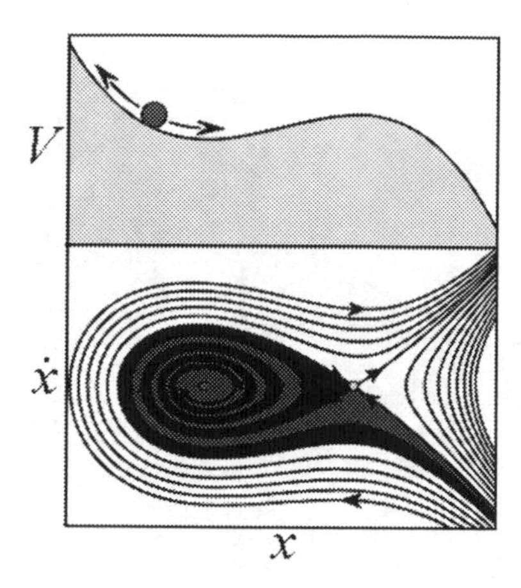

Fig. 21.11. Motion of a damped particle in a potential well, driven by a periodic force $F\cos(wt)$,. Up: potential $(x-V)$–plot, with $V=x^2/2-x^3/3$; down: the corresponding phase $(x-\dot{x})$–portrait, showing the safe basin of attraction – if the driving is switched off ($F=0$).

Now, if the driving is switched off, i.e., $F=0$, we have an autonomous 2D–system with the phase–portrait (and the safe basin of attraction) given in Figure 21.11 (below). The grey area of escape starts over the hilltop to infinity. Once we start driving, the system (21.1) becomes 3–dimensional, with its 3D phase–space. We need to see the basin in a *stroboscopic section* (see Figure 21.12). The hill–top solution still has an inset and and outset. As the driving increases, the inset and outset get tangled. They intersect one another an infinite number of times. The boundary of the safe basin becomes fractal. As the driving increases even more, the so–called fractal–fingers created by the homoclinic tangling, make a sudden incursion into the safe basin. At that point, the integrity of the in–well motions is lost [TS01].

21.1.3 Lorenz' Weather Prediction and Chaos

Recall that an *attractor* is a set of system's states (i.e., points in the system's phase–space), invariant under the dynamics, towards which neighboring states in a given *basin of attraction* asymptotically approach in the course of dynamic evolution.[15] An attractor is defined as the smallest unit which cannot be itself decomposed into two or more attractors with distinct basins of attraction. This restriction is necessary since a dynamical system may have multiple attractors, each with its own basin of attraction.

Conservative systems do not have attractors, since the motion is periodic. For dissipative dynamical systems, however, volumes shrink exponentially, so attractors have 0 volume in nD phase–space.

[15] A *basin of attraction* is a set of points in the system's phase–space, such that initial conditions chosen in this set dynamically evolve to a particular attractor.

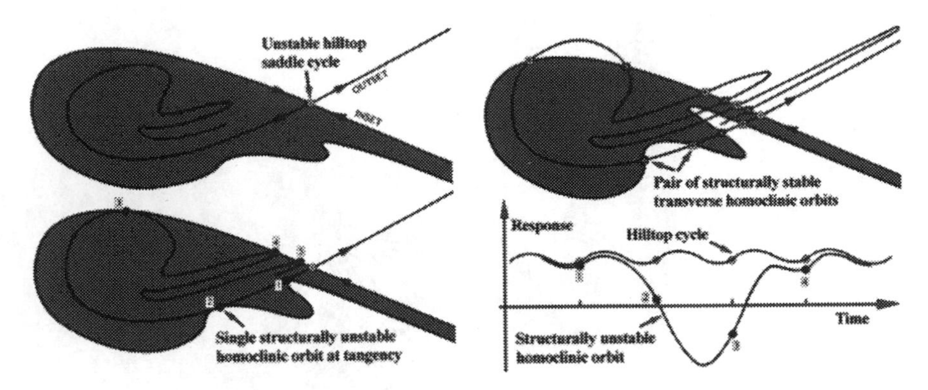

Fig. 21.12. Dynamics of a homoclinic tangle. The hill–top solution of a damped particle in a potential well driven by a periodic force. As the driving increases, the inset and outset get tangled.

In particular, a stable *fixed–point* surrounded by a dissipative region is an attractor known as a *map sink*.[16] Regular attractors (corresponding to 0 *Lyapunov exponents*) act as *limit cycles*, in which trajectories circle around a limiting trajectory which they asymptotically approach, but never reach. The so–called *strange attractors*[17] are bounded regions of phase–space (corresponding to positive Lyapunov characteristic exponents) having zero measure in the embedding phase–space and a *fractal dimension*. Trajectories within a strange attractor appear to skip around randomly.

In 1963, Ed Lorenz from MIT was trying to improve weather forecasting. Using a primitive computer of those days, he discovered the first *chaotic attractor*. Lorenz used three Cartesian variables, (x, y, z), to define *atmospheric convection*. Changing in time, these variables gave him a trajectory in a (Euclidean) 3D–space. From all starts, trajectories settle onto a chaotic, or *strange attractor*.[18]

[16] A *map sink* is a stable fixed–point of a map which, in a dissipative dynamical system, is an attractor.

[17] A strange attractor is an attracting set that has zero measure in the embedding phase–space and has fractal dimension. Trajectories within a strange attractor appear to skip around randomly.

[18] Edward Lorenz is a professor of meteorology at MIT who wrote the first clear paper on *deterministic chaos*. The paper was called 'Deterministic Nonperiodic Flow' and it was published in the Journal of Atmospheric Sciences in 1963. Before that, in 1960, Lorenz began a project to simulate weather patterns on a computer system called the Royal McBee. Lacking much memory, the computer was unable to create complex patterns, but it was able to show the interaction between major meteorological events such as tornados, hurricanes, easterlies and westerlies. A variety of factors was represented by a number, and Lorenz could use computer printouts to analyze the results. After watching his systems develop on the computer, Lorenz began to see patterns emerge, and was able to predict with some degree of accuracy what would happen next. While carrying out an experiment, Lorenz made an accidental discovery. He had completed a run, and wanted to recreate the pattern.

More precisely, Lorenz reduced the *Navier–Stokes equations* for *convective Bénard fluid flow* into three first order coupled nonlinear differential equations and demonstrated with these the idea of sensitive dependence upon initial conditions and chaos (see [Lor63, Spa82]).

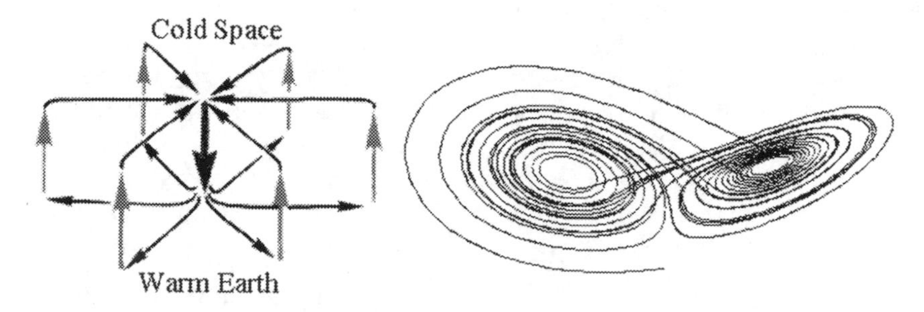

Fig. 21.13. Bénard cells, showing a typical vortex of a rolling air, with a warm air rising in a ring and a cool air descending in the center (left). A simple model of the Bénard cells provided by the celebrated 'Lorenz–butterfly' (or, 'Lorenz–mask') *strange attractor* (right).

We rewrite the celebrated Lorenz equations here as

$$\dot{x} = a(y - x), \qquad \dot{y} = bx - y - xz, \qquad \dot{z} = xy - cz, \tag{21.2}$$

Using a printout, Lorenz entered some variables into the computer and expected the simulation to proceed the same as it had before. To his surprise, the pattern began to diverge from the previous run, and after a few 'months' of simulated time, the pattern was completely different. Lorenz eventually discovered why seemingly identical variables could produce such different results. When Lorenz entered the numbers to recreate the scenario, the printout provided him with numbers to the thousandth position (such as 0.617). However, the computer's internal memory held numbers up to the millionth position (such as 0.617395); these numbers were used to create the scenario for the initial run. This small deviation resulted in a completely divergent weather pattern in just a few months. This discovery creates the groundwork of chaos theory: In a system, small deviations can result in large changes. This concept is now known as a *butterfly effect*.

Lorenz definition of chaos is: "The property that characterizes a dynamical system in which most orbits exhibit sensitive dependence." Dynamical systems (like the weather) are all around us. They have recurrent behavior (it is always hotter in summer than winter) but are very difficult to pin down and predict apart from the very short term. 'What will the weather be tomorrow?' – can be anticipated, but 'What will the weather be in a months time?' is an impossible question to answer.

Lorenz showed that with a set of simple differential equations seemingly very complex turbulent behavior could be created that would previously have been considered as random. He further showed that accurate longer range forecasts in any chaotic system were impossible, thereby overturning the previous orthodoxy. It had been believed that the more equations you add to describe a system, the more accurate will be the eventual forecast.

where x, y and z are dynamical variables, constituting the 3D *phase–space* of the *Lorenz system*; and a, b and c are the parameters of the system. Originally, Lorenz used this model to describe the unpredictable behavior of the weather, where x is the rate of convective overturning (convection is the process by which heat is transferred by a moving fluid), y is the horizontal temperature overturning, and z is the vertical temperature overturning; the parameters are: $a \equiv P-$proportional to the *Prandtl number* (ratio of the fluid viscosity of a substance to its thermal conductivity, usually set at 10), $b \equiv R-$proportional to the Rayleigh number (difference in temperature between the top and bottom of the system, usually set at 28), and $c \equiv K-$a number proportional to the physical proportions of the region under consideration (width to height ratio of the box which holds the system, usually set at 8/3). The Lorenz system (21.2) has the properties:

1. *Symmetry*: $(x, y, z) \to (-x, -y, z)$ for all values of the parameters, and
2. The $z-$axis ($x = y = 0$) is *invariant* (i.e., all trajectories that start on it also end on it).

Nowadays it is well–known that the Lorenz model is a paradigm for low–dimensional chaos in dynamical systems in synergetics and this model or its modifications are widely investigated in connection with modelling purposes in meteorology, hydrodynamics, laser physics, superconductivity, electronics, oil industry, chemical and biological kinetics, etc.

The 3D *phase–portrait* of the Lorenz system (21.13) shows the celebrated '*Lorenz mask*', a special type of *fractal attractor* (see Figure 21.13). It depicts the famous '*butterfly effect*', (i.e., sensitive dependence on initial conditions) – the popular idea in meteorology that 'the flapping of a butterfly's wings in Brazil can set off a tornado in Texas' (i.e., a tiny difference is amplified until two outcomes are totally different), so that the long term behavior becomes impossible to predict (e.g., long term weather forecasting). The Lorenz mask has the following characteristics:

1. Trajectory does not intersect itself in three dimensions;
2. Trajectory is not periodic or transient;
3. General form of the shape does not depend on initial conditions; and
4. Exact sequence of loops is very sensitive to the initial conditions.

21.1.4 Feigenbaum's Constant and Universality

Mitchell Jay Feigenbaum (born December 19, 1944; Philadelphia, USA) is a mathematical physicist whose pioneering studies in chaos theory led to the discovery of the *Feigenbaum constant*.

In 1964 he began graduate studies at the MIT. Enrolling to study electrical engineering, he changed to physics and was awarded a doctorate in 1970 for a thesis on dispersion relations under Francis Low. After short positions at Cornell

University and Virginia Polytechnic Institute, he was offered a longer–term post at Los Alamos National Laboratory to study turbulence. Although the group was ultimately unable to unravel the intractable theory of turbulent fluids, his research led him to study chaotic maps.

Many mathematical maps involving a single linear parameter exhibit apparently random behavior known as chaos when the parameter lies in a certain range. As the parameter is increased towards this region, the map undergoes bifurcations at precise values of the parameter. At first there is one stable point, then bifurcating to oscillate between two points, then bifurcating again to oscillate between four points and so on. In 1975 Feigenbaum, using the HP-65 computer he was given, discovered that the ratio of the difference between the values at which such successive *period–doubling bifurcations* (called the *Feigenbaum cascade*) occur tends to a constant of around 4.6692. He was then able to provide a mathematical proof of the fact, and showed that the same behavior and the same constant would occur in a wide class of mathematical functions prior to the onset of chaos. For the first time this universal result enabled mathematicians to take their first huge step to unravelling the apparently intractable 'random' behavior of chaotic systems. This 'ratio of convergence' is now known as the Feigenbaum constant.

More precisely, the Feigenbaum constant δ is a universal constant for functions approaching chaos via successive period doubling bifurcations. It was discovered by Feigenbaum in 1975, while studying the fixed–points of the iterated function $f(x) = 1 - \mu|x|^r$, and characterizes the geometric approach of the bifurcation parameter to its limiting value (see Figure 21.14) as the parameter μ is increased for fixed x [Fei79].

The Logistic map is a well known example of the maps that Feigenbaum studied in his famous Universality paper [Fei78].

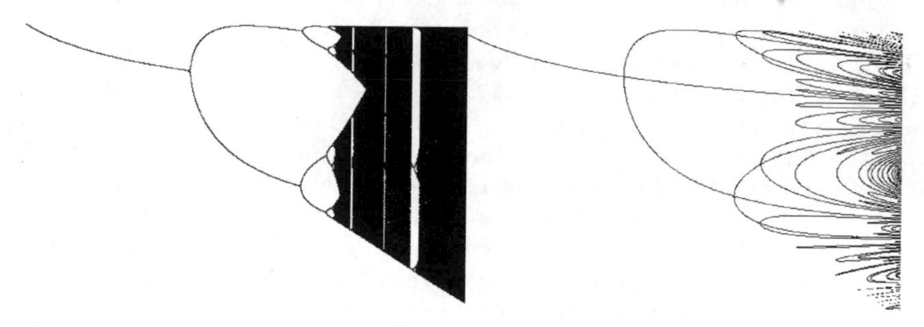

Fig. 21.14. Feigenbaum constant: approaching chaos via successive period doubling bifurcations. The plot on the left is made by iterating equation $f(x) = 1 - \mu|x|^r$ with $r = 2$ several hundred times for a series of discrete but closely spaced values of μ, discarding the first hundred or so points before the iteration has settled down to its fixed–points, and then plotting the points remaining. The plot on the right more directly shows the cycle may be constructed by plotting function $f^n(x) - x$ as a function of μ, showing the resulting curves for $n = 1, 2, 4$. Simulated in *Mathematica*^{TM}.

In 1986 Feigenbaum was awarded the Wolf Prize in Physics. He has been Toyota Professor at Rockefeller University since 1986.

For details on Feigenbaum universality, see [Gle87].

21.1.5 May's Population Modelling and Chaos

Let $x(t)$ be the population of the species at time t; then the *conservation law* for the population is conceptually given by (see [Mur02])

$$\dot{x} = births - deaths + migration, \tag{21.3}$$

where $\dot{x} = dx/dt$. The above conceptual equation gave rise to a series of *population models*. The simplest continuous–time model, due to Thomas Malthus from 1798 [Mal798],[19] has no migration, while the birth and death terms are proportional to x,

[19] The Rev. Thomas Robert Malthus, FRS (February, 1766–December 23, 1834), was an English demographer and political economist best known for his pessimistic but highly influential views. Malthus's views were largely developed in reaction to the optimistic views of his father, Daniel Malthus and his associates, notably Jean-Jacques Rousseau and William Godwin. Malthus's essay was also in response to the views of the Marquis de Condorcet. In An Essay on the Principle of Population, first published in 1798, Malthus made the famous prediction that population would outrun food supply, leading to a decrease in food per person: "The power of population is so superior to the power of the earth to produce subsistence for man, that premature death must in some shape or other visit the human race. The vices of mankind are active and able ministers of depopulation. They are the precursors in the great army of destruction; and often finish the dreadful work themselves. But should they fail in this war of extermination, sickly seasons, epidemics, pestilence, and plague, advance in terrific array, and sweep off their thousands and tens of thousands. Should success be still incomplete, gigantic inevitable famine stalks in the rear, and with one mighty blow levels the population with the food of the world." This Principle of Population was based on the idea that population if unchecked increases at a geometric rate, whereas the food supply grows at an arithmetic rate. Only natural causes (eg. accidents and old age), misery (war, pestilence, and above all famine), moral restraint and vice (which for Malthus included infanticide, murder, contraception and homosexuality) could check excessive population growth. Thus, Malthus regarded his Principle of Population as an explanation of the past and the present situation of humanity, as well as a prediction of our future. The eight major points regarding evolution found in his 1798 *Essay* are: (i) Population level is severely limited by subsistence. (ii) When the means of subsistence increases, population increases. (iii) Population pressures stimulate increases in productivity. (iv) Increases in productivity stimulates further population growth. (v) Since this productivity can never keep up with the potential of population growth for long, there must be strong checks on population to keep it in line with carrying capacity. (vi) It is through individual cost/benefit decisions regarding sex, work, and children that population and production are expanded or contracted. (vii) Positive checks will come into operation as population exceeds subsistence level. (viii) The nature of these checks will have significant effect on the rest of the sociocultural system.

$$\dot{x} = bx - dx \qquad \Longrightarrow \qquad x(t) = x_0 \mathrm{e}^{(b-d)t}, \tag{21.4}$$

where b, d are positive constants and $x_0 = x(0)$ is the initial population. Thus, according to the *Malthus model* (21.4), if $b > d$, the population grows exponentially, while if $b < d$, it dies out. Clearly, this approach is fairly oversimplified and apparently fairly unrealistic. (However, if we consider the past and predicted growth estimates for the total world population from the 1900, we see that it has actually grown exponentially.)

This simple example shows that it is difficult to make long–term predictions (or, even relatively short–term ones), unless we know sufficient facts to incorporate in the model to make it a *reliable predictor*. In the long run, clearly, there must be some adjustment to such exponential growth. François Verhulst [Ver838, Ver845][20] proposed that a *self–limiting process* should operate when a population becomes too large. He proposed the so–called *logistic growth* population model,

$$\dot{x} = rx(1 - x/K), \tag{21.5}$$

where r, K are positive constants. In the Verhulst logistic model (21.5), the constant K is the *carrying capacity* of the environment (usually determined by the available sustaining resources), while the per capita birth rate $rx(1-x/K)$ is dependent on x. There are two steady states (where $\dot{x} = 0$) for (21.5): (i) $x = 0$ (unstable, since linearization about it gives $\dot{x} \approx rx$); and (ii) $x = K$ (stable, since linearization about it gives $\frac{d}{dt}(x - K) \approx -r(x - K)$, so $\lim_{t\to\infty} x = K$). The carrying capacity K determines the size of the stable steady state population, while r is a measure of the rate at which it is reached (i.e., the measure of the dynamics) – thus $1/r$ is a representative timescale of the response of the model to any change in the population. The solution of (21.5) is

$$x(t) = \frac{x_0 K \mathrm{e}^{rt}}{[K + x_0(\mathrm{e}^{rt} - 1)]} \qquad \Longrightarrow \qquad \lim_{t\to\infty} x(t) = K.$$

In general, if we consider a population to be governed by

$$\dot{x} = f(x), \tag{21.6}$$

where typically $f(x)$ is a nonlinear function of x, then the equilibrium solutions x^* are solutions of $f(x) = 0$, and are linearly stable to small perturbations if $\dot{f}(x^*) < 0$, and unstable if $\dot{f}(x^*) > 0$ [Mur02].

Evolutionists John Maynard Smith and Ronald Fisher were both critical of Malthus' theory, though it was Fisher who referred to the *growth rate* r (used in *logistic equation*) as the *Malthusian parameter*. Fisher referred to "...a relic of creationist philosophy..." in observing the fecundity of nature and deducing (as Darwin did) that this therefore drove natural selection. Smith doubted that famine was the great leveller that Malthus insisted it was.

[20] François Verhulst (October 28, 1804–February 15, 1849, Brussels, Belgium) was a mathematician and a doctor in number theory from the University of Ghent in 1825. Verhulst published in 1838 the logistic demographic model (21.5).

In the mid 20th century, ecologists realized that many species had no overlap between successive generations and so population growth happens in discrete–time steps x_t, rather than in continuous–time $x(t)$ as suggested by the conservative law (21.3) and its Maltus–Verhulst derivations. This leads to study *discrete–time models* given by *difference equations*, or, *maps*, of the form

$$x_{t+1} = f(x_t), \tag{21.7}$$

where $f(x_t)$ is some generic nonlinear function of x_t. Clearly, (21.7) is a discrete–time version of (21.6). However, instead of solving differential equations, if we know the particular form of $f(x_t)$, it is a straightforward matter to evaluate x_{t+1} and subsequent generations by simple recursion of (21.7). The skill in modelling a specific population's growth dynamics lies in determining the appropriate form of $f(x_t)$ to reflect known observations or facts about the species in question.

In 1970s, Robert May, a physicist by training, won the Crafoord Prize for 'pioneering ecological research in theoretical analysis of the dynamics of populations, communities and ecosystems', by proposing a simple *logistic map* model for the generic population growth (21.7).[21] May's model of population growth is the celebrated *logistic map* [May76, May73, May76],

$$x_{t+1} = r\, x_t\, (1 - x_t), \tag{21.8}$$

where r is the *Malthusian parameter* that varies between 0 and 4, and the initial value of the population $x_0 = x(0)$ is restricted to be between 0 and 1. Therefore, in May's logistic map (21.8), the generic function $f(x_t)$ gets a specific quadratic form

$$f(x_t) = r\, x_t\, (1 - x_t).$$

For $r < 3$, the x_t have a single value. For $3 < r < 3.4$, the x_t oscillate between two values (see *bifurcation diagram*[22] on Figure 21.15). As r increases, bifurcations occur where the number of iterates doubles. These *period doubling bifurcations* continue to a limit point at $r_{lim} = 3.569944$ at which the period is 2^∞ and the dynamics become chaotic. The r values for the first two bifurcations can be found analytically, they are $r_1 = 3$ and $r_2 = 1 + \sqrt{6}$. We can label the successive values of r at which bifurcations occur as r_1, r_2, ... The universal number associated with such period doubling sequences is called the *Feigenbaum number*,

---

[21] Lord Robert May received his Ph.D. in theoretical physics from University of Sydney in 1959. He then worked at Harvard University and the University of Sydney before developing an interest in animal population dynamics and the relationship between complexity and stability in natural communities. He moved to Princeton University in 1973 and to Oxford and the Imperial College in 1988. May was able to make major advances in the field of population biology through the application of mathematics. His work played a key role in the development of *theoretical ecology* through the 1970s and 1980s. He also applied these tools to the study of disease and to the study of *bio–diversity*.

[22] A bifurcation diagram shows the possible long–term values a variable of a system can get in function of a parameter of the system.

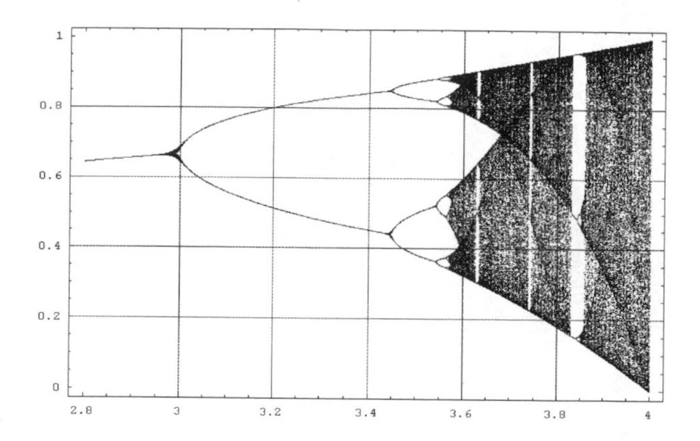

Fig. 21.15. Bifurcation diagram for the logistic map, simulated using $Mathematica^{TM}$.

$$\delta = \lim_{k\to\infty} \frac{r_k - r_{k-1}}{r_{k+1} - r_k} \approx 4.669.$$

This series of period–doubling bifurcations says that close enough to r_{lim} the distance between bifurcation points decreases by a factor of δ for each bifurcation. The complex *fractal pattern* got in this way shrinks indefinitely.

21.1.6 Hénon's 2D Map and Its Strange Attractor

Michel Hénon (born 1931 in Paris, France) is a mathematician and astronomer. He is currently at the Nice Observatory. In astronomy, Hénon is well known for his contributions to stellar dynamics, most notably the problem of *globular cluster* (see [Gle87]). In late 1960s and early 1970s he was involved in dynamical evolution of star clusters, in particular the globular clusters. He developed a numerical technique using *Monte Carlo methods*, to follow the dynamical evolution of a spherical star cluster much faster than the so–called n–body methods. In mathematics, he is well known for the Hénon map, a simple discrete dynamical system that exhibits chaotic behavior. Lately he has been involved in the restricted 3–body problem.

His celebrated *Hénon map* [Hen69] is a discrete–time dynamical system that is an extension of the *logistic map* (21.8) and exhibits a chaotic behavior. The map was introduced by Michel Hénon as a simplified model of the *Poincaré section* of the *Lorenz system* (21.2). This 2D–map takes a point (x, y) in the plane and maps it to a new point defined by equations

$$x_{n+1} = y_n + 1 - ax_n^2, \qquad y_{n+1} = bx_n,$$

The map depends on two parameters, a and b, which for the canonical Hénon map have values of $a = 1.4$ and $b = 0.3$ (see Figure 21.16). For the canonical values the Hénon map is chaotic. For other values of a and b the map may be chaotic, intermittent, or converge to a periodic orbit. An overview of the type

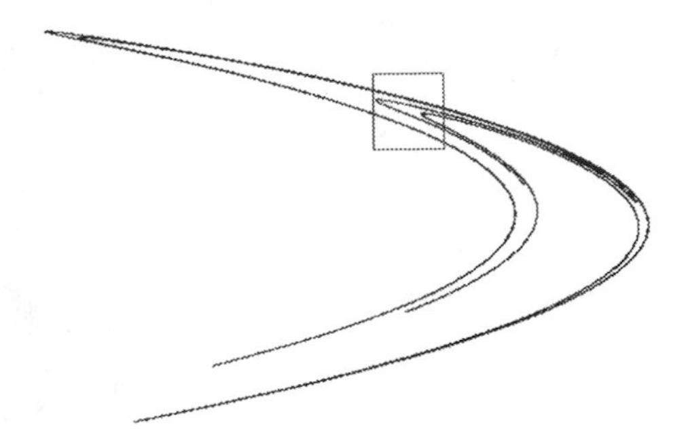

Fig. 21.16. *Hénon strange attractor* (see text for explanation), simulated using *Dynamics Solver*^{TM}.

of behavior of the map at different parameter values may be obtained from its orbit (or, bifurcation) diagram (see Figure 21.17). For the canonical map, an initial point of the plane will either approach a set of points known as the *Hénon strange attractor*, or diverge to infinity. The Hénon attractor is a fractal, smooth in one direction and a Cantor set in another. Numerical estimates yield a correlation dimension of 1.42 ± 0.02 and a Hausdorff dimension of 1.261 ± 0.003 for the Hénon attractor. As a dynamical system, the canonical Hénon map is interesting because, unlike the logistic map, its orbits defy a simple description. The Hénon map maps two points into themselves: these are the invariant points. For the canonical values of a and b, one of these points is on the attractor: $x = 0.631354477...$ and $y = 0.189406343...$ This point is unstable. Points close to this fixed–point and along the slope 1.924 will approach the fixed–point and points along the slope -0.156 will move away from the fixed–point. These slopes arise from the linearizations of the *stable manifold* and *unstable manifold* of the fixed–point. The unstable manifold of the fixed–point in the attractor is contained in the strange attractor of the Hénon map. The Hénon map does not have a strange attractor for all values of the parameters a and b. For example, by keeping b fixed at 0.3 the bifurcation diagram shows that for a = 1.25 the Hénon map has a stable periodic orbit as an attractor. Cvitanovic *et al.* [CGP88]

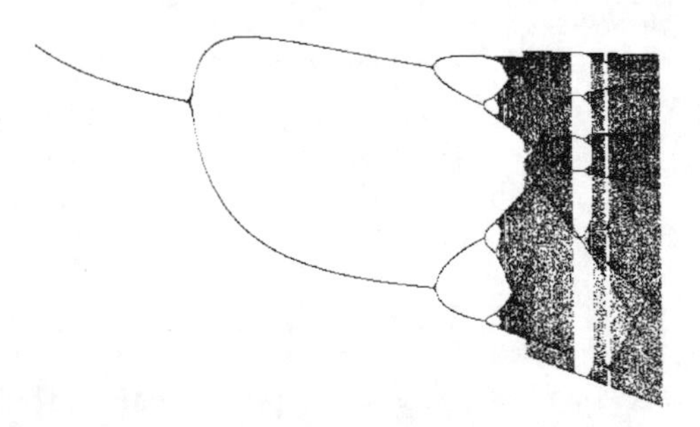

Fig. 21.17. Bifurcation diagram of the *Hénon strange attractor*, simulated using *Dynamics Solver*^{TM}.

showed how the structure of the Hénon strange attractor could be understood in terms of unstable periodic orbits within the attractor.

For the (slightly modified) Hénon map: $x_{n+1} = ay_n + 1 - x_n^2, \; y_{n+1} = bx_n$, there are three *basins of attraction* (see Figure 21.18).

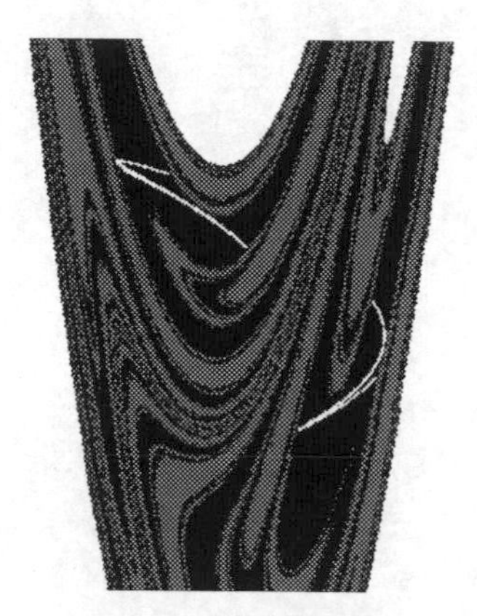

Fig. 21.18. Three basins of attraction for the Hénon map $x_{n+1} = ay_n + 1 - x_n^2$, $y_{n+1} = bx_n$, with $a = 0.475$.

The *generalized Hénon map* is a 3D–system (see Figure 21.19)

$$x_{n+1} = a\,x_n - z\,(y_n - x_n^2)), \qquad y_{n+1} = z\,x_n + a\,(y_n - x_n^2)), \qquad z_{n+1} = z_n,$$

where $a = 0.24$ is a parameter. It is an *area–preserving map*, and simulates the *Poincaré map* of period orbits in *Hamiltonian systems*. Repeated random initial conditions are used in the simulation and their gray–scale color is selected at random.

Other Famous 2D Chaotic Maps

1. The *standard map*:

$$x_{n+1} = x_n + y_{n+1}/2\pi, \qquad y_{n+1} = y_n + a\sin(2\pi x_n).$$

2. The *circle map*:

$$x_{n+1} = x_n + c + y_{n+1}/2\pi, \qquad y_{n+1} = by_n - a\sin(2\pi x_n).$$

3. The *Duffing map*:

$$x_{n+1} = y_n, \qquad y_{n+1} = -bx_n + ay_n - y_n^3.$$

4. The *Baker map*:

$$x_{n+1} = bx_n, \qquad y_{n+1} = y_n/a \qquad \text{if} \quad y_n \leq a,$$
$$x_{n+1} = (1-c) + cx_n, \qquad y_{n+1} = (y_n - a)/(1-a) \qquad \text{if} \quad y_n > a.$$

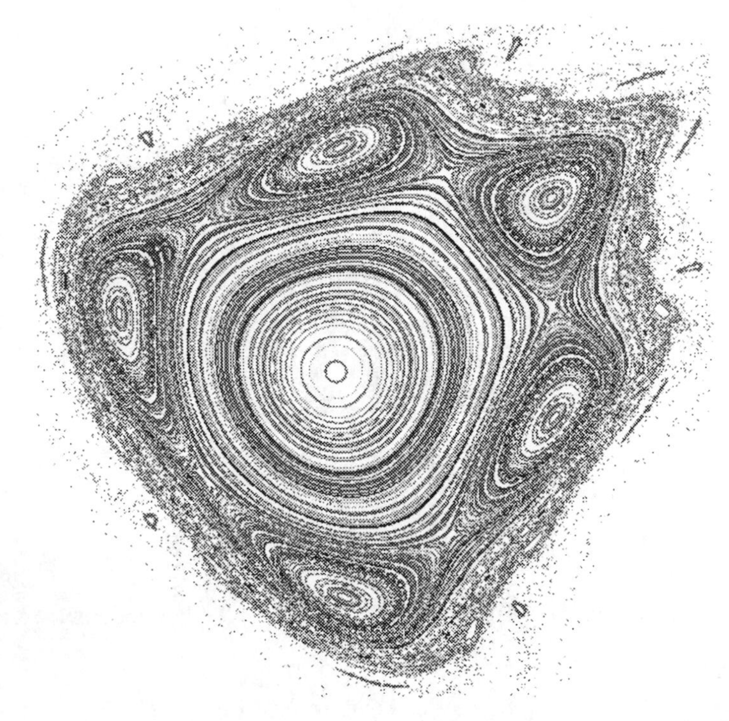

Fig. 21.19. Phase–plot of the *area–preserving generalized Hénon map*, simulated using *Dynamics Solver*^{TM}.

5. The *Kaplan–Yorke map*:

$$x_{n+1} = ax_n \bmod 1, \qquad y_{n+1} = -by_n + \cos(2\pi x_n).$$

6. The *Ott–Grebogi–Yorke map*:

$$x_{n+1} = x_n + w_1 + aP_1(x_n, y_n) \bmod 1,$$
$$y_{n+1} = y_n + w_2 + aP_2(x_n, y_n) \bmod 1,$$

where the nonlinear functions P_1, P_2 are sums of sinusoidal functions $A_{rs}^{(i)} \sin[2\pi(rx + sy + B_{rs}^{(i)})]$, with $(r, s) = (0, 1), (1, 0), (1, 1), (1, -1)$, while $A_{rs}^{(i)}, B_{rs}^{(i)}$ were selected randomly in the range $[0, 1]$.

21.1.7 Nonlinear Neurodynamics

The so–called *complex nonlinearity* [II08b] plays the key role in modern computational neuroscience. All standard neural models are nonlinear and for certain values of their system parameters they can exhibit unpredictable nonlinear behavior, most commonly strange attractors. In this chapter I will briefly review three especially popular fields within the *computational neuroscience*: (i) spiking & bursting neurons, (ii) synchronization in oscillatory neural networks, and (iii) neural attractor dynamics. There are many applications of this approach, including: brain science, medical diagnosis, autonomous robotics, pattern classification, as well as nonlinear and adaptive control theory.

A neuron is said to fire a burst of spikes when it fires two or more action potentials followed by a period of quiescence. A burst of two spikes is called a doublet, of three spikes is called a triplet, four – quadruplet, etc. Most spiking neurons can burst if stimulated with a current that slowly drives the neuron above and below the firing threshold. Such a current could be injected via an electrode or generated by the synaptic input.

Throughout this section, v denotes the membrane potential. All the parameters in the models are chosen so that v has mV scale and the time has ms scale. To compare computational cost, one assumes that each model, written as a dynamical system $\dot{x} = f(x)$, is implemented using the simplest, fixed–step first–order Euler method, with the integration time step chosen to achieve a reasonable numerical accuracy (see [IJP08, II07a, II07b] for more details).

Integrate–and–Fire Neuron

One of the most widely used models in computational neuroscience is the *leaky integrate–and–fire neuron*, (I&F neuron, for short) given by

$$\dot{v} = I + a - bv, \qquad \text{If} \quad v \geq v_{trsh} \quad \text{Then} \quad v \leftarrow c,$$

where v is the membrane potential, I is the input current, and a, b, c, and v_{trsh} are the parameters. When the membrane potential v reaches the threshold value v_{trsh}, the neuron is said to fire a *spike*, and v is reset to c. The I&F neuron can fire tonic spikes with constant frequency, and it is an integrator. The I&F neuron can fire tonic spikes with constant frequency, and it is an integrator. It is the simplest model to implement when the integration time step τ is 1 ms. Because I&F neuron has only one variable, it cannot have phasic spiking, bursting of any kind, rebound responses, threshold variability, bistability of attractors, or autonomous unpredictable nonlinear dynamics. Because of the fixed threshold, the spikes do not have latencies.

Integrate–and–Fire Neuron with Adaptation

The I&F model is $1D$, hence it cannot burst or have other properties of cortical neurons. One may think that having a second linear equation

$$\dot{v} = I + a - bv + g(d - v), \qquad \dot{g} = (e\delta(t) - g)/\tau,$$

describing activation dynamics of a high–threshold K–current, can make an improvement, e.g., endow the model with spike–frequency adaptation. Indeed, each firing increases the K–activation gate via Dirac δ–function and produces an outward current that slows down the frequency of tonic spiking. This model is fast, yet still lacks many important properties of cortical spiking neurons.

Integrate–and–Fire–or–Burst Neuron

The *integrate–and–fire–or–burst neuron* model is given by

$$\dot{v} = I + a - bv + gH(v - v_h)h(v_T - v),$$
$$\text{If } v \geq v_{trsh} \text{ Then } v \leftarrow c, \qquad \dot{h} = \begin{cases} \frac{-h}{T^-}, & \text{if } v > v_h, \\ \frac{1-h}{\tau^+}, & \text{if } v < v_h \end{cases}$$

to model thalamo–cortical neurons. Here h describes the inactivation of the calcium T–current, g, v_h, v_T, τ^+ and τ^- are parameters describing dynamics of the T–current, and H is the Heaviside step function. Having this kind of a second variable creates the possibility for bursting and other interesting regimes but is already a much slower (depending on the value of v).

Complex–Valued Resonate–and–Fire Neuron

The *resonate–and–fire neuron* is a complex–valued (i.e., $2D$) analogue of the I&F neuron, given by:

$$\dot{z} = I + (b + iw)z, \qquad \text{if } \text{Im}z = a_{trsh} \text{ then } z \longleftarrow z_0(z), \tag{21.9}$$

where $z = x + iy \in \mathbb{C}$ is a complex–valued variable that describes oscillatory activity of the neuron. Here b, w, and a_{trsh} are parameters, $i = \sqrt{-1}$, and $z_0(z)$ is an arbitrary function describing activity–dependent after–spike reset. (21.9) is equivalent to the linear system

$$\dot{x} = bx - wy, \qquad \dot{y} = wx + by,$$

where the real part x is the current–like variable, while the imaginary part y is the voltage–like variable. The resonate–and–fire model is simple and efficient. When the frequency of oscillation $w = 0$, it becomes an integrator.

Quadratic Integrate–and–Fire Neuron

An alternative to the leaky I&F neuron is the *quadratic I&F neuron*, also known as the *theta–neuron*, or the *Ermentrout canonical model*. It can be presented as

$$\dot{v} = I + a(v - v_{rest})(v - v_{trsh}), \qquad \text{If } v = v_{trsh} \text{ Then } v \leftarrow v_{rest},$$

where v_{rest} and v_{trsh} are the resting and threshold values of the membrane potential. This model is canonical in the sense that any *Class 1 excitable system* described by smooth ODEs can be transformed into this form by a continuous change of variables. It takes only seven operations to simulate 1 ms of the model, and this should be the model of choice when one simulates large–scale networks of integrators. Unlike its linear analogue, the quadratic I&F neuron has spike latencies, activity dependent threshold (which is v_{trsh} only when $I = 0$), and bistability of resting and tonic spiking modes.

Hindmarsh–Rose Neuron

The *Hindmarsh–Rose thalamic neuron* model can be written as a $3D$ ODE system:

$$\dot{v} = I + u - F(v) - w, \qquad \dot{u} = G(v) - u, \qquad \dot{w} = (H(v) - w)/\tau,$$

where F, G, and H are some functions. This model is quite expensive to implement as a large–scale spike simulator.

Morris–Lecar Neuron

Morris and Lecar suggested a simple $2D$ model to describe oscillations in barnacle giant muscle fiber. Because it has biophysically meaningful and measurable parameters, the *Morris–Lecar neuron* model became quite popular in computational neuroscience community. It consists of a membrane potential equation with instantaneous activation of Ca current and an additional equation describing slower activation of K current,

$$\begin{aligned}
&C\dot{V} = I - g_L(V - V_L) - g_{Ca}m_\infty(V)(V - V_{Ca}) - g_K n(V - V_K),\\
&\dot{n} = \lambda(V)(n_\infty(V) - n), \qquad \text{where}\\
&m_\infty(V) = \frac{1}{2}\left(1 + \tanh\left[\frac{V - V_1}{V_2}\right]\right), \qquad \text{and}\\
&n_\infty(V) = \frac{1}{2}\left(1 + \tanh\left[\frac{V - V_3}{V_4}\right]\right), \qquad \lambda(V) = \bar{\lambda}\cosh\left[\frac{V - V_3}{2V_4}\right],
\end{aligned}$$

with parameters: $C = 20\,\mu F/cm^2$, $g_L = 2\,mmho/cm^2$, $V_L = -50\,mV$, $g_{Ca} = 4\,mmho/cm^2$, $V_{Ca} = 10\,mV$, $g_K = 8\,mmho/cm^2$, $V_K = -70\,mV$, $V_1 = 0\,mV$, $V_2 = 15\,mV$, $V_3 = 10\,mV$, $V_4 = 10\,mV$, $\bar{\lambda} = 0.1\,s^{-1}$, and applied current $I(\mu A/cm^2)$. The model can exhibit various types of spiking, but could exhibit tonic bursting only when an additional equation is added, e.g., slow inactivation of Ca current. In this case, the model becomes equivalent to the classical *Hodgkin–Huxley neural model*.

Wilson-Cowan Neuronal Population Model

Wilson-Cowan model describes the dynamics of interactions between populations of excitatory $E(t)$ and inhibitory $I(t)$ neuronal populations. The model includes the following components:

- Cells in refractory period, given by

$$\int_{t-r}^{t} E(t')dt',$$

- Sensitive cells, given by
$$1-\int_{t-r}^{t} E(t')dt',$$
- Subpopulation response function based on the distribution of neuronal thresholds, given by
$$S(x)=\int_{0}^{x(t)} D(\theta)d\theta,$$
- Subpopulation response function based on the distribution of afferent synapses per cell, given by
$$S(x)=\int_{\frac{\theta}{x(t)}}^{\infty} C(w)dw,$$
- Average excitation level, given by
$$\int_{-\infty}^{t} \alpha(t-t')[c_1E(t)-c_2I(t')+P(t')]dt',$$
and
- Excitatory subpopulation expression, given by
$$[1-\int_{t-r}^{t} E(t')dt']S(x)dt.$$

The complete Wilson-Cowan model reads:

$$E(t+\tau)=[1-\int_{t-r}^{t} E(t')dt']S\left\{\int_{-\infty}^{t} \alpha(t-t')[c_1E(t)-c_2I(t')+P(t')]dt'\right\},$$

$$I(t+\tau)=[1-\int_{t-r}^{t} I(t')dt']S\left\{\int_{-\infty}^{t} \alpha(t-t')[c_3E(t)-c_4I(t')+Q(t')]dt'\right\}.$$

Time course graining of the Wilson-Cowan model is given by:

$$\tau\frac{d\bar{E}}{dt}=-\bar{E}+(1-r\bar{E})S_e[kc_1\bar{E}(t)+kP(t)],$$

where τ is the constant relaxation rate. The corresponding isocline equation reads:

$$c_2I=c_1E-S_e^{-1}\left(\frac{E}{k_e-r_eE}\right)+P,$$

while the sigmoid function is defined as:

$$S(x)=\frac{1}{1+exp[-a(x-\theta)]}-\frac{1}{1+\exp(a\theta)}.$$

Classical Neural Models

Hodgkin–Huxley Neural Model

Among spiking neuron models, the oldest one, celebrated *Hodgkin–Huxley model* (or, HH–model), is expected to be the most realistic in the biological sense [HH52, Hod64, HH52]. Nobel Laureates Alan Hodgkin and Andrew Huxley performed experiments on the giant axon of the squid and found three different types of ion current, viz., sodium, potassium, and a leak current that consists mainly of Cl^- ions. Specific voltage–dependent ion channels, one for sodium and another one for potassium, control the flow of those ions through the cell membrane. The leak current takes care of other channel types which are not described explicitly. From a biophysical point of view, action potentials are the result of currents that pass through ion channels in the cell membrane. In an extensive series of experiments on the giant axon of the squid, Hodgkin and Huxley succeeded to measure these currents and to describe their dynamics in the form of an analogous circuit, which in turn can be characterized by a set of nonlinear ODEs. These equations contain an empirically adequate description of the feedback system which changes the conductances (or equivalently, the permeabilities) of the membrane as a function of time and potential difference.

The HH–neuron model is described by the nonlinear coupled ODEs for the four variables, V for the membrane potential, and m, h and n for the gating variables of sodium and potassium channels,

$$C\dot{V} = -g_{\text{Na}} m^3 h(V - V_{\text{Na}}) - g_{\text{K}} n^4 (V - V_{\text{K}}) - g_{\text{L}}(V - V_{\text{L}}) + I_j^{\text{ext}},$$

$$\dot{m} = -(a_m + b_m)\, m + a_m, \quad \dot{h} = -(a_h + b_h)\, h + a_h, \quad \dot{n} = -(a_n + b_n)\, n + a_n,$$

where

$$a_m = 0.1\,(V + 40)/[1 - e^{-(V+40)/10}], \qquad b_m = 4\, e^{-(V+65)/18},$$
$$a_h = 0.01\,(V + 55)/[1 - e^{-(V+55)/10}], \qquad b_h = 0.125\, e^{-(V+65)/80},$$
$$a_n = 0.07\, e^{-(V+65)/20}, \qquad b_n = 1/[1 + e^{-(V+35)/10}].$$

Here the reversal potentials of Na, an K channels and leakage are $V_{\text{Na}} = 50$ mV, $V_{\text{K}} = -77$ mV and $V_{\text{L}} = -54.5$ mV; the maximum values of corresponding conductivities are $g_{\text{Na}} = 120$ mS/cm^2, $g_{\text{K}} = 36$ mS/cm^2 and $g_{\text{L}} = 0.3$ mS/cm^2; the capacity of the membrane is $C = 1\ \mu$F/cm^2. The external, input current is given by

$$I_j^{\text{ext}} = g_{syn}(V_a - V_c) \sum_n \alpha(t - t_{in}), \tag{21.10}$$

which is induced by the pre–synaptic spike–train input applied to the neuron i, given by

$$U_{\text{i}}(t) = V_{\text{a}} \sum_n \delta(t - t_{in}). \tag{21.11}$$

In (21.10) and (21.11), t_{in} is the nth firing time of the spike–train inputs, g_{syn} and V_c denote the conductance and the reversal potential, respectively, of the synapse, $\tau_{\rm s}$ is the time constant relevant to the synapse conduction, and $\alpha(t)$ is the alpha function given by

$$\alpha(t) = (t/\tau_{\rm s})\, e^{-t/\tau_{\rm s}}\Theta(t),$$

where $\Theta(t)$ is the Heaviside function. The HH model was originally proposed to account for the property of squid giant axons and it has been generalized with modifications of ion conductances. The HH–type models have been widely adopted for a study on activities of *transducer neurons* such as motor and thalamus relay neurons, which transform the amplitude–modulated input to spike–train outputs. In this section, one pays attention to *data–processing neurons* which receive and emit the spike–train pulses.

To make a network of HH–neurons, the interaction term was added to the HH–equation

$$C\dot{V} = -g_{\rm Na}m^3h(V - V_{\rm Na}) - g_{\rm K}n^4(V - V_{\rm K}) - g_{\rm L}(V - V_{\rm L}) + I_j^{\rm ext} + I_j^{\rm int},$$

such that

$$I_j^{\rm int} = G(\sum_{k(\neq j)}\sum_m (g_{jk}^{exc} - g_{jk}^{inh})(V_a - V_c)\,\alpha(t - \tau_{jk} - t_{okm})),$$

where g_{jk}^{exc} and g_{jk}^{inh} denote respectively conductances of excitatory and inhibitory synapses of Hebbian or generalized Hebbian type.

FitzHugh–Nagumo Neural Model

Another plausible biological NN–model is *FitzHugh–Nagumo model* (FHM, [Fit61, NAY60]), which is actually a lower–dimensional descendant of the HH–model. The FHN equations correspond to an excitable threshold model but, as will be seen briefly, due to their cubic nonlinearity, they exhibit the characteristic behavior of a bistable system. Our main objective is to show analytically the appearance of SR in this model under a periodic external forcing.

The reduced (without diffusion) 2D non–dimensional FHN equations read:

$$\dot{v} = v(a - v)(v - 1) - w + I_a, \tag{21.12}$$

$$\dot{w} = bv - \gamma w, \tag{21.13}$$

where $0 < a < 1$ is essentially the threshold value, b and γ are positive constants and I_a is the applied current. The drift field for this model is given by

$$u_1(v, w) = v(a - v)(v - 1) - w, \qquad u_2(v, w) = bv - \gamma w.$$

As can be seen from (21.13) the null cline of the deterministic dynamics of this equations is the line $v = \frac{\gamma}{b}w$. By substitution on the right-hand-side (r.h.s) of equation (21.12) the following equation for steady states is obtained

$$v(a - v)(v - 1) - \frac{b}{\gamma}v = 0.$$

When this system is in a noisy environment, in the limit of weak noise, one can approximate the dynamics of the fluctuations by the 1D *Langevin equation*

$$\dot{v} = v(a - v)(v - 1) - \frac{b}{\gamma}v + \xi(t),$$

that is, the fluctuations run along the line $v = \frac{\gamma}{b}w$.

Phase plane analysis was used effectively by R. FitzHugh to understand various aspects of the HH equations and the two-variable FitzHugh–Nagumo model. His review also defines some basic mathematical terminology of nonlinear dynamics applied to neurodynamics.

21.2 Basics of Quantum Mechanics

21.2.1 History of Quantum Mechanics

Quantum mechanics was born on December 14, 1900, when Max Planck announced his famous formula on the black-body spectrum. He succeeded in deriving the correct shape of the black-body spectrum which now bears his name, eliminating the ultraviolet catastrophe. However, this involved an assumption so bizarre that even he distanced himself from it for many years afterwards: that energy was only emitted in certain finite units, or "energy quanta". In defining his black-body radiation law, Max Plank introduced the idea of electric oscillators in the walls of the cavity, vibrating back and forth under thermal agitation. He assumed that all possible frequencies would be present, as well as the average frequency to increase at higher temperatures as heating the walls caused the oscillators to vibrate faster and faster until thermal equilibrium was reached. Therefore, this was ultimately the thermodynamics problem and thus Planck applied Boltzmann's statistical formulation of the Second Law of Thermodynamics, in particular Boltzmann's expression for the entropy (measure of disorder) of a collection of particles, or oscillators. He found that he had to choose energy units proportional to the oscillator frequencies, namely: $e = n\,h\,f$, where e is the energy, n is the number of oscillators (that equally share the energy e), f is the frequency and h is a nonzero constant that would later get Planck's name. This formula implies that it is not possible for an oscillator to absorb and emit energy in a continuous range — it must gain and lose energy discontinuously, in small indivisible units of the form: $e = h\,f$, which Planck called "energy quanta". This discovery by Planck showed that at the microscopic level, energy and matter are discontinuous and for this he won the Nobel Prize.

In 1905, Albert Einstein took Planck's bold idea one step further. Assuming that radiation could only transport energy in such units, called "photons", he was able to explain the so-called photoelectric effect, which is related to the processes used in present-day solar cells and the image sensors in digital cameras. In 1905, Albert Einstein published three papers in a single volume of the top physics journal Annalen der Physik (in which Planck was the Editor), as an unknown physicist, he instantly became the world-leader. The first paper explained the photoelectric effect, based on Planck's energy-quantization formula: $e = h\ f$. Einstein showed that illuminating radiation was actually a collection of light-particles. If one assumes that the incident light consists of photons (light energy quanta) of magnitude$h\ f$, it is possible to conceive of the ejection of electrons (matter energy quanta) as follows. Energy quanta penetrate the surface layer of the metal of the target electrode. Their energy is partly transferred into the kinetic energy of the electrons: $h\ f - P$, where P is the characteristics of the radiating metal; also some electrons are ejected. Einstein came up with the photoelectric formula: $q\ V = h\ f - P$, q is the electron charge, and V is the threshold voltage to which the plate must be charged in order to offset the kinetic energy (and reduce the electron current to zero). This tangible discovery finally won for Einstein the long expected Nobel Prize, which he could not win for his Relativity Theory.

In 1911, Ernest Rutherford had convincingly argued that atoms consisted of electrons orbiting a positively charged nucleus much like a miniature solar system. This planetary atom model won for Rutherford the Nobel Prize.[23] However, Rutherford's solar model was unstable, which does not correspond to our everyday reality which consists purely of stable atoms. If the electrons are like planets in a microscopic solar system, moving in circular orbits about the nucleus, and thus accelerating, they would radiate continuously as the classical electrodynamics would predict. They would lose all their energy in a fraction of a second. Clearly, additional assumptions would be needed to complete the picture, in particular with regard to the details of atomic structure.

In 1913, Niels Bohr entered the micro-world scene. He was the grandfather of quantum physics and the founder of the mainstream Copenhagen Interpretation of quantum mechanics. Bohr made a real breakthrough in this new exciting field of science by postulating that the amount of angular momentum in an atom was quantized, while the electrons were confined to a discrete set of orbits, each with a definite energy. If the electron jumped from one orbit to a lower one, the energy difference was sent off in the form of a photon. If the electron was in the innermost allowed orbit, then there were no orbits with less energy to jump to, and therefore the atom was stable. Niels Bohr quantized the angular momentum of an electron in units of Planck's constant h. The angular momentum is defined as $L = m\ v\ r$, where m is the mass of a particle, v is its speed around the orbit and r is the radius of the orbit. Constant angular momentum means that there is no torque acting on the particle. Bohr's first postulate is the quantum orbital condition: an atom can exist only in one of several special orbits with no emission of radiation, which is contrary to the expectations of classical physics. These orbits are called *stationary states* and have the quantized values of the orbital angular momentum: $L = m\ v\ r = n\ h/(2\pi)$. And when an electron is in a stationary state, it is ready to make a *quantum jump* from one orbit to another. In other words, the angular momentum L of an electron cannot take any arbitrary value (as it is the case in classical mechanics) but only certain integer values: $L = 1 \cdot\ h/(2\pi)$ in the first orbit ($n = 1$);$L = 2 \cdot\ h/(2\pi)$ in the second

[23] Note that the popular solar model is still in use in chemistry.

orbit ($n = 2$) etc. This integer n is called the principal quantum number. This discovery won Bohr the Nobel Prize. It was also a foundation for today's view of the micro-world. In particular, Bohr defined the popular "quantum jump" or "quantum leap".

The famous *wave–particle duality of matter* was proposed by French prince Louis de Broglie in 1923 in his Ph.D. thesis: that electrons and other particles acted like standing waves. Such waves, like vibrations of a guitar string, can only occur with certain discrete (or quantized) frequencies. The idea was so new that the examining committee went outside its circle for advice on the acceptability of the thesis. Einstein gave a favorable opinion and the thesis was accepted.

In November 1925, Erwin Schrödinger gave a seminar on de Broglie's work in Zurich. When he was finished, P. Debye said in effect, "You speak about waves. But where is the wave equation?" Schrödinger went on to produce and publish his famous wave equation, the master key for so much of modern physics.

Another formulation of quantum mechanics, involving infinite matrices, was proposed by Werner Heisenberg and his co-workers, Max Born and Pasquale Jordan, around the same time. Next year, Paul Dirac showed that both Schrödinger's wave mechanics and Heisenberg's matrix mechanics were equivalent and subsumed by his own vector mechanics. All of these people won the Nobel Prize for physics. With this new powerful mathematical underpinning, quantum theory made explosive progress. Within a few years, a host of hitherto unexplained measurements had been successfully explained, including spectra of more complicated atoms and various numbers describing properties of chemical reactions. But what was this quantity, the "wave–function", which Schrödinger's equation described? This central puzzle of quantum mechanics remains a potent and controversial issue to this day. Max Born had the dramatic insight that the wave–function should be interpreted in terms of probabilities, which won him the Nobel Prize. If we measure the location of an electron, the probability of finding it in a given region depends on the intensity of its wave–function in that region. This interpretation suggested that a fundamental randomness was built into the laws of nature. Einstein was deeply unhappy with this interpretation, and expressed his preference for a deterministic Universe with the oft-quoted remark "I can't believe that God plays dice". Schrödinger was also uneasy. Wave functions could describe combinations of different states, so-called superpositions. For example, an electron could be in a superposition of several different locations. Schrödinger pointed out that if microscopic objects like atoms could be in strange superpositions, so could macroscopic objects, since they are made of atoms.

The Copenhagen Interpretation of quantum mechanics is still the dominant interpretation as taught at the majority of universities. This interpretation evolved from discussions between Bohr and Heisenberg (mostly as answers to Einstein's questions) in the late 1920s, and addresses the mystery by asserting that observations, or measurements, are special. So long as a quantum system is unobserved, its wave–function evolves by obeying the Schrödinger equation – a continuous and smooth "unitary" evolution, which produces the superposition of all possible states of a quantum system. However, the act of observing the quantum system triggers an abrupt change in its wave function, commonly called a "collapse": the observer sees the card in one definite classical state (face up or face down) and from then onward only that portion of the wave–function survives. Nature supposedly decided which particular state to collapse into at random, with the probabilities determined by the wave function. In this way, the Copenhagen Interpretation makes a distinction between the observer and the observed; when no one is watching, a system evolves deterministically according to a wave equation, but when someone is watching, the wave–function of the system "collapses" to the observed state, which is why the act of observing changes the system. Although this provided a strikingly successful calculation recipe, there was a lingering feeling that there ought to be some equation describing when and how this collapse occurred. Many physicists took this to mean that there is something fundamentally wrong with quantum mechanics, and that it would soon be replaced by some even more fundamental theory that provided such an equation.[24]

21.2.2 Double–Slit Experiment and the Riemann Sphere

Light consists of quantum particles called *photons*, and the Figure 21.20 shows a photon source which we assume emits photons one at a time. There are two slits, A and B, and a screen behind them. The photons arrive at the screen as individual events, where they are detected separately, just as if they were ordinary particles. The curious quantum behavior arise in the following way, as nicely described by Sir Roger Penrose. If only slit A were open and the other closed, there would be many places on the screen which the photon could reach. If we now close the slit A and open the slit B, we may again find that the photon could reach the same spot on the screen. However, if we open *both slits*, and if we have chosen the point on the screen carefully, we may now find that the photon cannot reach that spot, even though it could have done so if either slit alone were open. Somehow, the two possible things which the photon *might* do cancel each other out. This type of behavior does not take place in classical physics. Either one thing happens or another thing happens – we do not get two possible things which might happen, somehow conspiring to cancel each other out (see [II08a] and references therein).

The way we understand the outcome of this experiment in quantum mechanics is to say that when the photon is *en route* from the source to the screen, the state of the photon is not that of having gone through one slit or the other, but is some mysterious combination of the two, weighted by complex numbers. That is, we can write the state of the photon as a wave $\psi-$function,[25] which is

[24] See Max Tegmark, John Archibald Wheeler "100 Years of Quantum Mystery," Scientific American no. 284, pp. 68-75, (2001).

[25] In the *Schrödinger picture*, the *unitary evolution* U of a quantum system is described by the *Schrödinger equation*, which provides the time rate of change of the quantum state or wave function $\psi = \psi(t)$.

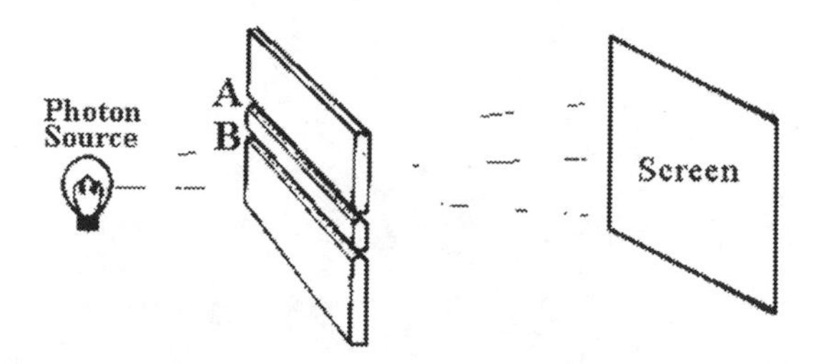

Fig. 21.20. The two–slit experiment, with individual photons of monochromatic light (see text for explanation).

the *linear superposition* of the two states, $|A\rangle$ and $|B\rangle$,[26] corresponding to the A–slot and B–slot alternatives,

$$|\psi\rangle = z_1|A\rangle + z_2|B\rangle,$$

where z_1 and z_2 are complex numbers (not both zero), while $|\cdot\rangle$ denotes the *quantum state ket–vector.*

Now, in quantum mechanics, we are not so interested in the sizes of the complex numbers z_1 and z_2 themselves as we are in their ratio – it is only the ratio of these numbers which has direct physical meaning (as multiplying a quantum state with a nonzero complex number does not change the physical situation). The *Riemann sphere* is a way of representing complex numbers (plus ∞) and their ratios on a sphere on unit radius, whose equatorial plane is the complex–plane, whose center is the origin of that plane and the equator of this sphere is the unit circle in the complex–plane. We can project each point on the equatorial complex–plane onto the Riemann sphere, projecting from its south pole S, which corresponds to the *point at infinity* in the complex–plane. To represent a particular complex ratio, say $u = z/w$ (with $w \neq 0$), we take the stereographic projection from the sphere onto the plane.

The Riemann sphere plays a fundamental role in the quantum picture of two–state systems. If we have a spin–$\frac{1}{2}$ particle, such as an electron, a proton, or a neutron, then the various combinations of their spin states can be realised geometrically on the Riemann sphere. Spin –$\frac{1}{2}$ particles can have two spin states: (i) spin–up (with the rotation vector pointing upwards), and (ii) spin–down (with the rotation vector pointing downwards). The superposition of the two spin–states can be represented symbolically as

$$|\nearrow\rangle = w|\uparrow\rangle + z|\downarrow\rangle.$$

[26] We are using here the standard Dirac 'bra–ket' notation for quantum states. Paul Dirac was one of the outstanding physicists of the 20th century. Among his achievements was a general formulation of quantum mechanics (having Heisenberg matrix mechanics and Shrödinger wave mechanics as special cases) and also its relativistic generalization involving the 'Dirac equation', which he discovered, for the electron. He had an unusual ability to 'smell out' the truth, judging his equations, to a large degree, by their aesthetic qualities!

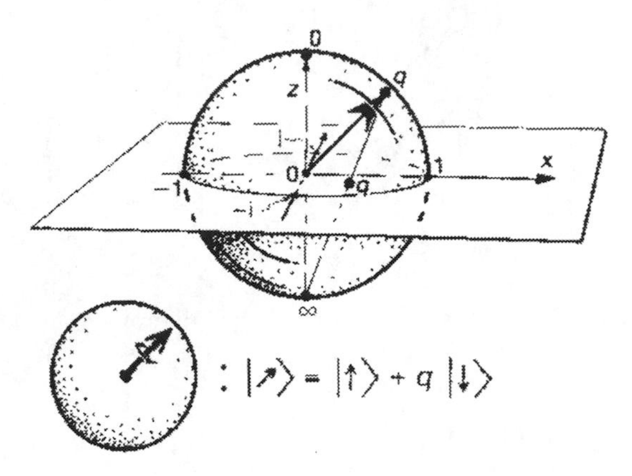

Fig. 21.21. The quantum Riemann sphere, represented as the space of physically distinct spin–states of a spin–$\frac{1}{2}$ particle (e.g., electron, proton, neutron): $|\nearrow\rangle = |\uparrow\rangle + q|\downarrow\rangle$. The sphere is projected stereographically from its south pole (∞) to the complex–plane through its equator (see text for explanation).

Different combinations of these spin states give us rotation about some other axis and, if we want to know where that axis is, we take the ratio of complex numbers $u = z/w$. We place this new complex number u on the Riemann sphere and the direction of u from the center is the direction of the spin axis (see Figure 21.21).

More general quantum state vectors might have a form such as [II08a]:

$$|\psi\rangle = z_1|A_1\rangle + z_2|A_2\rangle + ... + z_n|A_n\rangle,$$

where $z_1... z_n$ are complex numbers (not all zero) and the state vectors $|A_1\rangle, ..., |A_n\rangle$ might represent various possible locations for a particle (or perhaps some other property of a particle, such as its state of spin). Even more generally, infinite sums would be allowed for a wave ψ–function or quantum state vector.

Now, the most basic feature of *unitary quantum evolution* U[27] is that it is *linear*. This means that, if we have two states, say $|\psi\rangle$ and $|\phi\rangle$, and if the Schrödinger equation would tell us that, after some time t, the states $|\psi\rangle$ and $|\phi\rangle$ would each

[27] Unitary quantum evolution U is governed by the time–dependent *Schrödinger equation*,

$$i\hbar\,\partial_t|\psi(t)\rangle = H|\psi(t)\rangle,$$

where $\partial_t \equiv \partial/\partial t$, $\hbar$ is the *Planck's constant*, and H is the Hamiltonian (total energy) operator. Given the quantum state $|\psi(t)\rangle$ at some initial time ($t = 0$), we can integrate the Schrödinger equation to get the state at any subsequent time. In particular, if H is independent of time, then

$$|\psi(t)\rangle = \exp\left(-\frac{\mathrm{i}Ht}{\hbar}\right)|\psi(0)\rangle.$$

individually evolve to new states $|\psi'\rangle$ and $|\phi'\rangle$, respectively then any linear superposition $z_1|\psi\rangle + z_2|\phi\rangle$, must evolve, after some time t, to the corresponding superposition $z_1|\psi'\rangle + z_2|\phi'\rangle$. Let us use the symbol $\rightsquigarrow$ to denote the evolution after time t, Then linearity asserts that if

$$|\psi\rangle \rightsquigarrow |\psi'\rangle \qquad \text{and} \qquad |\phi\rangle \rightsquigarrow |\phi'\rangle,$$

then the evolution

$$z_1|\psi\rangle + z_2|\phi\rangle \quad \rightsquigarrow \quad z_1|\psi'\rangle + z_2|\phi'\rangle$$

would also hold. This would consequently apply also to linear superpositions of more than two individual quantum states. For example, $z_1|\psi\rangle + z_2|\phi\rangle + z_3|\chi\rangle$ would evolve, after time t, to $z_1|\psi'\rangle + z_2|\phi'\rangle + z_3|\chi'\rangle$, if $|\psi\rangle, |\phi\rangle$, and $|\chi\rangle$ would each individually evolve to $|\psi'\rangle, |\phi'\rangle$, and $|\chi'\rangle$, respectively. Thus, the evolution always proceeds as though each different component of a superposition were oblivious to the presence of the others.

As a second experiment, consider a situation in which light impinges on a half–silvered mirror, that is a semi-transparent mirror that reflects just half the light (composed of a *stream of photons*) falling upon it and transmits the remaining half. We might well have imagined that for a stream of photons impinging on our half–silvered mirror, half the photons would be reflected and half would be transmitted. Not so! Quantum theory tells us that, instead, each individual photon, as it impinges on the minor, is separately put into a superposed state of reflection and transmission. If the photon before its encounter with the minor is in state $|A\rangle$, then afterwards it evolves according to U to become a state that can be written $|B\rangle + i|C\rangle$, where $|B\rangle$ represents the state in which the photon is transmitted through the mirror and $|C\rangle$ the state where the photon is reflected from it (see Figure 21.22). Let us write this as

$$|A\rangle \rightsquigarrow |B\rangle + i|C\rangle.$$

The imaginary factor 'i' arises here because of a net phase shift by a quarter of a wavelength, which occurs between the reflected and transmitted beams at such a mirror.

Although, from the classical picture of a particle, we would have to imagine that $|B\rangle$ and $|C\rangle$ just represent alternative things that the photon *might* do, in quantum mechanics we have to try to believe that the photon is now actually

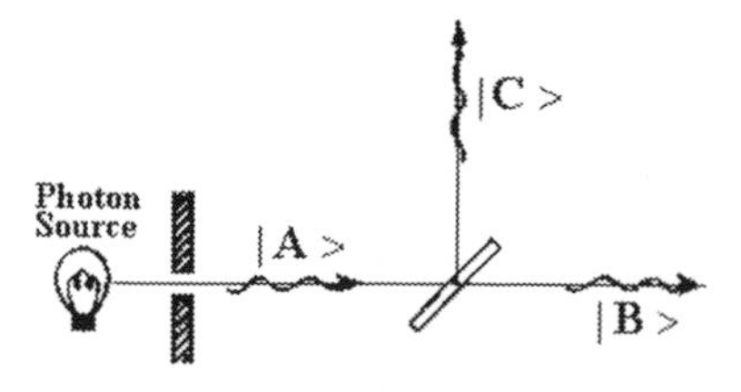

Fig. 21.22. A photon in state $|A\rangle$ impinges on a half–silvered mirror and its state evolves according to U into a a superposition $|B\rangle + i|C\rangle$ (see text for explanation).

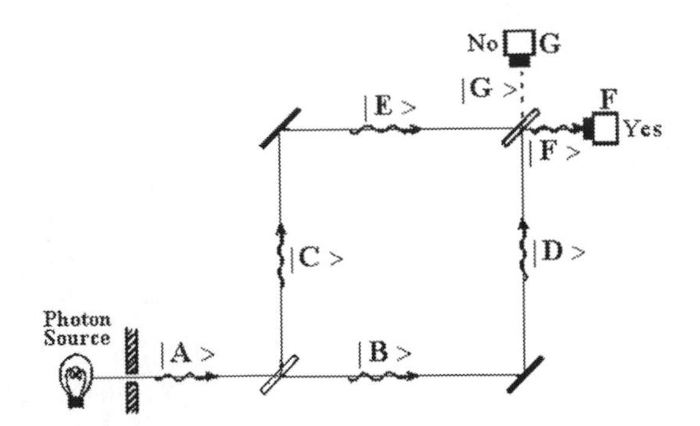

Fig. 21.23. *Mach–Zehnder interferometer*: the two parts of the photon state are brought together by two fully silvered mirrors (black), so as to encounter each other at a final half–silvered mirror (white). They interfere in such a way that the entire state emerges in state $|F\rangle$, and the detector at G cannot receive the photon (see text for explanation).

doing *both things at once* in this strange, complex superposition. To see that it cannot just be a matter of classical probability–weighted alternatives, let us take this example a little further and try to bring the two parts of the photon state, i.e., the two photon beams, back together again. We can do this by first reflecting each beam with a fully silvered mirror. After reflection, the photon state $|B\rangle$ would evolve according to U, into another state $i|D\rangle$, whilst $|C\rangle$ would evolve into $i|E\rangle$,

$$|B\rangle \rightsquigarrow i|D\rangle \qquad \text{and} \qquad |C\rangle \rightsquigarrow i|E\rangle.$$

Thus the entire state $|B\rangle + i|C\rangle$ evolves by U into

$$|B\rangle + i|C\rangle \rightsquigarrow i|D\rangle + i(i|E\rangle) = i|D\rangle - |E\rangle,$$

since $i^2 = -1$. Now, suppose that these two beams come together at a fourth mirror, which is now half silvered (see Figure 21.23). The state $|D\rangle$ evolves into a combination $|G\rangle + i|F\rangle$, where $|G\rangle$ represents the transmitted state and $|F\rangle$ the reflected one. Similarly, $|E\rangle$ evolves into $|F\rangle + i|G\rangle$, since it is now the state $|F\rangle$ that is the transmitted state and $|G\rangle$ the reflected one,

$$|D\rangle \rightsquigarrow |G\rangle + i|F\rangle \qquad \text{and} \qquad |E\rangle \rightsquigarrow |F\rangle + i|G\rangle.$$

Our entire state $i|D\rangle - |E\rangle$ is now seen (because of the linearity of U) to evolve as:

$$\begin{aligned} i|D\ \rangle\ - |E\rangle &\rightsquigarrow i(|G\rangle + i|F\rangle) - (|F\rangle + i|G\rangle) \\ &= i|G\rangle - |F\rangle - |F\rangle - i|G\rangle = -2|F\rangle. \end{aligned}$$

As mentioned above, the multiplying factor -2 appearing here plays no physical role, thus we see that the possibility $|G\rangle$ is *not* open to the photon; the two beams together combine to produce just a *single* possibility $|F\rangle$. This curious outcome arises because *both* beams are present *simultaneously* in the physical state of the photon, between its encounters with the first and last mirrors. We say that the two beams *interfere* with one another[28] [II08a].

21.2.3 Feynman Path Integral in Quantum Mechanics

Dirac's *amplitude* $\langle q_F|\mathrm{e}^{-\mathrm{i}HT}|q_I\rangle$ of a quantum-mechanical system (in natural units $\hbar = c = 1$) with Hamiltonian H to propagate from a point q_I to a point q_F in time T is governed by the *unitary operator* $\mathrm{e}^{-\mathrm{i}HT}$, or more completely, by the complex *Feynman path integral* (in real time):

$$\langle q_F|\mathrm{e}^{-\mathrm{i}HT}|q_I\rangle = \int \mathcal{D}[q(t)]\,\mathrm{e}^{\mathrm{i}\int_0^T dt L(\dot{q},q)}. \tag{21.14}$$

For example, a quantum particle in a potential $V(q)$ has the *Hamiltonian operator*: $\hat{H} = \hat{p}^2/2m + V(\hat{q})$, with the corresponding *Lagrangian*: $L = \frac{1}{2}m\dot{q}^2 - V(q)$, so the path integral reads:

$$\langle q_F|\mathrm{e}^{-\mathrm{i}HT}|q_I\rangle = \int \mathcal{D}[q(t)]\,\mathrm{e}^{\mathrm{i}\int_0^T dt[\frac{1}{2}m\dot{q}^2 - V(q)]}.$$

It is somewhat more rigorous to perform a so-called *Wick rotation* to Euclidean time, which means substituting $t \to -it$ and rotating the integration contour in the complex t–plane, so that we obtain the real path integral in complex time:

$$\langle q_F|\mathrm{e}^{-\mathrm{i}HT}|q_I\rangle = \int \mathcal{D}[q(t)]\,\mathrm{e}^{-\int_0^T dt L(\dot{q},q)} = \int \mathcal{D}[q(t)]\,\mathrm{e}^{-(\mathrm{i}/\hbar)\int_0^T dt[\frac{1}{2}m\dot{q}^2 - V(q)]},$$

known as the *Euclidean path integral*.

One particularly nice feature of the path-integral formalism is that the classical limit of quantum mechanics can be recovered easily. We simply restore Planck's constant $\hbar$ in (21.14)

$$\langle q_F|\mathrm{e}^{-(\mathrm{i}/\hbar)HT}|q_I\rangle = \int \mathcal{D}[q(t)]\,\mathrm{e}^{(\mathrm{i}/\hbar)\int_0^T dt L(\dot{q},q)},$$

and take the $\hbar \to 0$ limit. Applying the stationary phase (or steepest descent) method, see [Zee03, II08a, II08b]), we obtain $\mathrm{e}^{(\mathrm{i}/\hbar)\int_0^T dt L(\dot{q}_c,q_c)}$,[29] where $q_c(t)$ is the recovered *classical path* determined by solving the *Euler-Lagrangian equation*: $(d/dt)(\delta L/\delta \dot{q}) - (\delta L/\delta q) = 0$, with appropriate boundary conditions.

[28] This is a property of single photons: each individual photon must be considered to feel out both routes that are open to it, but it remains one photon; it does not split into two photons in the intermediate stage, but its location undergoes the strange kind of complex–number–weighted *co–existence of alternatives* that is characteristic of quantum theory.

[29] To do an exponential integral of the form: $I = \int_{-\infty}^{+\infty} dq\,\mathrm{e}^{-(\mathrm{i}/\hbar)f(q)}$, we often have to resort to the following steepest-descent approximation [Zee03]. In the limit of $\hbar$ small, this integral is dominated by the minimum of $f(q)$. Expanding: $f(q) = f(a) + \frac{1}{2}f''(a)(q-a)^2 + O[(q-a)^3]$, and applying the *Gaussian integral* rule:

$$\int_{-\infty}^{+\infty} dq\,\mathrm{e}^{-\frac{1}{2}ax^2} = \sqrt{\frac{2\pi}{a}}, \qquad \text{we obtain:} \qquad I = \mathrm{e}^{-(1/\hbar)f(a)}\left(\frac{2\pi\hbar}{f''(a)}\right)\mathrm{e}^{-O(\hbar^{\frac{1}{2}})}.$$

22

Modern Tennis Biomechanics

22.1 Jet Methods in Time–Dependent Lagrangian Biomechanics

Most of dynamics in contemporary human biomechanics is *autonomous* (see [II06a, II06b, II06c, II07e, II07a, II08c, II08b]). This approach works fine for most individual movement simulations and predictions, in which the total human energy dissipations are insignificant. However, if we analyze a 100 m-dash sprinting motion, which is in case of top athletes finished under 10 s, we can recognize a significant slow-down after about 70 m in *all* athletes – despite of their strong intention to finish and win the race, which is an obvious sign of the total energy dissipation. This can be seen, for example, in a current record-braking speed-distance curve of Usain Bolt, the world-record holder with 9.69 s, or in a former record-braking speed–distance curve of Carl Lewis, the former world-record holder (and 9 time Olympic gold medalist) with 9.86 s (see Figure 22.1). In other words, the *total mechanical energy* of a sprinter *cannot be conserved* even for 10 s. So, if we want to develop a realistic model of intensive human motion that is longer than 7–8 s, we necessarily need to use the more advanced formalism of time-dependent mechanics.

Similarly, if we analyze individual movements of gymnasts, we can clearly see that the high speed of these movements is based on quickly-varying mass-inertia distribution of various body segments (mostly arms and legs). Similar is the case of pirouettes in ice skating. As the total mass-inertia matrix M_{ij} of a biomechanical system corresponds to the Riemannian metric tensor g_{ij} of its configuration manifold, we can formulate this problem in terms of time-dependent Riemannian geometry [II06c, II07e].

The purpose of this section is to introduce a general framework for time-dependent biomechanics [Iva09, Iva10], consisting of the following four steps:

1. Human biomechanical configuration manifold and its (co)tangent bundles;
2. Biomechanical configuration bundle, as a time–extension of the configuration manifold;

T.T. Ivancevic et al.: Paradigm Shift for Future Tennis, COSMOS 12, pp. 227–258.
springerlink.com

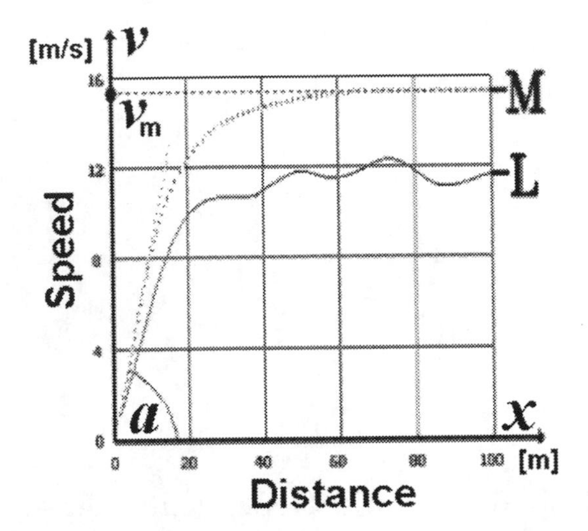

Fig. 22.1. Sprint speed–distance curve of Carl Lewis for 100m-dash. Bottom line L: Carl Lewis' result (9.86s). Top line M: theoretical-model curve, as an exponential solution of the linear first-order differential equation: $Mdv/dx + Bv = E$ (with mass M, damping B, input E and output v), with an initial condition: $v(0) = 0$ (adapted from [IJD09]). Also, v_m denotes maximal speed of the model curve, while a is the initial running acceleration. The L-curve maximum occurs at 73 m, after which there is a wavy decline in speed.

3. Biomechanical jet spaces and prolongation of locomotion vector-fields developed on the configuration bundle; and
4. Time–dependent Lagrangian dynamics using biomechanical jet spaces.

In addition, we will show that Riemannian geometrical basis of this framework is defined by the Ricci flow. In particular, we will show that the exponential–like decay of total biomechanical energy (due to exhaustion of biochemical resources [IJD09]) is closely related to the Ricci flow on the configuration manifold of human motion.

22.1.1 Configuration Manifold for Autonomous Biomechanics

Recall from [II08c] that representation of an ideal humanoid–robot motion is rigorously defined in terms of rotational constrained $SO(3)$–groups in all main robot joints. Therefore, the configuration manifold Q_{rob} for humanoid dynamics is defined as a topological product of all included $SO(3)$ groups, $Q_{rob} = \prod_i SO(3)^i$. Consequently, the natural stage for autonomous Lagrangian dynamics of robot motion is the tangent bundle TQ_{rob}, defined as follows. To each $n-$dimensional (nD) configuration manifold Q there is associated its $2n$D *velocity phase–space manifold*, denoted by TQ and called the tangent bundle of Q. The original smooth manifold Q is called the *base* of TQ. There is an onto map $\pi : TQ \to Q$, called the *projection*. Above each point $x \in Q$ there is a tangent space $T_xQ = \pi^{-1}(x)$ to Q at x, which is called a fibre. The fibre $T_xQ \subset TQ$ is the subset of TQ, such that the total tangent bundle, $TQ = \bigsqcup_{m \in Q} T_xQ$, is a disjoint union of tangent spaces T_xQ to Q for all points $x \in Q$. From dynamical perspective, the most important quantity in the tangent bundle concept is the smooth map $v : Q \to TQ$, which is an inverse to the projection π, i.e, $\pi \circ v = \mathrm{Id}_Q$, $\pi(v(x)) = x$. It is called the *velocity vector–field*. Its graph $(x, v(x))$ represents the cross–section of the tangent bundle TQ. This explains the dynamical term *velocity phase–space*, given to the tangent bundle TQ of the manifold Q. The tangent bundle is where tangent vectors live, and is itself a smooth manifold. Vector–fields are cross-sections of the tangent bundle. Robot's *Lagrangian* (energy function) is a natural energy function on the tangent bundle TQ.[1]

On the other hand, human joints are more flexible than robot joints. Namely, every rotation in all synovial human joints is followed by the corresponding micro–translation, which occurs after the rotational amplitude is reached [II08c]. So, representation of human motion is rigorously defined in terms of Euclidean $SE(3)$–groups of full rigid–body motion [II06a, II06c, II07e] in all main human joints (see Figure 22.5). Therefore, the configuration manifold Q for human dynamics is defined as a topological product of all included constrained $SE(3)$–groups,[2] $Q = \prod_i SE(3)^i$. Consequently, the natural stage for autonomous Lagrangian dynamics of human motion is the tangent bundle TQ (and for the corresponding autonomous Hamiltonian dynamics is the cotangent bundle T^*Q).

Therefore, the *configuration manifold* Q for human musculo-skeletal dynamics is defined as a Cartesian product of all included constrained $SE(3)$ groups, $Q = \prod_j SE(3)^j$ where j labels the active joints. The configuration manifold Q is coordinated by local joint coordinates $x^i(t)$, $i = 1, ..., n$ = total number of active DOF. The corresponding joint velocities $\dot{x}^i(t)$ live in the *velocity phase space* TQ, which is the *tangent bundle* of the configuration manifold Q.

The velocity phase space TQ has the Riemannian geometry with the *local metric form:*

$$\langle g \rangle \equiv ds^2 = g_{ij} dx^i dx^j, \qquad \text{(Einstein's summation convention is in always use)}$$

[1] The corresponding autonomous Hamiltonian robot dynamics takes place in the cotangent bundle T^*Q_{rob}, defined as follows. A dual notion to the tangent space T_mQ to a smooth manifold Q at a point m is its cotangent space T_m^*Q at the same point m. Similarly to the tangent bundle, for a smooth manifold Q of dimension n, its cotangent bundle T^*Q is the disjoint union of all its cotangent spaces T_m^*Q at all points $m \in Q$, i.e., $T^*Q = \bigsqcup_{m \in Q} T_m^*Q$. Therefore, the cotangent bundle of an n–manifold Q is the vector bundle $T^*Q = (TQ)^*$, the (real) dual of the tangent bundle TQ. The cotangent bundle is where 1–forms live, and is itself a smooth manifold. Covector–fields (1–forms) are cross-sections of the cotangent bundle. Robot's *Hamiltonian* is a natural energy function on the cotangent bundle.

[2] Briefly, the Euclidean SE(3)–group is defined as a semidirect (noncommutative) product (denoted by $\triangleright$) of 3D rotations and 3D translations: $SE(3) := SO(3) \triangleright \mathbb{R}^3$. Its most important subgroups are the following (for technical details see [II06c,

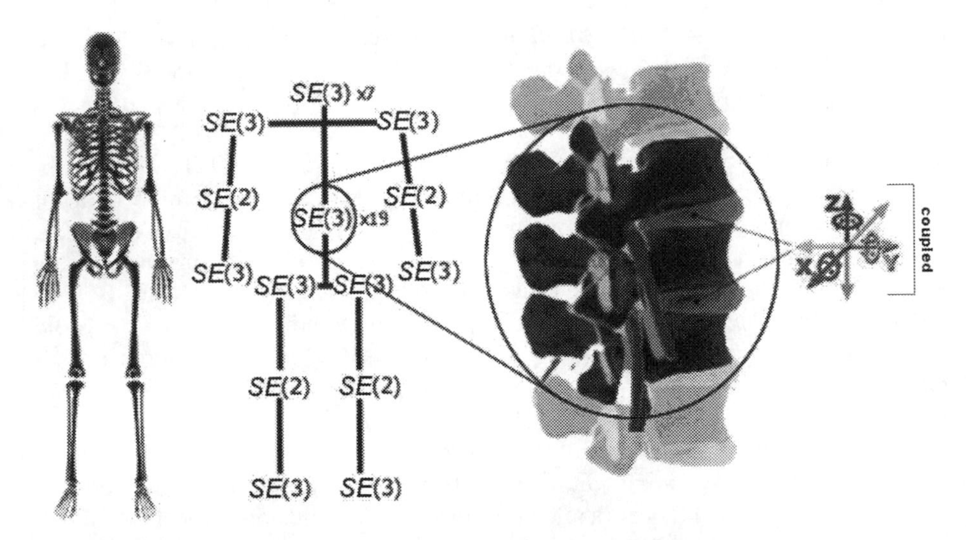

Fig. 22.2. The configuration manifold Q of the human musculo-skeletal dynamics is defined as a topological product of constrained $SE(3)$ groups acting in all major (synovial) human joints, $Q = \prod_i SE(3)^i$.

where $g_{ij}(x)$ is the material metric tensor defined by the biomechanical system's *mass-inertia matrix* and dx^i are differentials of the local joint coordinates x^i on Q. Besides giving the local distances between the points on the manifold Q, the Riemannian metric form $\langle g \rangle$ defines the system's kinetic energy:

$$T = \frac{1}{2} g_{ij} \dot{x}^i \dot{x}^j,$$

giving the *Lagrangian equations* of the conservative skeleton motion with kinetic-minus-potential energy Lagrangian $L = T - V$, with the corresponding *geodesic form* [II06c]

$$\frac{d}{dt} L_{\dot{x}^i} - L_{x^i} = 0 \qquad \text{or} \qquad \ddot{x}^i + \Gamma^i_{jk} \dot{x}^j \dot{x}^k = 0, \tag{22.1}$$

II07e]):

Subgroup	Definition
$SO(3)$, group of rotations in 3D (a spherical joint)	Set of all proper orthogonal 3×3 – rotational matrices
$SE(2)$, special Euclidean group in 2D (all planar motions)	Set of all 3×3 – matrices: $\begin{bmatrix} \cos\theta & \sin\theta & r_x \\ -\sin\theta & \cos\theta & r_y \\ 0 & 0 & 1 \end{bmatrix}$
$SO(2)$, group of rotations in 2D subgroup of $SE(2)$–group (a revolute joint)	Set of all proper orthogonal 2×2 – rotational matrices included in $SE(2)$ – group
$\mathbb{R}^3$, group of translations in 3D (all spatial displacements)	Euclidean 3D vector space

where subscripts denote partial derivatives, while Γ^i_{jk} are the Christoffel symbols of the affine Levi-Civita connection of the biomechanical manifold Q.

This is the basic geometrical structure for *autonomous Lagrangian biomechanics.* In the next sections will extend this basic structure to embrace the time-dependent Lagrangian biomechanics.

22.1.2 Biomechanical Bundle, Sections and Connections

While standard autonomous Lagrangian biomechanics is developed on the configuration manifold X, the *time–dependent biomechanics* necessarily includes also the real time axis $\mathbb{R}$, so we have an *extended configuration manifold* $\mathbb{R} \times X$. Slightly more generally, the fundamental geometrical structure is the so-called *configuration bundle* $\pi : X \to \mathbb{R}$. Time-dependent biomechanics is thus formally developed either on the *extended configuration manifold* $\mathbb{R} \times X$, or on the configuration bundle $\pi : X \to \mathbb{R}$, using the concept of *jets*, which are based on the idea of *higher–order tangency*, or higher–order contact, at some designated point (i.e., certain joint) on a biomechanical configuration manifold X.

In general, tangent and cotangent bundles, TM and T^*M, of a smooth manifold M, are special cases of a more general geometrical object called *fibre bundle*, denoted $\pi : Y \to X$, where the word *fiber* V of a map $\pi : Y \to X$ is the *preimage* $\pi^{-1}(x)$ of an element $x \in X$. It is a space which *locally* looks like a product of two spaces (similarly as a manifold locally looks like Euclidean space), but may possess a different *global* structure. To get a visual intuition behind this fundamental geometrical concept, we can say that a fibre bundle Y is a *homeomorphic generalization* of a *product space* $X \times V$ (see Figure 22.6), where X and V are called the *base* and the *fibre*, respectively. $\pi : Y \to X$ is called the *projection*, $Y_x = \pi^{-1}(x)$ denotes a fibre over a point x of the base X, while the map $f = \pi^{-1} : X \to Y$ defines the *cross–section*, producing the *graph* $(x, f(x))$ in the bundle Y (e.g., in case of a tangent bundle, $f = \dot{x}$ represents a velocity vector–field).

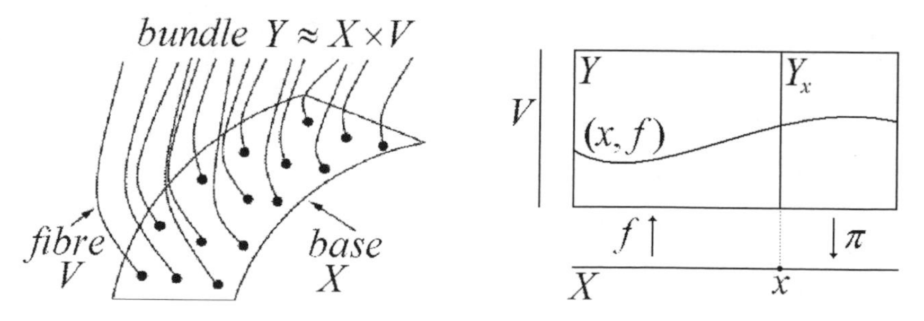

Fig. 22.3. A sketch of a locally trivial fibre bundle $Y \approx X \times V$ as a generalization of a product space $X \times V$; left – main components; right – a few details (see text for explanation).

More generally, a biomechanical configuration bundle, $\pi : Y \longrightarrow X$, is a locally trivial fibred (or, projection) manifold over the base X. It is endowed with an atlas of fibred bundle coordinates (x^λ, y^i), where (x^λ) are coordinates of X.

All dynamical objects in time–dependent biomechanics (including vectors, tensors, differential forms and gauge potentials) are *cross–sections* of biomechanical bundles, representing generalizations of graphs of continuous functions.

An *exterior differential form* α of order p (or, a p*–form* α) on a base manifold X is a section of the bundle $\overset{p}{\wedge} T^*X \longrightarrow X$ [II07e]. It has the following expression in local coordinates on X

$$\alpha = \alpha_{\lambda_1 \dots \lambda_p} dx^{\lambda_1} \wedge \dots \wedge dx^{\lambda_p} \qquad \text{(such that } |\alpha| = p),$$

where summation is performed over all ordered collections $(\lambda_1, ..., \lambda_p)$. $\Omega^p(X)$ is the vector space of p–forms on a biomechanical manifold X. In particular, the 1–forms are called the *Pfaffian forms.*

The *contraction* $\rfloor$ of any vector-field $u = u^\mu \partial_\mu$ and a p–form $\alpha = \alpha_{\lambda_1 \dots \lambda_p} dx^{\lambda_1} \wedge \dots \wedge dx^{\lambda_p}$ on a biomechanical manifold X is given in local coordinates on X by

$$u \rfloor \alpha = u^\mu \alpha_{\mu \lambda_1 \dots \lambda_{p-1}} dx^{\lambda_1} \wedge \dots \wedge dx^{\lambda_{p-1}}.$$

It satisfies the following relation

$$u \rfloor (\alpha \wedge \beta) = u \rfloor \alpha \wedge \beta + (-1)^{|\alpha|} \alpha \wedge u \rfloor \beta.$$

The *Lie derivative* $\mathfrak{L}_u \alpha$ of p–form α along a vector-field u is defined by Cartan's 'magic' formula (see [II06c, II07e]):

$$\mathfrak{L}_u \alpha = u \rfloor d\alpha + d(u \rfloor \alpha).$$

It satisfies the *Leibnitz relation*

$$\mathfrak{L}_u (\alpha \wedge \beta) = \mathfrak{L}_u \alpha \wedge \beta + \alpha \wedge \mathfrak{L}_u \beta.$$

A linear connection $\bar{\Gamma}$ on a biomechanical bundle $Y \longrightarrow X$ is given in local coordinates on Y by [II07e]

$$\bar{\Gamma} = dx^\lambda \otimes [\partial_\lambda - \Gamma^i{}_{j\lambda}(x) y^j \partial_i]. \tag{22.2}$$

An affine connection Γ on a biomechanical bundle $Y \longrightarrow X$ is given in local coordinates on Y by

$$\Gamma = dx^\lambda \otimes [\partial_\lambda + (-\Gamma^i{}_{j\lambda}(x) y^j + \Gamma^i{}_\lambda(x)) \partial_i].$$

Clearly, a linear connection $\bar{\Gamma}$ is a special case of an affine connection Γ.

22.1.3 Biomechanical Jets

A pair of smooth manifold maps, $f_1, f_2 : M \to N$ (see Figure 22.7), are said to be $k-$*tangent* (or *tangent of order* k, or have a kth *order contact*) at a point x on a domain manifold M, denoted by $f_1 \sim f_2$, iff

$$
\begin{aligned}
f_1(x) &= f_2(x) \qquad \text{called} \quad 0-\text{tangent},\\
\partial_x f_1(x) &= \partial_x f_2(x), \qquad \text{called} \quad 1-\text{tangent},\\
\partial_{xx} f_1(x) &= \partial_{xx} f_2(x), \qquad \text{called} \quad 2-\text{tangent},\\
&\ldots \qquad \text{etc. to the order } k
\end{aligned}
$$

In this way defined $k-$*tangency* is an *equivalence relation.*

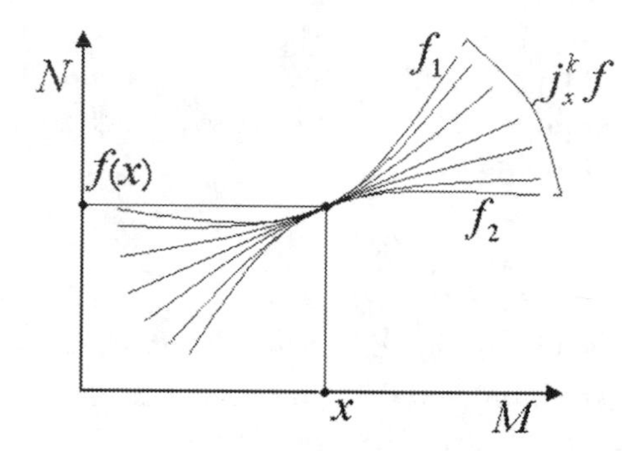

Fig. 22.4. An intuitive geometrical picture behind the $k-$jet concept, based on the idea of a higher–order tangency (or, higher–order contact).

A $k-$*jet* (or, a *jet of order* k), denoted by $j_x^k f$, of a smooth map $f : Q \to N$ at a point $x \in Q$ (see Figure 22.7), is defined as an *equivalence class* of $k-$tangent maps at x,

$$j_x^k f : Q \to N = \{f' : f' \text{ is } k-\text{tangent to } f \text{ at } x\}.$$

For example, consider a simple function $f : X \to Y, x \mapsto y = f(x)$, mapping the $X-$axis into the $Y-$axis in $\mathbb{R}^2$. At a chosen point $x \in X$ we have:
a $0-$jet is a graph: $(x, f(x))$;
a $1-$jet is a triple: $(x, f(x), f'(x))$;
a $2-$jet is a quadruple: $(x, f(x), f'(x), f''(x))$,
and so on, up to the order k (where $f'(x) = \frac{df(x)}{dx}$, etc).
The set of all $k-$jets from $j_x^k f : X \to Y$ is called the $k-$jet manifold $J^k(X, Y)$.

Formally, given a biomechanical bundle $Y \longrightarrow X$, its first order *jet manifold* J^1Y comprises the set of equivalence classes $j_x^1 s$, $x \in X$, of sections $s : X \longrightarrow Y$ so that sections s and s' belong to the same class iff

$$Ts\mid_{T_xX} = Ts'\mid_{T_xX}.$$

Intuitively, sections $s, s' \in j^1_x s$ are identified by their values $s^i(x) = s'^i(x)$ and the values of their partial derivatives $\partial_\mu s^i(x) = \partial_\mu s'^i(x)$ at the point x of X. There are the natural fibrations [II07e]

$$\pi_1 : J^1Y \ni j^1_x s \mapsto x \in X, \qquad \pi_{01} : J^1Y \ni j^1_x s \mapsto s(x) \in Y.$$

Given bundle coordinates (x^λ, y^i) of Y, the associated jet manifold J^1Y is endowed with the adapted coordinates

$$(x^\lambda, y^i, y^i_\lambda), \qquad (y^i, y^i_\lambda)(j^1_x s) = (s^i(x), \partial_\lambda s^i(x)), \qquad y'^i_\lambda = \frac{\partial x^\mu}{\partial x'^\lambda}(\partial_\mu + y^j_\mu \partial_j) y'^i.$$

In particular, given the biomechanical configuration bundle $Q \to \mathbb{R}$ over the time axis $\mathbb{R}$, the 1*–jet space* $J^1(\mathbb{R}, Q)$ is the set of equivalence classes $j^1_t s$ of sections $s^i : \mathbb{R} \to Q$ of the configuration bundle $Q \to \mathbb{R}$, which are identified by their values $s^i(t)$, as well as by the values of their partial derivatives $\partial_t s^i = \partial_t s^i(t)$ at time points $t \in \mathbb{R}$. The 1–jet manifold $J^1(\mathbb{R}, Q)$ is coordinated by $(t, q^i, \dot{q}^i)$, that is by *(time, coordinates and velocities)* at every active human joint, so the 1–jets are local joint coordinate maps

$$j^1_t s : \mathbb{R} \to Q, \qquad t \mapsto (t, q^i, \dot{q}^i).$$

The *repeated jet manifold* J^1J^1Y is defined to be the jet manifold of the bundle $J^1Y \longrightarrow X$. It is endowed with the adapted coordinates $(x^\lambda, y^i, y^i_\lambda, y^i_{(\mu)}, y^i_{\lambda\mu})$.

The *second order jet manifold* J^2Y of a bundle $Y \longrightarrow X$ is the subbundle of $\widehat{J}^2Y \longrightarrow J^1Y$ defined by the coordinate conditions $y^i_{\lambda\mu} = y^i_{\mu\lambda}$. It has the local coordinates $(x^\lambda, y^i, y^i_\lambda, y^i_{\lambda\le\mu})$ together with the transition functions [II07e]

$$y'^i_{\lambda\mu} = \frac{\partial x^\alpha}{\partial x'^\mu}(\partial_\alpha + y^j_\alpha \partial_j + y^j_{\nu\alpha}\partial^\nu_j) y'^i_\lambda.$$

The second order jet manifold J^2Y of Y comprises the equivalence classes $j^2_x s$ of sections s of
$Y \longrightarrow X$ such that

$$y^i_\lambda(j^2_x s) = \partial_\lambda s^i(x), \qquad y^i_{\lambda\mu}(j^2_x s) = \partial_\mu \partial_\lambda s^i(x).$$

In other words, two sections $s, s' \in j^2_x s$ are identified by their values and the values of their first and second order derivatives at the point $x \in X$.

In particular, given the biomechanical configuration bundle $Q \to \mathbb{R}$ over the time axis $\mathbb{R}$, the 2*–jet space* $J^2(\mathbb{R}, Q)$ is the set of equivalence classes $j^2_t s$ of sections $s^i : \mathbb{R} \to Q$ of the configuration bundle $\pi : Q \to \mathbb{R}$, which are identified by their values $s^i(t)$, as well as the values of their first and second partial derivatives, $\partial_t s^i = \partial_t s^i(t)$ and $\partial_{tt} s^i = \partial_{tt} s^i(t)$, respectively, at time points $t \in \mathbb{R}$. The 2–jet manifold $J^2(\mathbb{R}, Q)$ is coordinated by $(t, q^i, \dot{q}^i, \ddot{q}^i)$, that is by *(time, coordinates, velocities and accelerations)* at every active human joint, so the 2–jets are local joint coordinate maps:

$$j^2_t s : \mathbb{R} \to Q, \qquad t \mapsto (t, q^i, \dot{q}^i, \ddot{q}^i).$$

22.1.4 Lagrangian Time–Dependent Biomechanics

Jet Dynamics and Quadratic Equations

The general form of time-dependent Lagrangian biomechanics with *time-dependent Lagrangian* function $L(t;q^i;\dot{q}^i)$ defined on the jet space $X = J^1(\mathbb{R},Q) \cong \mathbb{R}\times TQ$, with local canonical coordinates: $(t;q^i;\dot{q}^i) =$ (time, coordinates and velocities) in active local joints, can be formulated as [II06c, II07e]

$$\frac{d}{dt}L_{\dot{q}^i} - L_{q^i} = \mathcal{F}_i\left(t,q,\dot{q}\right), \qquad (i=1,...,n), \tag{22.3}$$

where the coordinate and velocity partial derivatives of the Lagrangian are respectively denoted by L_{q^i} and $L_{\dot{q}^i}$.

The most interesting instances of (22.3) are quadratic biomechanical equations, of the general form

$$\xi^i \equiv \ddot{q}^i = a^i_{jk}(q^\mu)\dot{q}^j\dot{q}^k + b^i_j(q^\mu)\dot{q}^j + f^i(q^\mu). \tag{22.4}$$

They are coordinate–independent due to the affine transformation law of coordinates $\dot{q}^i$. Then, it is clear that the corresponding dynamical connection Γ_ξ is affine [II06c, II07e]:

$$\Gamma_\xi = dq^\alpha\otimes[\partial_\alpha + (\Gamma^i_{\lambda 0}(q^\nu) + \Gamma^i_{\lambda j}(q^\nu)\dot{q}^j)\partial^t_i].$$

This connection is symmetric iff $\Gamma^i_{\lambda\mu} = \Gamma^i_{\mu\lambda}$.

There is 1–1 correspondence between the affine connections Γ on the affine jet bundle $J^1(\mathbb{R},Q)\to Q$ and the linear connections K on the tangent bundle $TQ\to Q$ of the autonomous biomechanical manifold Q. This correspondence is given by the relation

$$\Gamma^i_\mu = \Gamma^i_{\mu 0} + \Gamma^i_{\mu j}\dot{q}^j, \qquad \Gamma^i_{\mu\lambda} = K_\mu{}^i{}_\alpha.$$

Any quadratic biomechanical equation (22.4) is equivalent to the geodesic equation [II07e]

$$\dot{t}=1, \qquad \ddot{t}=0, \qquad \ddot{q}^i = a^i_{jk}(q^\mu)\dot{q}^i\dot{q}^j + b^i_j(q^\mu)\dot{q}^j\dot{t} + f^i(q^\mu)\dot{t}\dot{t},$$

for the symmetric linear connection

$$K = dq^\alpha\otimes(\partial_\alpha + K^\mu_{\alpha\nu}(t,q^i)\dot{q}^\nu\dot{\partial}_\mu)$$

on the tangent bundle $TQ\to Q$, given by the components

$$K^0_{\alpha\nu}=0, \quad K_0{}^i{}_0 = f^i, \quad K_0{}^i{}_j = K_j{}^i{}_0 = \frac{1}{2}b^i_j, \quad K_j{}^i{}_k = a^i_{jk}.$$

Conversely, any linear connection K on the tangent bundle $TQ\to Q$ defines the quadratic dynamical equation

$$\ddot{q}^i = K_j{}^i{}_k\dot{q}^j\dot{q}^k + (K_0{}^i{}_j + K_j{}^i{}_0)\dot{q}^j + K_0{}^i{}_0,$$

written with respect to a given reference frame $(t,q^i)\equiv q^\mu$ (see [II07e] for technical details).

22.1.5 Local Muscle–Joint Mechanics

The right–hand side terms $\mathcal{F}_i(t, q, \dot{q})$ of (22.3) denote any type of external torques and forces, including excitation and contraction dynamics of muscular–actuators and rotational dynamics of hybrid robot actuators, as well as (nonlinear) dissipative joint torques and forces and external stochastic perturbation torques and forces. In particular, we have [II06a, II06b]):

1. Synovial joint mechanics, giving the first stabilizing effect to the conservative skeleton dynamics, is described by the $(q, \dot{q})$–form of the Rayleigh–Van der Pol's dissipation function

$$R = \frac{1}{2}\sum_{i=1}^{n}(\dot{q}^i)^2\,[\alpha_i\,+\,\beta_i(q^i)^2],$$

where α_i and β_i denote dissipation parameters. Its partial derivatives give rise to the viscous–damping torques and forces in the joints

$$\mathcal{F}_i^{joint} = \partial R/\partial \dot{q}^i,$$

which are linear in $\dot{q}^i$ and quadratic in q^i.

2. Muscular mechanics, giving the driving torques and forces $\mathcal{F}_i^{muscle} = \mathcal{F}_i^{muscle}(t, q, \dot{q})$ with $(i = 1, \dots, n)$ for human biomechanics, describes the internal excitation and contraction dynamics of equivalent muscular actuators [Hat78].

(a) *Excitation dynamics* can be described by an impulse force–time relation

$$\begin{aligned} F_i^{imp} &= F_i^0(1\,-\,e^{-t/\tau_i}) && \text{if stimulation} > 0 \\ F_i^{imp} &= F_i^0 e^{-t/\tau_i} && \text{if stimulation } = 0, \end{aligned}$$

where F_i^0 denote the maximal isometric muscular torques and forces, while τ_i denote the associated time characteristics of particular muscular actuators. This relation represents a solution of the Wilkie's muscular active–state element equation [Wil56]

$$\dot{\mu}\,+\,\Gamma\,\mu\,=\,\Gamma\,S\,A,\quad \mu(0)\,=\,0,\quad 0 < S < 1,$$

where $\mu = \mu(t)$ represents the active state of the muscle, Γ denotes the element gain, A corresponds to the maximum tension the element can develop, and $S = S(r)$ is the 'desired' active state as a function of the motor unit stimulus rate r. This is the basis for biomechanical force controller.

(b) *Contraction dynamics* has classically been described by the Hill's hyperbolic force–velocity relation [Hil38]

$$F_i^{Hill}\;=\;\frac{(F_i^0 b_i\,-\,\delta_{ij}a_i\dot{q}^j\,)}{(\delta_{ij}\dot{q}^j\,+\,b_i)},$$

where a_i and b_i denote the Hill's parameters, corresponding to the energy dissipated during the contraction and the phosphagenic energy conversion rate, respectively, while δ_{ij} is the Kronecker's δ–tensor.

In this way, human biomechanics describes the excitation/contraction dynamics for the ith equivalent muscle–joint actuator, using the simple impulse–hyperbolic product relation

$$\mathcal{F}_i^{muscle}(t,q,\dot{q}) = F_i^{imp} \times F_i^{Hill}.$$

Now, for the purpose of biomedical engineering and rehabilitation, human biomechanics has developed the so–called *hybrid rotational actuator*. It includes, along with muscular and viscous forces, the D.C. motor drives, as used in robotics [II06a]

$$\mathcal{F}_k^{robo} = i_k(t) - J_k\ddot{q}_k(t) - B_k\dot{q}_k(t), \qquad \text{with}$$
$$l_k i_k(t) + R_k i_k(t) + C_k\dot{q}_k(t) = u_k(t),$$

where $k = 1,\dots,n$, $i_k(t)$ and $u_k(t)$ denote currents and voltages in the rotors of the drives, R_k, l_k and C_k are resistances, inductances and capacitances in the rotors, respectively, while J_k and B_k correspond to inertia moments and viscous dampings of the drives, respectively.

Finally, to make the model more realistic, we need to add some stochastic torques and forces [II07b]

$$\mathcal{F}_i^{stoch} = B_{ij}[q^i(t),t]\,dW^j(t),$$

where $B_{ij}[q(t),t]$ represents continuous stochastic diffusion fluctuations, and $W^j(t)$ is an $N-$variable Wiener process (i.e., generalized Brownian motion), with

$$dW^j(t) = W^j(t+dt) - W^j(t), \qquad (\text{for } \; j = 1,\dots,N).$$

22.1.6 Time–Dependent Geometry of Biomechanics

As illustrated in the introduction, the mass-inertia matrix of human body, defining the Riemannian metric tensor $g_{ij}(q)$ need not be time-constant, as in case of fast gymnastic movements and pirouettes in ice skating, which are based on quick variations of inertia moments and products constituting the material metric tensor $g_{ij}(q)$. In particular, in the geodesic framework (22.10), the (in)stability of the biomechanical joint and center-of-mass trajectories is the (in)stability of the geodesics, and it is completely determined by the curvature properties of the underlying manifold according to the *Jacobi equation* of *geodesic deviation* [II06c, II07e]

$$\frac{D^2J^i}{ds^2} + R^i{}_{jkm}\frac{dq^j}{ds}J^k\frac{dq^m}{ds} = 0,$$

whose solution J, usually called *Jacobi variation field*, locally measures the distance between nearby geodesics; D/ds stands for the *covariant derivative* along a geodesic and $R^i{}_{jkm}$ are the components of the *Riemann curvature tensor*.

In general, the biomechanical metric tensor g_{ij} is both time and joint dependent, $g_{ij} = g_{ij}(t,q)$. This time-dependent Riemannian geometry can be formalized in terms of the *Ricci flow* [II07e], the nonlinear heat–like evolution metric equation:

$$\partial_t g_{ij} = -R_{ij}, \tag{22.5}$$

for a time–dependent Riemannian metric $g = g_{ij}(t)$ on a smooth n–manifold Q with the Ricci curvature tensor R_{ij}. This equation roughly says that we can deform any metric on the configuration manifold Q by the negative of its curvature; after *normalization*, the final state of such deformation will be a metric with constant curvature. The negative sign in (22.5) insures a kind of global *volume exponential decay*,[3] since the Ricci flow equation (22.5) is a kind of nonlinear geometric generalization of the standard linear *heat equation*

$$\partial_t u = \Delta u.$$

In a suitable local coordinate system, the Ricci flow equation (22.5) on a biomechanical configuration manifold Q has a nonlinear heat–type form, as follows. At any time t, we can choose local harmonic coordinates so that the coordinate functions are locally defined harmonic functions in the metric $g(t)$. Then the Ricci flow takes the general form [II07e]

$$\partial_t g_{ij} = \Delta_Q g_{ij} + G_{ij}(g, \partial g), \qquad \text{where} \tag{22.6}$$
$$\Delta_Q \equiv \frac{1}{\sqrt{\det(g)}} \frac{\partial}{\partial q^i} \left(\sqrt{\det(g)} g^{ij} \frac{\partial}{\partial q^j} \right)$$

is the *Laplace–Beltrami operator* of the configuration manifold Q and $G_{ij}(g, \partial g)$ is a lower–order term quadratic in g and its first order partial derivatives ∂g. From the analysis of nonlinear heat PDEs, one obtains existence and uniqueness of forward–time solutions to the Ricci flow on some time interval, starting at any smooth initial metric g_0 on Q.

The exponentially-decaying geometrical diffusion (22.6) is a formal description for pirouettes in ice skating and fast rotational movements in gymnastics.

22.2 Jet Spaces in Modern Hamiltonian Biomechanics

In this section we present our recent work on et methods in modern time-dependent Hamiltonian biomechanics [Iva10]. This work is a sequel to our previous paper on time-dependent Lagrangian biomechanics [Iva09]. In the Hamiltonian paper we have used the covariant force law in conjunction with the

---

[3] This complex geometric process is globally similar to a generic exponential decay ODE:

$$\dot{q} = -\lambda f(q),$$

for a positive function $f(q)$. We can get some insight into its solution from the simple exponential decay ODE,

$$\dot{q} = -\lambda q \qquad \text{with the solution} \qquad q(t) = q_0 \mathrm{e}^{-\lambda t},$$

where $q = q(t)$ is the observed quantity with its initial value q_0 and λ is a positive decay constant.

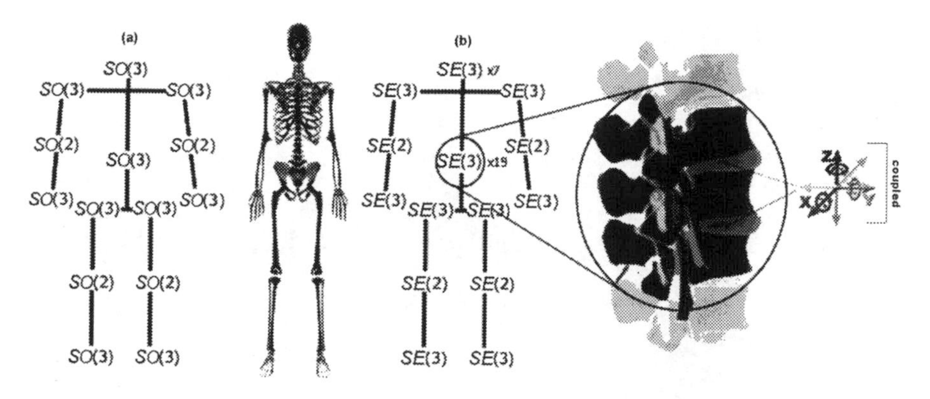

Fig. 22.5. Biomechanical configuration manifold for a humanoid robot (a) and human (loco)motion system (b), both defined as anthropomorphic product trees of constrained motion symmetry groups, SO(3) and SE(3), respectively. Each humanoid joint consists of a pair of coupled segments with only Eulerian rotational degrees of freedom, defined as a constrained SO(3) rotational group. On the other hand, in each human (synovial) joint, besides gross Eulerian rotational movements, we also have some hidden and restricted translations along Eulerian/Cartesian $(X, Y, Z)-$axes. For example, in the knee joint, patella (knee cap) moves for about 7–10 cm from maximal extension to maximal flexion). Even greater are translational amplitudes in the shoulder joint. In other words, within the realm of rigid body mechanics, a segment of a human arm or leg is not properly represented as a rigid body fixed at a certain point, but rather as a rigid body hanging on rope–like ligaments. More generally, the whole skeleton mechanically represents a system of flexibly coupled rigid bodies, each of them defined by the Euclidean group of (rotational + translational) motions in $\mathbb{R}^3$ (see text below for explanation). This implies more complex kinematics, dynamics and control then in the case of humanoid robots (see [II08c] for technical details).

modern geometric formalism of jet manifolds to develop general Hamiltonian approach to time-dependent human biomechanics in which *total mechanical energy is not conserved.*

Before we get into specific techniques of (time-dependent) Hamiltonian biomechanics, we remark here the three essential differences between a human (loco)motion system and a humanoid robot (or a mechanical multi-body system in general):

1. Human joints are more flexible than robot joints (see Figure 22.5);
2. Human (or animal) muscular actuators are more complex then any robot actuators; they have their own excitation-contraction dynamics (defined in section 22.2.3 below);
3. In a human (loco)motion system, total mechanical energy is not conserved but dissipated.

22.2.1 The Covariant Force Law

Autonomous Hamiltonian biomechanics (as well as autonomous Lagrangian biomechanics), based on the postulate of conservation of the total mechanical energy, can be derived from the *covariant force law* [II06a, II06b, II06c, II07e], which in 'plain English' states:

Force 1–form = Mass distribution × Acceleration vector-field,

and formally reads (using Einstein's summation convention over repeated indices):

$$F_i = m_{ij} a^j. \tag{22.7}$$

Here, the force 1–form $F_i = F_i(t, q, p) = F_i'(t, q, \dot{q})$, $(i = 1, ..., n)$ denotes any type of torques and forces acting on a human skeleton, including excitation and contraction dynamics of muscular–actuators [Hil38, Wil56, Hat78] and rotational dynamics of hybrid robot actuators, as well as (nonlinear) dissipative joint torques and forces and external stochastic perturbation torques and forces [II07a]. m_{ij} is the material (mass–inertia) metric tensor, which gives the total mass distribution of the human body, by including all segmental masses and their individual inertia tensors. a^j is the total acceleration vector-field, including all segmental vector-fields, defined as the absolute (Bianchi) derivative $\dot{\bar{v}}^i$ of all the segmental angular and linear velocities $v^i = \dot{x}^i$, $(i = 1, ..., n)$, where n is the total number of active degrees of freedom (DOF) with local coordinates (x^i).

More formally, this *central Law of biomechanics* represents the *covariant force functor* $\mathcal{F}_*$ defined by the commutative diagram:

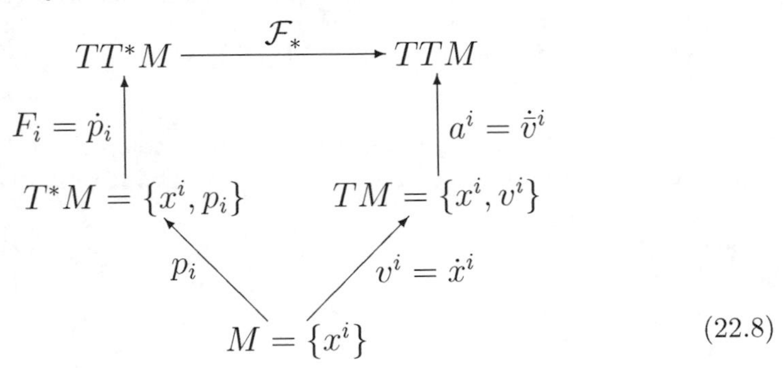

(22.8)

Here, $M \equiv M^n = \{x^i, \ (i = 1, ..., n)\}$ is the biomechanical configuration $n-$manifold (Figure 22.2), that is the set of all active DOF of the biomechanical system under consideration (in general, human skeleton), with local coordinates (x^i).

The right-hand branch of the fundamental covariant force functor $\mathcal{F}_* : TT^*M \longrightarrow TTM$ depicted in (22.8) is Lagrangian dynamics with its Riemannian geometry. To each $n-$dimensional (nD) smooth manifold M there is associated its $2n$D *velocity phase-space manifold*, denoted by TM and called the tangent bundle of M. The original configuration manifold M is called the *base* of TM. There is an onto map $\pi : TM \rightarrow M$, called the *projection*. Above each point

$x \in M$ there is a tangent space $T_xM = \pi^{-1}(x)$ to M at x, which is called a fibre. The fibre $T_xM \subset TM$ is the subset of TM, such that the total tangent bundle, $TM = \bigsqcup_{m \in M} T_xM$, is a disjoint union of tangent spaces T_xM to M for all points $x \in M$. From dynamical perspective, the most important quantity in the tangent bundle concept is the smooth map $v : M \to TM$, which is an inverse to the projection π, i.e, $\pi \circ v = \mathrm{Id}_M$, $\pi(v(x)) = x$. It is called the *velocity vector-field* $v^i = \dot{x}^i$.[4] Its graph $(x, v(x))$ represents the cross–section of the tangent bundle TM. Velocity vector-fields are cross-sections of the tangent bundle. Biomechanical *Lagrangian* (that is, kinetic minus potential energy) is a natural energy function on the tangent bundle TM. The tangent bundle is itself a smooth manifold. It has its own tangent bundle, TTM. Cross-sections of the second tangent bundle TTM are the acceleration vector-fields.

The left-hand branch of the fundamental covariant force functor $\mathcal{F}_* : TT^*M \longrightarrow TTM$ depicted in (22.8) is Hamiltonian dynamics with its symplectic geometry. It takes place in the *cotangent bundle* T^*M_{rob}, defined as follows. A *dual* notion to the tangent space T_xM to a smooth manifold M at a point $x = (x^i)$ with local is its cotangent space T_x^*M at the same point x. Similarly to the tangent bundle TM, for any smooth nD manifold M, there is associated its $2n$D *momentum phase-space manifold*, denoted by T^*M and called the *cotangent bundle*. T^*M is the disjoint union of all its cotangent spaces T_x^*M at all points $x \in M$, i.e., $T^*M = \bigsqcup_{x \in M} T_x^*M$. Therefore, the cotangent bundle of an $n-$manifold M is the vector bundle $T^*M = (TM)^*$, the (real) dual of the tangent bundle TM. Momentum 1–forms (or, covector-fields) p_i are cross-sections of the cotangent bundle. Biomechanical *Hamiltonian* (that is, kinetic plus potential energy) is a natural energy function on the cotangent bundle. The cotangent bundle T^*M is itself a smooth manifold. It has its own tangent bundle, TT^*M. Cross-sections of the mixed-second bundle TT^*M are the force 1–forms $F_i = \dot{p}_i$.

There is a unique smooth map from the right-hand branch to the left-hand branch of the diagram (22.8):

$$TM \ni (x^i, v^i) \mapsto (x^i, p^i) \in T^*M.$$

It is called the *Legendre transformation*, or *fiber derivative* (for details see, e.g. [II06c, II07e]).

The fundamental covariant force functor $\mathcal{F}_* : TT^*M \longrightarrow TTM$ states that the force 1–form $F_i = \dot{p}_i$, defined on the mixed tangent–cotangent bundle TT^*M, causes the acceleration vector-field $a^i = \dot{\bar{v}}^i$, defined on the second tangent bundle TTM of the configuration manifold M. The corresponding *contravariant acceleration functor* is defined as its inverse map, $\mathcal{F}^* : TTM \longrightarrow TT^*M$.

Representation of human motion is rigorously defined in terms of Euclidean $SE(3)$–groups of full rigid–body motion in all main human joints [II08c]. The configuration manifold M for human musculo-skeletal dynamics is defined as a

[4] This explains the dynamical term *velocity phase–space*, given to the tangent bundle TM of the manifold M.

Cartesian product of all included constrained $SE(3)$ groups, $M = \prod_j SE(3)^j$ where j labels the active joints. The configuration manifold M is coordinated by local joint coordinates $x^i(t)$, $i = 1, ..., n =$ total number of active DOF. The corresponding joint velocities $\dot{x}^i(t)$ live in the *velocity phase space* TM, which is the *tangent bundle* of the configuration manifold M.

The velocity phase-space TM has the Riemannian geometry with the *local metric form*:

$$\langle g \rangle \equiv ds^2 = g_{ij} dx^i dx^j, \tag{22.9}$$

where $g_{ij} = g_{ij}(m, x)$ is the material metric tensor defined by the biomechanical system's *mass-inertia matrix* and dx^i are differentials of the local joint coordinates x^i on M. Besides giving the local distances between the points on the manifold M, the Riemannian metric form $\langle g \rangle$ defines the system's kinetic energy:

$$T = \frac{1}{2} g_{ij} \dot{x}^i \dot{x}^j,$$

giving the *Lagrangian equations* of the conservative skeleton motion with Lagrangian $L = T - V$, with the corresponding *geodesic form* of the free motion:

$$\frac{d}{dt} L_{\dot{x}^i} - L_{x^i} = 0, \qquad \text{or} \qquad \ddot{x}^i + \Gamma^i_{jk} \dot{x}^j \dot{x}^k = 0, \tag{22.10}$$

where subscripts denote partial derivatives, while Γ^i_{jk} are the Christoffel symbols of the affine Levi-Civita connection of the biomechanical manifold M.

The corresponding momentum phase-space $P = T^*M$ provides a natural *symplectic structure* that can be defined as follows. As the biomechanical configuration space M is a smooth $n-$manifold, we can pick local coordinates $\{dx^1, ..., dx^n\} \in M$. Then $\{dx^1, ..., dx^n\}$ defines a basis of the cotangent space $T^*_x M$, and by writing $\theta \in T^*_x M$ as $\theta = p_i dx^i$, we get local coordinates $\{x^1, ..., x^n, p_1, ..., p_n\}$ on T^*M. We can now define the canonical symplectic form ω on $P = T^*M$ as:

$$\omega = dp_i \wedge dx^i,$$

where '$\wedge$' denotes the wedge or exterior product of exterior differential forms.[5] This $2-$form ω is obviously independent of the choice of coordinates $\{x^1, ..., x^n\}$ and independent of the base point $\{x^1, ..., x^n, p_1, ..., p_n\} \in T^*_x M$. Therefore, it is locally constant, and so $d\omega = 0$.[6]

---

[5] Recall that an *exterior differential form* α of order p (or, a $p-$*form* α) on a base manifold X is a section of the exterior product bundle $\overset{p}{\wedge} T^*X \to X$. It has the following expression in local coordinates on X

$$\alpha = \alpha_{\lambda_1 \dots \lambda_p} dx^{\lambda_1} \wedge \dots \wedge dx^{\lambda_p} \qquad \text{(such that } |\alpha| = p),$$

where summation is performed over all ordered collections $(\lambda_1, ..., \lambda_p)$. $\mathbf{\Omega}^p(X)$ is the vector space of $p-$forms on a biomechanical manifold X. In particular, the 1–forms are called the *Pfaffian forms*.

[6] The canonical $1-$form θ on T^*M is the unique $1-$form with the property that, for any $1-$form β which is a section of T^*M we have $\beta^*\theta = \theta$.

If (P, ω) is a $2n$D symplectic manifold then about each point $x \in P$ there are local coordinates $\{x^1, ..., x^n, p_1, ..., p_n\}$ such that $\omega = dp_i \wedge dx^i$. These coordinates are called canonical or symplectic. By the Darboux theorem, ω is constant in this local chart, i.e., $d\omega = 0$.

22.2.2 Autonomous Hamiltonian Biomechanics

We develop autonomous Hamiltonian biomechanics on the configuration biomechanical manifold M in three steps, following the standard symplectic geometry prescription (see [II06a, II06c, II07e, II08b]):

Step A. Find a symplectic *momentum phase–space* (P, ω).

A symplectic structure on a smooth manifold M is a nondegenerate closed 2–form ω on M, i.e., for each $x \in M$, $\omega(x)$ is nondegenerate, and $d\omega = 0$. The cotangent bundle $P = T^*M$ is our momentum phase–space. On P there is a nondegenerate symplectic 2–form ω is defined in local joint coordinates $x^i, p_i \in U$, U open in P, as $\omega = dx^i \wedge dp_i$. In that case the coordinates $x^i, p_i \in U$ are called canonical. In a usual procedure the canonical 1–form θ is first defined as $\theta = p_i dx^i$, and then the canonical 2–form ω is defined as $\omega = -d\theta$. A *symplectic phase–space manifold* is a pair (P, ω).

Step B. Find a *Hamiltonian vector-field* X_H on (P, ω).

Let (P, ω) be a symplectic manifold. A vector-field $X : P \to TP$ is called *Hamiltonian* if there is a smooth function $F : P \longrightarrow \mathbb{R}$ such that $i_X\omega = dF$ ($i_X\omega$ denotes the *interior product* or *contraction* of the vector-field X and the 2–form ω). X is *locally Hamiltonian* if $i_X\omega$ is closed.

Let the smooth real–valued *Hamiltonian function* $H : P \to \mathbb{R}$, representing the total biomechanical energy $H(x, p) = T(p) + V(x)$ (T and V denote kinetic and potential energy of the system, respectively), be given in local canonical coordinates $x^i, p_i \in U$, U open in P. The *Hamiltonian vector-field* X_H, condition by $i_{X_H}\omega = dH$, is actually defined via symplectic matrix J, in a local chart U, as

$$X_H = J\nabla H = (\partial_{p_i} H, -\partial_{x^i} H), \qquad J = \begin{pmatrix} 0 & I \\ -I & 0 \end{pmatrix}, \tag{22.11}$$

where I denotes the $n \times n$ identity matrix and ∇ is the gradient operator.

Step C. Find a *Hamiltonian phase–flow* ϕ_t of X_H.

Let (P, ω) be a symplectic phase–space manifold and $X_H = J\nabla H$ a Hamiltonian vector-field corresponding to a smooth real–valued Hamiltonian function $H : P \to \mathbb{R}$, on it. If a unique one–parameter group of diffeomorphisms $\phi_t : P \to P$ exists so that $\frac{d}{dt}|_{t=0}\, \phi_t x = J\nabla H(x)$, it is called the *Hamiltonian phase–flow*.

Let $f : M \to M$ be a diffeomorphism. Then T^*f preserves the canonical 1–form θ on T^*M, i.e., $(T^*f)^*\theta = \theta$. Thus T^*f is symplectic diffeomorphism.

A smooth curve $t \mapsto (x^i(t), p_i(t))$ on (P,ω) represents an *integral curve* of the Hamiltonian vector-field $X_H = J\nabla H$, if in the local canonical coordinates $x^i, p_i \in U$, U open in P, *Hamiltonian canonical equations* hold:

$$\dot{q}^i = \partial_{p_i} H, \qquad \dot{p}_i = -\partial_{q^i} H. \tag{22.12}$$

An integral curve is said to be *maximal* if it is not a restriction of an integral curve defined on a larger interval of $\mathbb{R}$. It follows from the standard theorem on the *existence* and *uniqueness* of the solution of a system of ODEs with smooth r.h.s, that if the manifold (P,ω) is Hausdorff, then for any point $x = (x^i, p_i) \in U$, U open in P, there exists a maximal integral curve of $X_H = J\nabla H$, passing for $t = 0$, through point x. In case X_H is complete, i.e., X_H is C^p and (P,ω) is compact, the maximal integral curve of X_H is the Hamiltonian phase–flow $\phi_t : U \to U$.

The phase–flow ϕ_t is *symplectic* if ω is constant along ϕ_t, i.e., $\phi_t^*\omega = \omega$, (where $\phi_t^*\omega$ denotes the *pull–back* of ω by ϕ_t), if and only if ('iff' for short) $\mathfrak{L}_{X_H}\omega = 0$ (where $\mathfrak{L}_{X_H}\omega$ denotes the *Lie derivative* of ω upon X_H). Symplectic phase–flow ϕ_t consists of canonical transformations on (P,ω), i.e., diffeomorphisms in canonical coordinates $x^i, p_i \in U$, U open on all (P,ω) which leave ω invariant. In this case the *Liouville theorem* is valid: ϕ_t *preserves* the *phase volume* on (P,ω).

The general form of autonomous Hamiltonian biomechanics is given by dissipative, driven Hamiltonian equations on T^*M:

$$\dot{x}^i = \frac{\partial H}{\partial p_i} + \frac{\partial R}{\partial p_i}, \tag{22.13}$$

$$\dot{p}_i = F_i - \frac{\partial H}{\partial x^i} + \frac{\partial R}{\partial x^i}, \tag{22.14}$$

$$x^i(0) = x_0^i, \qquad p_i(0) = p_i^0, \tag{22.15}$$

including *contravariant equation* (22.13) – the *velocity vector-field*, and *covariant equation* (22.14) – the *force 1–form* (field), together with initial joint angles and momenta (22.15). Here $R = R(x,p)$ denotes the Raileigh nonlinear (biquadratic) dissipation function, and $F_i = F_i(t,x,p)$ are covariant driving torques of *equivalent muscular actuators*, resembling muscular excitation and contraction dynamics in rotational form. The velocity vector-field (22.13) and the force 1–form (22.14) together define the generalized Hamiltonian vector-field X_H; the Hamiltonian energy function $H = H(x,p)$ is its generating function.

As a Lie group, the biomechanical configuration manifold $M = \prod_j SE(3)^j$ is *Hausdorff*. Therefore, for $x = (x^i, p_i) \in U_p$, where U_p is an open coordinate chart in T^*M, there exists a unique one–parameter group of diffeomorphisms $\phi_t : T^*M \to T^*M$, that is the *autonomous Hamiltonian phase–flow:*

$$\phi_t : T^*M \to T^*M : (p(0), x(0)) \mapsto (p(t), x(t)), \tag{22.16}$$
$$(\phi_t \circ \phi_s = \phi_{t+s}, \quad \phi_0 = \text{identity}),$$

given by (22.13–22.15) such that

$$\frac{d}{dt}|_{t=0}\, \phi_t x = J\nabla H(x).$$

22.2.3 Time–Dependent Hamiltonian Biomechanics

In this subsection we develop time-dependent Hamiltonian biomechanics. For this, we first need to extend our autonomous Hamiltonian machinery, using the general concepts of bundles, jets and connections.

Biomechanical Bundles and Jets

While standard autonomous Lagrangian biomechanics is developed on the configuration manifold X, the *time–dependent biomechanics* necessarily includes also the real time axis $\mathbb{R}$, so we have an *extended configuration manifold* $\mathbb{R} \times X$. Slightly more generally, the fundamental geometrical structure is the so-called *configuration bundle* $\pi : X \to \mathbb{R}$. Time-dependent biomechanics is thus formally developed either on the *extended configuration manifold* $\mathbb{R} \times X$, or on the configuration bundle $\pi : X \to \mathbb{R}$, using the concept of *jets*, which are based on the idea of *higher–order tangency*, or higher–order contact, at some designated point (i.e., certain anatomical joint) on a biomechanical configuration manifold X (see [Iva09, Iva10]).

In general, tangent and cotangent bundles, TM and T^*M, of a smooth manifold M, are special cases of a more general geometrical object called *fibre bundle*, denoted $\pi : Y \to X$, where the word *fiber* V of a map $\pi : Y \to X$ is the *preimage* $\pi^{-1}(x)$ of an element $x \in X$. It is a space which *locally* looks like a product of two spaces (similarly as a manifold locally looks like Euclidean space), but may possess a different *global* structure. To get a visual intuition behind this fundamental geometrical concept, we can say that a fibre bundle Y is a *homeomorphic generalization* of a *product space* $X \times V$ (see Figure 22.6), where X and V are called the *base* and the *fibre*, respectively. $\pi : Y \to X$ is called the *projection*, $Y_x = \pi^{-1}(x)$ denotes a fibre over a point x of the base X, while the map $f = \pi^{-1} : X \to Y$ defines the *cross–section*, producing the *graph* $(x, f(x))$ in the bundle Y (e.g., in case of a tangent bundle, $f = \dot{x}$ represents a velocity vector–field).

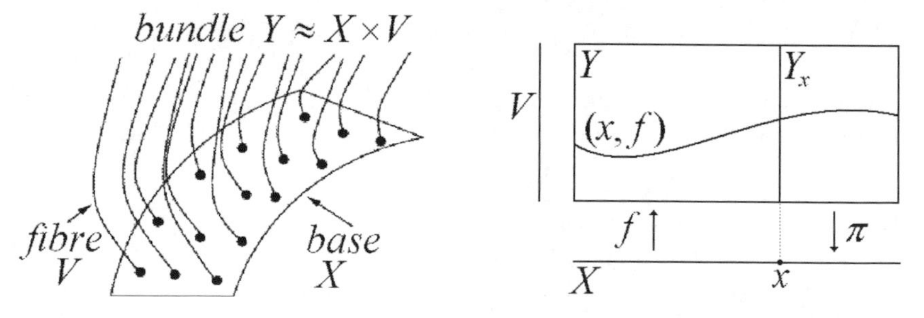

Fig. 22.6. A sketch of a locally trivial fibre bundle $Y \approx X \times V$ as a generalization of a product space $X \times V$; left – main components; right – a few details (see text for explanation).

More generally, a biomechanical configuration bundle, $\pi : Y \longrightarrow X$, is a locally trivial fibred (or, projection) manifold over the base X. It is endowed with an atlas of fibred bundle coordinates (x^λ, y^i), where (x^λ) are coordinates of X.

Now, a pair of smooth manifold maps, $f_1, f_2 : M \to N$ (see Figure 22.7), are said to be $k-$*tangent* (or *tangent of order* k, or have a kth *order contact*) at a point x on a domain manifold M, denoted by $f_1 \sim f_2$, iff

$$\begin{aligned} f_1(x) &= f_2(x) \qquad \text{called} \quad 0-\text{tangent},\\ \partial_x f_1(x) &= \partial_x f_2(x), \qquad \text{called} \quad 1-\text{tangent},\\ \partial_{xx} f_1(x) &= \partial_{xx} f_2(x), \qquad \text{called} \quad 2-\text{tangent},\\ &\ldots \qquad \text{etc. to the order } k \end{aligned}$$

In this way defined $k-$*tangency* is an *equivalence relation.*

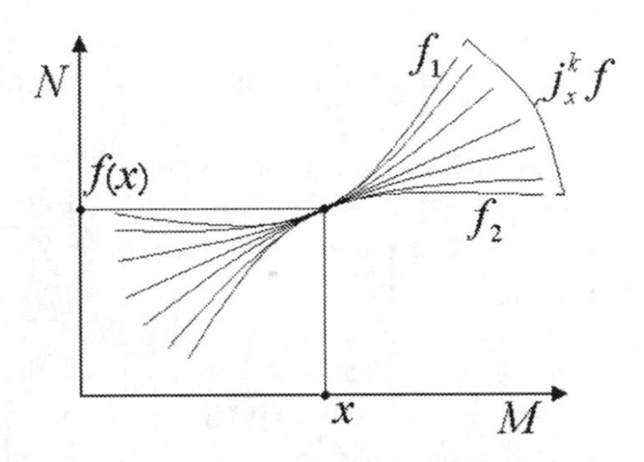

Fig. 22.7. An intuitive geometrical picture behind the $k-$jet concept, based on the idea of a higher–order tangency (or, higher–order contact).

A $k-$*jet* (or, a *jet of order* k), denoted by $j_x^k f$, of a smooth map $f : M \to N$ at a point $x \in M$ (see Figure 22.7), is defined as an *equivalence class* of $k-$tangent maps at x,

$$j_x^k f : M \to N = \{f' : f' \text{ is } k-\text{tangent to } f \text{ at } x\}.$$

For example, consider a simple function $f : X \to Y, x \mapsto y = f(x)$, mapping the $X-$axis into the $Y-$axis in $\mathbb{R}^2$. At a chosen point $x \in X$ we have:

a $0-$jet is a graph: $(x, f(x))$;
a $1-$jet is a triple: $(x, f(x), f'(x))$;
a $2-$jet is a quadruple: $(x, f(x), f'(x), f''(x))$,
and so on, up to the order k (where $f'(x) = \frac{df(x)}{dx}$, etc).

The set of all $k-$jets from $j_x^k f : X \to Y$ is called the $k-$jet manifold $J^k(X, Y)$.

Formally, given a biomechanical bundle $Y \longrightarrow X$, its first–order *jet manifold* J^1Y comprises the set of equivalence classes $j_x^1 s$, $x \in X$, of sections $s : X \longrightarrow Y$ so that sections s and s' belong to the same class iff

$$Ts\,|_{T_xX} = Ts'\,|_{T_xX}\,.$$

Intuitively, sections $s, s' \in j^1_x s$ are identified by their values $s^i(x) = s'^i(x)$ and the values of their partial derivatives $\partial_\mu s^i(x) = \partial_\mu s'^i(x)$ at the point x of X. There are the natural fibrations [II07e]

$$\pi_1 : J^1Y \ni j^1_x s \mapsto x \in X, \qquad \pi_{01} : J^1Y \ni j^1_x s \mapsto s(x) \in Y.$$

Given bundle coordinates (x^λ, y^i) of Y, the associated jet manifold J^1Y is endowed with the adapted coordinates

$$(x^\lambda, y^i, y^i_\lambda), \qquad (y^i, y^i_\lambda)(j^1_x s) = (s^i(x), \partial_\lambda s^i(x)), \qquad y'^i_\lambda = \frac{\partial x^\mu}{\partial x'^\lambda}(\partial_\mu + y^j_\mu \partial_j) y'^i.$$

In particular, given the biomechanical configuration bundle $M \to \mathbb{R}$ over the time axis $\mathbb{R}$, the 1*–jet space* $J^1(\mathbb{R}, M)$ is the set of equivalence classes $j^1_t s$ of sections $s^i : \mathbb{R} \to M$ of the configuration bundle $M \to \mathbb{R}$, which are identified by their values $s^i(t)$, as well as by the values of their partial derivatives $\partial_t s^i = \partial_t s^i(t)$ at time points $t \in \mathbb{R}$. The 1–jet manifold $J^1(\mathbb{R}, M)$ is coordinated by $(t, x^i, \dot{x}^i)$, that is by *(time, coordinates and velocities)* at every active human joint, so the 1–jets are local joint coordinate maps

$$j^1_t s : \mathbb{R} \to M, \qquad t \mapsto (t, x^i, \dot{x}^i).$$

The *second–order jet manifold* J^2Y of a bundle $Y \to X$ is the subbundle of $\widehat{J}^2Y \to J^1Y$ defined by the coordinate conditions $y^i_{\lambda\mu} = y^i_{\mu\lambda}$. It has the local coordinates $(x^\lambda, y^i, y^i_\lambda, y^i_{\lambda \leq \mu})$ together with the transition functions [II07e]

$$y'^i_{\lambda\mu} = \frac{\partial x^\alpha}{\partial x'^\mu}(\partial_\alpha + y^j_\alpha \partial_j + y^j_{\nu\alpha} \partial^\nu_j) y'^i_\lambda.$$

The second–order jet manifold J^2Y of Y comprises the equivalence classes $j^2_x s$ of sections s of
$Y \to X$ such that

$$y^i_\lambda(j^2_x s) = \partial_\lambda s^i(x), \qquad y^i_{\lambda\mu}(j^2_x s) = \partial_\mu \partial_\lambda s^i(x).$$

In other words, two sections $s, s' \in j^2_x s$ are identified by their values and the values of their first and second–order derivatives at the point $x \in X$.

In particular, given the biomechanical configuration bundle $M \to \mathbb{R}$ over the time axis $\mathbb{R}$, the 2*–jet space* $J^2(\mathbb{R}, M)$ is the set of equivalence classes $j^2_t s$ of sections $s^i : \mathbb{R} \to M$ of the configuration bundle $\pi : M \to \mathbb{R}$, which are identified by their values $s^i(t)$, as well as the values of their first and second partial derivatives, $\partial_t s^i = \partial_t s^i(t)$ and $\partial_{tt} s^i = \partial_{tt} s^i(t)$, respectively, at time points $t \in \mathbb{R}$. The 2–jet manifold $J^2(\mathbb{R}, M)$ is coordinated by $(t, x^i, \dot{x}^i, \ddot{x}^i)$, that is by *(time, coordinates, velocities and accelerations)* at every active human joint, so the 2–jets are local joint coordinate maps:

$$j^2_t s : \mathbb{R} \to M, \qquad t \mapsto (t, x^i, \dot{x}^i, \ddot{x}^i).$$

Nonautonomous Dissipative Hamiltonian Dynamics

We can now formulate the time-dependent biomechanics in which the biomechanical phase space is the Legendre manifold Π (see [II06c, II07e]), endowed with the holonomic coordinates (t, y^i, p_i) with the transition functions

$$p_i' = \frac{\partial y^j}{\partial y'^i} p_j.$$

Π admits the canonical form Λ given by

$$\Lambda = dp_i \wedge dy^i \wedge dt \otimes \partial_t.$$

We say that a connection

$$\gamma = dt \otimes (\partial_t + \gamma^i \partial_i + \gamma_i \partial^i) \qquad \text{(where} \quad \partial_i = \partial^i = \frac{\partial}{\partial x^i}),$$

on the bundle $\Pi \to X$ is *locally Hamiltonian* if the exterior form $\gamma \rfloor \Lambda$ is closed and Hamiltonian if the form $\gamma \rfloor \Lambda$ is exact [Iva10]. A connection γ is locally Hamiltonian iff it obeys the conditions:

$$\partial^i \gamma^j - \partial^j \gamma^i = 0, \quad \partial_i \gamma_j - \partial_j \gamma_i = 0, \quad \partial_j \gamma^i + \partial^i \gamma_j = 0.$$

Note that every connection $\Gamma = dt \otimes (\partial_t + \Gamma^i \partial_i)$ on the bundle $Y \longrightarrow X$ gives rise to the Hamiltonian connection $\widetilde{\Gamma}$ on $\Pi \longrightarrow X$, given by

$$\widetilde{\Gamma} = dt \otimes (\partial_t + \Gamma^i \partial_i - \partial_j \Gamma^i p_i \partial^j).$$

The corresponding Hamiltonian form H_Γ is given by

$$H_\Gamma = p_i dy^i - p_i \Gamma^i dt.$$

Let H be a *dissipative Hamiltonian form* on Π, which reads:

$$H = p_i dy^i - \mathcal{H} dt = p_i dy^i - p_i \Gamma^i dt - \widetilde{\mathcal{H}}_\Gamma dt. \tag{22.17}$$

We call $\mathcal{H}$ and $\widetilde{\mathcal{H}}$ in the decomposition (22.17) the *Hamiltonian* and the *Hamiltonian function* respectively. Let γ be a Hamiltonian connection on $\Pi \longrightarrow X$ associated with the Hamiltonian form (22.17). It satisfies the relations [II06c, II07e]

$$\begin{aligned} \gamma \rfloor \Lambda &= dp_i \wedge dy^i + \gamma_i dy^i \wedge dt - \gamma^i dp_i \wedge dt = dH, \\ \gamma^i &= \partial^i \mathcal{H}, \qquad \gamma_i = -\partial_i \mathcal{H}. \end{aligned} \tag{22.18}$$

From equations (22.18) we see that, in the case of biomechanics, one and only one Hamiltonian connection is associated with a given Hamiltonian form.

Every connection γ on $\Pi \longrightarrow X$ yields the system of first–order differential equations:

$$\dot{y}^i = \gamma^i, \qquad \dot{p}_i = \gamma_i. \tag{22.19}$$

They are called the *evolution equations*. If γ is a Hamiltonian connection associated with the Hamiltonian form H (22.17), the evolution equations (22.19) become the *dissipative time-dependent Hamiltonian equations*:

$$\dot{y}^i = \partial^i \mathcal{H}, \qquad \dot{p}_i = -\partial_i \mathcal{H}. \tag{22.20}$$

In addition, given any scalar function f on Π, we have the *dissipative Hamiltonian evolution equation*

$$d_H f = (\partial_t + \partial^i \mathcal{H} \partial_i - \partial_i \mathcal{H} \partial^i)\, f, \tag{22.21}$$

relative to the Hamiltonian $\mathcal{H}$. On solutions s of the Hamiltonian equations (22.20), the evolution equation (22.21) is equal to the total time derivative of the function f:

$$s^* d_H f = \frac{d}{dt}(f \circ s).$$

Time–Dependent Hamiltonian Biomechanics

The dissipative Hamiltonian system (22.20)–(22.21) is the basis for our time & fitness-dependent biomechanics. The scalar function f in (22.21) on the biomechanical Legendre phase-space manifold Π is now interpreted as an *individual neuro-muscular fitness function.* This fitness function is a 'determinant' for the performance of muscular drives for the driven, dissipative Hamiltonian biomechanics. These muscular drives, for all active DOF, are given by time & fitness-dependent Pfaffian form: $F_i = F_i(t, y, p, f)$. In this way, we obtain our final model for time & fitness-dependent Hamiltonian biomechanics:

$$\begin{aligned} \dot{y}^i &= \partial^i \mathcal{H}, \\ \dot{p}_i &= F_i - \partial_i \mathcal{H}, \\ d_H f &= (\partial_t + \partial^i \mathcal{H} \partial_i - \partial_i \mathcal{H} \partial^i)\, f. \end{aligned}$$

Physiologically, the active muscular drives $F_i = F_i(t, y, p, f)$ consist of [II06a, II06b]):

1. Synovial joint mechanics, giving the first stabilizing effect to the conservative skeleton dynamics, is described by the $(y, \dot{y})$–form of the *Rayleigh–Van der Pol's dissipation function*

$$R = \frac{1}{2} \sum_{i=1}^{n} (\dot{y}^i)^2 \, [\alpha_i \, + \, \beta_i (y^i)^2],$$

where α_i and β_i denote dissipation parameters. Its partial derivatives give rise to the viscous–damping torques and forces in the joints

$$\mathcal{F}_i^{joint} = \partial R / \partial \dot{y}^i,$$

which are linear in $\dot{y}^i$ and quadratic in y^i.

2. Muscular mechanics, giving the driving torques and forces $\mathcal{F}_i^{musc} = \mathcal{F}_i^{musc}(t, y, \dot{y})$ with $(i = 1, \dots, n)$ for human biomechanics, describes the internal excitation and contraction dynamics of *equivalent muscular actuators* [Hat78].

(a) The *excitation dynamics* can be described by an impulse force–time relation

$$F_i^{imp} = F_i^0(1 - e^{-t/\tau_i}) \qquad \text{if stimulation} > 0$$
$$F_i^{imp} = F_i^0 e^{-t/\tau_i} \qquad \text{if stimulation} = 0,$$

where F_i^0 denote the maximal isometric muscular torques and forces, while τ_i denote the associated time characteristics of particular muscular actuators. This relation represents a solution of the Wilkie's muscular *active–state element* equation [Wil56]

$$\dot{\mu} + \Gamma \mu = \Gamma S A, \quad \mu(0) = 0, \quad 0 < S < 1,$$

where $\mu = \mu(t)$ represents the active state of the muscle, Γ denotes the element gain, A corresponds to the maximum tension the element can develop, and $S = S(r)$ is the 'desired' active state as a function of the motor unit stimulus rate r. This is the basis for biomechanical force controller.

(b) The *contraction dynamics* has classically been described by the Hill's *hyperbolic force–velocity* relation [Hil38]

$$F_i^{Hill} = \frac{(F_i^0 b_i - \delta_{ij} a_i \dot{y}^j)}{(\delta_{ij} \dot{y}^j + b_i)},$$

where a_i and b_i denote the Hill's parameters, corresponding to the energy dissipated during the contraction and the phosphagenic energy conversion rate, respectively, while δ_{ij} is the Kronecker's δ–tensor.

In this way, human biomechanics describes the excitation/contraction dynamics for the ith equivalent muscle–joint actuator, using the simple impulse–hyperbolic product relation

$$\mathcal{F}_i^{musc}(t, y, \dot{y}) = F_i^{imp} \times F_i^{Hill}.$$

Now, for the purpose of biomedical engineering and rehabilitation, human biomechanics has developed the so–called *hybrid rotational actuator*. It includes, along with muscular and viscous forces, the D.C. motor drives, as used in robotics

$$\mathcal{F}_k^{robo} = i_k(t) - J_k \ddot{y}_k(t) - B_k \dot{y}_k(t), \qquad \text{with}$$
$$l_k i_k(t) + R_k i_k(t) + C_k \dot{y}_k(t) = u_k(t),$$

where $k = 1, \dots, n$, $i_k(t)$ and $u_k(t)$ denote currents and voltages in the rotors of the drives, R_k, l_k and C_k are resistances, inductances and capacitances in the rotors, respectively, while J_k and B_k correspond to inertia moments and viscous dampings of the drives, respectively.

Finally, to make the model more realistic, we need to add some *stochastic torques and forces*:

$$\mathcal{F}_i^{stoch} = B_{ij}[y^i(t), t]\, dW^j(t),$$

where $B_{ij}[y(t), t]$ represents continuous stochastic *diffusion fluctuations*, and $W^j(t)$ is an $N-$variable *Wiener process* (i.e., generalized Brownian motion) [II07a], with

$$dW^j(t) = W^j(t + dt) - W^j(t), \qquad (\text{for } \; j = 1, \ldots, n = \text{no. of active DOF}).$$

22.2.4 Summary

In this section we have proposed the time-dependent Hamiltonian form of human biomechanics. Starting with the Covariant Force Law: $F_i = m_{ij}a^j$ on the biomechanical configuration manifold M, we have first developed the autonomous Hamiltonian biomechanics:

$$\dot{x}^i = \frac{\partial H}{\partial p_i} + \frac{\partial R}{\partial p_i}, \qquad \dot{p}_i = F_i - \frac{\partial H}{\partial x^i} + \frac{\partial R}{\partial x^i},$$

on the symplectic phase space that is the cotangent bundle TM of M. Then we have introduced powerful geometrical machinery consisting of fibre bundles and jet manifolds associated to the biomechanical manifold M. Using the jet formalism, we derived time-dependent, dissipative, Hamiltonian equations:

$$\dot{y}^i = \partial^i \mathcal{H}, \qquad \dot{p}_i = F_i - \partial_i \mathcal{H},$$

together with the fitness evolution equation:

$$d_H f = (\partial_t + \partial^i \mathcal{H} \partial_i - \partial_i \mathcal{H} \partial^i)\, f.$$

for the general time-dependent human biomechanical system.

This is the time-dependent generalization of an 'ordinary' autonomous human biomechanics, in which total *mechanical + biochemical energy is not conserved.* In our view, this time-dependent energetic approach is much more realistic than the autonomous one, as in living systems energy is never conserved.

22.3 Cerebellum Motion Controller

Recall (see [II06a, II06b, II06c]) that realistic human biodynamics (RHB) is a science of human (and humanoid robot) motion in its full complexity. It is governed by both Newtonian dynamics and biological control laws.

There are over 200 bones in the human skeleton driven by about 640 muscular actuators (see, e.g. [Mar98, II06b]). While the muscles generate driving torques in the moving joints,[7] subcortical neural system performs both local and global (loco)motion control: first reflexly controlling contractions of individual muscles, and then orchestrating all the muscles into synergetic actions in order to produce efficient movements. While the local reflex control of individual muscles is performed on the *spinal control level*, the global integration of all the muscles into coordinated movements is performed within the *cerebellum* [II06a, II06b].

All hierarchical subcortical neuro–muscular physiology, from the bottom level of a single muscle fiber, to the top level of cerebellar muscular synergy, acts as a *temporal* $\langle \text{out}|\text{in} \rangle$ *reaction*, in such a way that the higher level acts as a command/control space for the lower level, itself representing an abstract image of the lower one:

1. At the *muscular level*, we have *excitation–contraction dynamics* [Hat77a, Hat78, Hat77b], in which $\langle \text{out}|\text{in} \rangle$ is given by the following sequence of nonlinear diffusion processes [II06a, II06b]:

$$\textit{neural action potential} \rightsquigarrow \textit{synaptic potential} \rightsquigarrow \textit{muscular action potential}$$
$$\rightsquigarrow \textit{excitation contraction coupling} \rightsquigarrow \textit{muscle tension generating}.$$

   Its purpose is the generation of muscular forces, to be transferred into driving torques within the joint anatomical geometry.
2. At the *spinal level*, $\langle \text{out}|\text{in} \rangle$ is given by *autogenetic–reflex stimulus–response control* [Hou79]. Here we have a neural image of all individual muscles. The main purpose of the spinal control level is to give both positive and negative feedbacks to stabilize generated muscular forces within the 'homeostatic' (or, more appropriately, 'homeokinetic') limits. The individual muscular actions are combined into flexor–extensor (or agonist–antagonist) pairs, mutually controlling each other. This is the mechanism of *reciprocal innervation of agonists and inhibition of antagonists*. It has a purely mechanical purpose to form the so–called *equivalent muscular actuators* (EMAs), which would generate driving torques $T_i(t)$ for all movable joints.
3. At the *cerebellar level*, $\langle \text{out}|\text{in} \rangle$ is given by *sensory–motor integration* [HBB96]. Here we have an abstracted image of all autogenetic reflexes. The main purpose of the cerebellar control level is integration and fine tuning of

[7] Here we need to emphasize that human joints are significantly more flexible than humanoid robot joints. Namely, each humanoid joint consists of a pair of coupled segments with only Eulerian rotational degrees of freedom. On the other hand, in each human synovial joint, besides gross Eulerian rotational movements (roll, pitch and yaw), we also have some hidden and restricted translations along (X, Y, Z)–axes. For example, in the knee joint, patella (knee cap) moves for about 7–10 cm from maximal extension to maximal flexion). It is well–known that even greater are translational amplitudes in the shoulder joint. In other words, within the realm of rigid body mechanics, a segment of a human arm or leg is not properly represented as a rigid body fixed at a certain point, but rather as a rigid body hanging on rope–like ligaments. More generally, the whole skeleton mechanically represents a system of flexibly coupled rigid bodies. This implies the more complex kinematics, dynamics and control then in the case of humanoid robots.

the action of all active EMAs into a synchronized movement, by *supervising* the individual autogenetic reflex circuits. At the same time, to be able to perform in new and unknown conditions, the cerebellum is continuously adapting its own neural circuitry by unsupervised (self–organizing) learning. Its action is subconscious and automatic, both in humans and in animals.

Naturally, we can ask the question: Can we assign a single $\langle \text{out}|\text{in} \rangle$ measure to all these neuro–muscular stimulus–response reactions? We think that we can do it; so in this Letter, we propose the concept of *adaptive sensory–motor transition amplitude* as a unique measure for this temporal $\langle \text{out}|\text{in} \rangle$ relation. Conceptually, this $\langle \text{out}|\text{in} \rangle -$*amplitude* can be formulated as the '*neural path integral*':

$$\langle \text{out}|\text{in} \rangle \equiv \underset{\text{amplitude}}{\langle \text{motor}|\text{sensory} \rangle} = \int \mathcal{D}[w,x]\, \mathrm{e}^{\mathrm{i}\, S[x]}. \tag{22.22}$$

Here, the integral is taken over all *activated* (or, 'fired') *neural pathways* $x^i = x^i(t)$ of the cerebellum, connecting its input *sensory*–state with its output *motor*–state, symbolically described by *adaptive neural measure* $\mathcal{D}[w,x]$, defined by the weighted product (of discrete time steps)

$$\mathcal{D}[w,x] = \lim_{n \longrightarrow \infty} \prod_{t=1}^{n} w_i(t)\, dx^i(t), \tag{22.23}$$

in which the *synaptic weights* $w_i = w_i(t)$, included in all active neural pathways $x^i = x^i(t)$, are updated by the standard learning rule

$$new\ value(t+1) \;=\; old\ value(t) \;+\; innovation(t).$$

More precisely, the weights w_i in (23.49) are updated according to one of the two standard neural learning schemes, in which the micro–time level is traversed in discrete steps, i.e., if $t = t_0, t_1, ..., t_n$ then $t+1 = t_1, t_2, ..., t_{n+1}$:[8]

1. A *self–organized, unsupervised* (e.g., Hebbian–like [Heb49]) learning rule:

$$w_i(t+1) = w_i(t) + \frac{\sigma}{\eta}(w_i^d(t) - w_i^a(t)), \tag{22.24}$$

   where $\sigma = \sigma(t)$, $\eta = \eta(t)$ denote *signal* and *noise*, respectively, while superscripts d and a denote *desired* and *achieved* micro–states, respectively; or

2. A certain form of a *supervised gradient descent learning*:

$$w_i(t+1) \;=\; w_i(t) - \eta \nabla J(t), \tag{22.25}$$

[8] Note that we could also use a reward–based, reinforcement learning rule (see [II07b] and references therein), in which system learns its optimal policy:

$$innovation(t) = |reward(t) - penalty(t)|.$$

where η is a small constant, called the *step size*, or the *learning rate*, and $\nabla J(n)$ denotes the gradient of the 'performance hyper–surface' at the t–th iteration.

Theoretically, equations (22.22–23.55) define an $\infty-$*dimensional neural network* [II08c]. Practically, in a computer simulation we can use $10^7 \leq n \leq 10^8$, roughly corresponding to the number of neurons in the cerebellum [II07b, II07d].

The exponent term $S[x]$ in equation (22.22) represents the *autogenetic–reflex action*, describing reflexly–induced motion of all active EMAs, from their initial $stimulus-$state to their final $response-$state, along the family of extremal (i.e., Euler–Lagrangian) paths $x^i_{\min}(t)$. ($S[x]$ is properly derived in (22.26–22.27) below.)

22.3.1 Cerebellum as an Adaptive Path–Integral Comparator

Having, thus, defined the spinal reflex control level, we proceed to model the top subcortical commander/controller, the *cerebellum*. The cerebellum is responsible for coordinating precisely timed $\langle \text{out}|\text{in} \rangle$ activity by integrating motor output with ongoing sensory feedback (see Figure 22.8). It receives extensive projections from sensory–motor areas of the cortex and the periphery and directs it back to premotor and motor cortex. This suggests a role in sensory–motor integration and the timing and execution of human movements. The cerebellum stores patterns of motor control for frequently performed movements, and therefore, its circuits are changed by experience and training [II06b]. It was termed the *adjustable pattern generator* in the work of James Houk and collaborators [HBB96]. Also, it has become the inspiring 'brain–model' in robotic research [II06a].

The cerebellum is known to be involved in the production and learning of smooth coordinated movements (see [II06b] and references therein). Two classes of inputs carry information into the cerebellum: the mossy fibers (MFs) and the climbing fibers (CFs). The MFs provide both plant state and contextual information. The CFs, on the other hand, are thought to provide information that reflect errors in recently generated movements. This information is used to adjust the programs encoded by the cerebellum. The MFs carry plant state, motor efference, and other contextual signals into the cerebellum. These fibers impinge on granule cells, whose axons give rise to parallel fibers (PFs). Through the combination of inputs from multiple classes of MFs and local inhibitory interneurons, the granule cells are thought to provide a sparse expansive encoding of the incoming state information. The large number of PFs converge on a much smaller set of Purkinje cells (PCs), while the PCs, in turn, provide inhibitory signals to a single cerebellar nuclear cell. Using this principle, the Cerebellar Model Arithmetic Computer, or CMAC–neural network has been built and implemented in robotics [II06b, II06a], using trial-and-error learning to produce bursts of muscular activity for controlling robot arms.

So, this 'cerebellar control' works for simple robotic problems, like non-redundant manipulation. However, comparing the number of its neurons (10^7 – 10^8), to the size of conventional neural networks (including CMAC), suggests

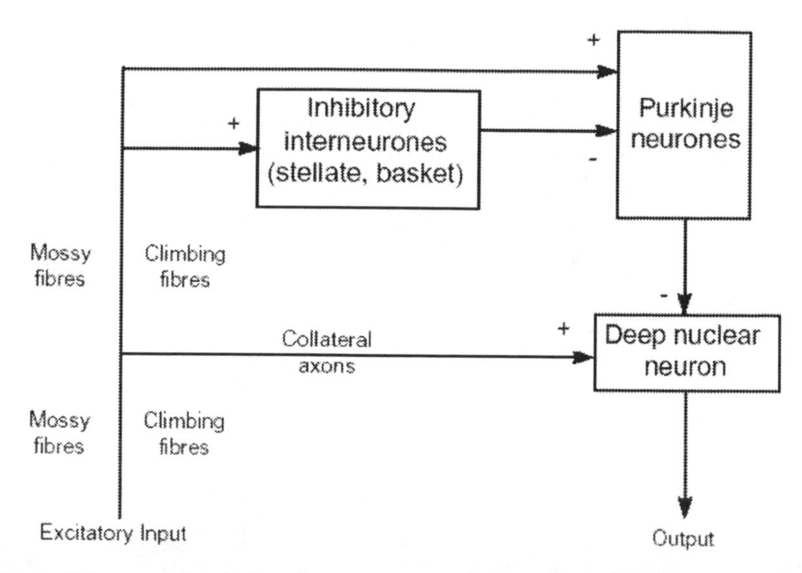

Fig. 22.8. Schematic ⟨out|in⟩ organization of the primary cerebellar circuit. In essence, excitatory inputs, conveyed by collateral axons of Mossy and Climbing fibers activate directly neurones in the Deep cerebellar nuclei. The activity of these latter is also modulated by the inhibitory action of the cerebellar cortex, mediated by the Purkinje cells.

that artificial neural nets *cannot* satisfactorily model the function of this sophisticated 'super–bio–computer', as its dimensionality is virtually infinite. Despite a lot of research dedicated to its structure and function (see [HBB96] and references there cited), the real nature of the cerebellum still remains a 'mystery'.

The main function of the cerebellum as a motor controller is depicted in Figure 22.9. A coordinated movement is easy to recognize, but we know little about how it is achieved. In search of the neural basis of coordination, a model of spinocerebellar interactions was presented in which the structural and functional

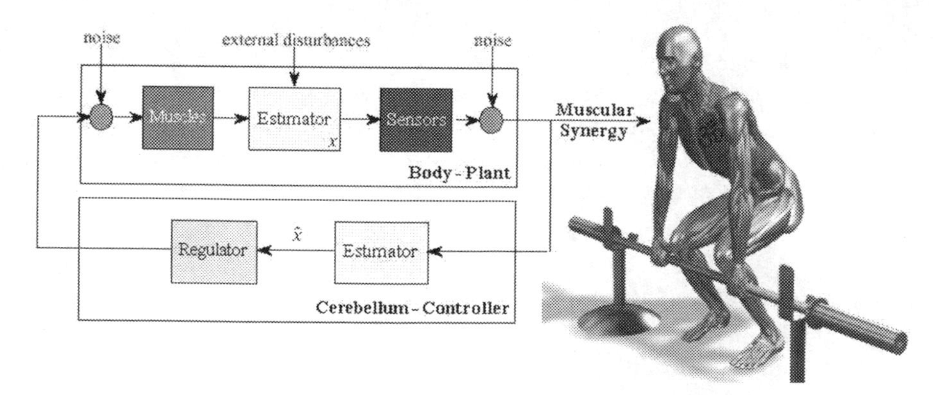

Fig. 22.9. The cerebellum as a motor controller.

organizing principle is a division of the cerebellum into discrete micro–complexes. Each micro–complex is the recipient of a specific motor error signal, that is, a signal that conveys information about an inappropriate movement. These signals are encoded by spinal reflex circuits and conveyed to the cerebellar cortex through climbing fibre afferents. This organization reveals salient features of cerebellar information processing, but also highlights the importance of systems level analysis for a fuller understanding of the neural mechanisms that underlie behavior.

Hamiltonian Action and Neural Path Integral

Here, we propose a *quantum–like adaptive control* approach to modeling the 'cerebellar mystery'. We define the *affine Hamiltonian control action*,

$$S_{aff}[q,p] = \int_{t_{in}}^{t_{out}} d\tau \left[p_i\dot{q}^i - H_{aff}(q,p)\right]. \tag{22.26}$$

From the affine Hamiltonian action (22.26) we further derive the associated expression for the *neural phase–space path integral* (in normal units), representing the *cerebellar sensory–motor amplitude* $\langle \text{out}|\text{in}\rangle$,

$$\begin{aligned} \langle q^i_{out}, p^{out}_i | q^i_{in}, p^{in}_i \rangle &= \int \mathcal{D}[w,q,p]\, \mathrm{e}^{\mathrm{i}\, S_{aff}[q,p]} \\ &= \int \mathcal{D}[w,q,p] \exp\left\{ \mathrm{i} \int_{t_{in}}^{t_{out}} d\tau \left[p_i\dot{q}^i - H_{aff}(q,p)\right]\right\}, \end{aligned} \tag{22.27}$$

$$\text{with} \qquad \int \mathcal{D}[w,q,p] = \int \prod_{\tau=1}^{n} \frac{w^i(\tau)dp_i(\tau)dq^i(\tau)}{2\pi},$$

where $w_i = w_i(t)$ denote the cerebellar synaptic weights positioned along its neural pathways, being continuously updated using the Hebbian–like self–organizing learning rule (22.24). Given the transition amplitude *out*|*in* (22.27), the *cerebellar sensory–motor transition probability* is defined as its absolute square, $|\langle \text{out}|\text{in}\rangle|^2$.

In the phase–space path integral (22.27), $q^i_{in} = q^i_{in}(t)$, $q^i_{out} = q^i_{out}(t)$; $p^{in}_i = p^{in}_i(t)$, $p^{out}_i = p^{out}_i(t)$; $t_{in} \le t \le t_{out}$, for all discrete time steps, $t = 1,...,n \longrightarrow \infty$, and we are allowing for the affine Hamiltonian $H_{aff}(q,p)$ to depend upon all the $(M \le N)$ EMA–angles and angular momenta collectively. Here, we actually systematically took a discretized differential time limit of the form $t_\sigma - t_{\sigma-1} \equiv d\tau$ (both σ and τ denote discrete time steps) and wrote $\frac{(q^i_\sigma - q^i_{\sigma-1})}{(t_\sigma - t_{\sigma-1})} \equiv \dot{q}^i$. For technical details regarding the path integral calculations on Riemannian and symplectic manifolds (including the standard regularization procedures).

Now, motor learning occurring in the cerebellum can be observed using functional MR imaging, showing changes in the cerebellar action potential, related to the motor tasks. To account for these electro–physiological currents, we need to add the *source* term $J_i(t)q^i(t)$ to the affine Hamiltonian action (22.26), (the current $J_i = J_i(t)$ acts as a source $J_i A^i$ of the *cerebellar electrical potential* $A^i = A^i(t)$),

$$S_{aff}[q,p,J] = \int_{t_{in}}^{t_{out}} d\tau \left[p_i \dot{q}^i - H_{aff}(q,p) + J_i q^i \right],$$

which, subsequently gives the cerebellar path integral with the action potential source, coming either from the motor cortex or from other subcortical areas.

Note that the standard *Wick rotation*: $t \mapsto t$ makes our path integral real [II06c, II08c], i.e.,

$$\int \mathcal{D}[w,q,p]\, \mathrm{e}^{\mathrm{i}\, S_{aff}[q,p]} \ \underrightarrow{Wick} \ \int \mathcal{D}[w,q,p]\, \mathrm{e}^{-\, S_{aff}[q,p]},$$

while their subsequent discretization gives the standard thermodynamic *partition function*,

$$Z = \sum_j {}^{-w_j E^j / T}, \tag{22.28}$$

where E^j is the energy eigenvalue corresponding to the affine Hamiltonian $H_{aff}(q,p)$, T is the temperature–like environmental control parameter, and the sum runs over all energy eigenstates (labelled by the index j). From (23.54), we can further calculate all statistical and thermodynamic system properties, as for example, *transition entropy* $S = k_B \ln Z$, etc.

Entropy and Motor Control

Our cerebellar path integral controller is closely related to *entropic motor control* [HN08a, HN08b], which deals with neuro-physiological feedback information and environmental uncertainty. The probabilistic nature of human motor action can be characterized by entropies at the level of the organism, task, and environment. Systematic changes in motor adaptation are characterized as task–organism and environment–organism tradeoffs in entropy. Such compensatory adaptations lead to a view of goal–directed motor control as the product of an underlying conservation of entropy across the task–organism–environment system. In particular, an experiment conducted in [HN08b] examined the changes in entropy of the coordination of isometric force output under different levels of task demands and feedback from the environment. The goal of the study was to examine the hypothesis that human motor adaptation can be characterized as a process of entropy conservation that is reflected in the compensation of entropy between the task, organism motor output, and environment. Information entropy of the coordination dynamics relative phase of the motor output was made conditional on the idealized situation of human movement, for which the goal was always achieved. Conditional entropy of the motor output decreased as the error tolerance and feedback frequency were decreased. Thus, as the likelihood of meeting the task demands was decreased increased task entropy and/or the amount of information from the environment is reduced increased environmental entropy, the subjects of this experiment employed fewer coordination patterns in the force output to achieve the goal. The conservation of entropy supports the view that context dependent adaptations in human goal–directed action are guided fundamentally by natural law and provides a novel means of examining human motor behavior. This is fundamentally related to the *Heisenberg uncertainty principle* [II08a] and further supports the argument for the primacy of a probabilistic approach toward the study of biodynamic cognition systems.

The action–amplitude formalism represents a kind of a generalization of the Haken-Kelso-Bunz (HKB) model of self-organization in the individual's motor system [HKB85], including: multi-stability, phase transitions and hysteresis effects, presenting a contrary view to the purely feedback driven systems. HKB uses the concepts of synergetics (order parameters, control parameters, instability, etc) and the mathematical tools of nonlinearly coupled (nonlinear) dynamical systems to account for self-organized behavior both at the cooperative, coordinative level and at the level of the individual coordinating elements. The HKB model stands as a building block upon which numerous extensions and elaborations have been constructed. In particular, it has been possible to derive it from a realistic model of the cortical sheet in which neural areas undergo a reorganization that is mediated by intra- and inter-cortical connections. Also, the HKB model describes phase transitions ('switches') in coordinated human movement as follows: (i) when the agent begins in the anti-phase mode and speed of movement is increased, a spontaneous switch to symmetrical, in-phase movement occurs; (ii) this transition happens swiftly at a certain critical frequency; (iii) after the switch has occurred and the movement rate is now decreased the subject remains in the symmetrical mode, i.e. she does not switch back; and (iv) no such transitions occur if the subject begins with symmetrical, in-phase movements. The HKB dynamics of the order parameter relative phase as is given by a nonlinear first-order ODE:

$$\dot{\phi} = (\alpha + 2\beta r^2) \sin \phi - \beta r^2 \sin 2\phi,$$

where ϕ is the phase relation (that characterizes the observed patterns of behavior, changes abruptly at the transition and is only weakly dependent on parameters outside the phase transition), r is the oscillator amplitude, while α, β are coupling parameters (from which the critical frequency where the phase transition occurs can be calculated).

23

Modern Tennis Psychology: Nonlinear and Quantum

23.1 Cognition

Today it is well known that disembodied cognition is a myth, albeit one that has had profound influence in Western science since Rene Descartes and others gave it credence during the Scientific Revolution. In fact, the mind-body separation had much more to do with explanation of method than with explanation of the mind and cognition, yet it is with respect to the latter that its impact is most widely felt. We find it to be an unsustainable assumption in the realm of crowd behavior. Mental intention is (almost immediately) followed by a physical action, that is, a human or animal movement [Sch07]. In animals, this physical action would be jumping, running, flying, swimming, biting or grabbing. In humans, it can be talking, walking, driving, or kicking, etc. Mathematical description of human/animal movement in terms of the corresponding neuro-musculo-skeletal equations of motion, for the purpose of prediction and control, is formulated within the realm of biodynamics (see [II06a, II06b, II06c]).

In this section we will focus on intelligence and problem solving.

23.1.1 Intelligence and Factor Analysis

Intelligence is a property of *human mind* that encompasses many related *mental abilities*, such as the capacities to *reason*, *plan*, solve problems, think abstractly, comprehend ideas and *language*, and learn. According to Encyclopedia Britannica, *intelligence* is the *ability to adapt effectively to the environment, either by making a change in oneself or by changing the environment or finding a new one.* Different investigators have emphasized different aspects of intelligence in their definitions. For example, in a 1921 symposium on the definition of intelligence, the American psychologist L. Terman emphasized the *ability to think abstractly*, while another American psychologist, E. Thorndike, emphasized *learning* and the ability to give good responses to questions.

T.T. Ivancevic et al.: Paradigm Shift for Future Tennis, COSMOS 12, pp. 259–355.
springerlink.com

Reason

In the philosophy of arguments, *reason* is the ability of the human mind to form and operate on concepts in abstraction, in varied accordance with rationality and logic —terms with which reason shares heritage. Reason is thus a very important word in Western intellectual history, to describe a type or aspect of mental thought which has traditionally been claimed as distinctly human, and not to be found elsewhere in the animal world. Discussion and debate about the nature, limits and causes of reason could almost be said to define the main lines of historical philosophical discussion and debate. Discussion about reason especially concerns:

(a) its relationship to several other related concepts: language, logic, consciousness etc,

(b) its ability to help people decide what is true, and

(c) its origin.

The concept of reason is connected to the concept of language, as reflected in the meanings of the Greek word 'logos', later to be translated by Latin 'ratio' and then French 'raison', from which the English word derived. As reason, rationality, and logic are all associated with the ability of the human mind to predict effects as based upon presumed causes, the word 'reason' also denotes a ground or basis for a particular argument, and hence is used synonymously with the word 'cause'.

It is sometimes said that the contrast between reason and logic extends back to the time of Plato and Aristotle. Indeed, although they had no separate Greek word for logic as opposed to language and reason, Aristotle's *syllogism* (Greek 'syllogismos') identified logic clearly for the first time as a distinct field of study: the most peculiarly reasonable ('logikê') part of reasoning, so to speak.

No philosopher of any note has ever argued that logic is the same as reason. They are generally thought to be distinct, although logic is one important aspect of reason. But the tendency to the preference for 'hard logic', or 'solid logic', in modern times has incorrectly led to the two terms occasionally being seen as essentially synonymous or perhaps more often logic is seen as the defining and pure form of reason.

However machines and animals can unconsciously perform logical operations, and many animals (including humans) can unconsciously, associate different perceptions as causes and effects and then make decisions or even plans. Therefore, to have any distinct meaning at all, 'reason' must be the type of thinking which links language, consciousness and logic, and at this time, only humans are known to combine these things.

However, note that reasoning is defined very differently depending on the context of the understanding of reason as a form of knowledge. The logical definition is the act of using reason to derive a conclusion from certain premises using a given methodology, and the two most commonly used explicit methods to reach a conclusion are deductive reasoning and inductive reasoning. However, within idealist philosophical contexts, reasoning is the mental process which informs our imagination, perceptions, thoughts, and feelings with whatever intelligibility these appear to contain; and thus links our experience with universal meaning. The specifics of the methods of reasoning are of interest to such disciplines as philosophy, logic, psychology, and artificial intelligence.

In deductive reasoning, given true premises, the conclusion must follow and it cannot be false. In this type of reasoning, the conclusion is inherent in the premises. Deductive reasoning therefore does not increase one's knowledge base and is said to be non–ampliative. Classic examples of deductive reasoning are found in such syllogisms as the following:

1. One must exist/live to perform the act of thinking.
2. I think.
3. Therefore, I am.

In inductive reasoning, on the other hand, when the premises are true, then the conclusion follows with some degree of *probability*. This method of reasoning is ampliative, as it gives more information than what was contained in the premises themselves. A classical example comes from David Hume:

1. The sun rose in the east every morning up until now.
2. Therefore the sun will also rise in the east tomorrow.

A third method of reasoning is called abductive reasoning, or inference to the best explanation. This method is more complex in its structure and can involve both inductive and deductive arguments. The main characteristic of abduction is that it is an attempt to favor one conclusion above others by either attempting to falsify alternative explanations, or showing the likelihood of the favored conclusion given a set of more or less disputable assumptions.

A fourth method of reasoning is analogy. Reasoning by analogy goes from a particular to another particular. The conclusion of an analogy is only plausible. Analogical reasoning is very frequent in common sense, science, philosophy and the humanities, but sometimes it is accepted only as an auxiliary method. A refined approach is *case–based reasoning*.

Plan

A *plan* represents a proposed or intended method of getting from one set of circumstances to another. They are often used to move from the present situation, towards the achievement of one or more objectives or goals.

Informal or ad–hoc plans are created by individual humans in all of their pursuits. Structured and formal plans, used by multiple people, are more likely to occur in projects, diplomacy, careers, economic development, military campaigns, combat, or in the conduct of other business.

It is common for less formal plans to be created as abstract ideas, and remain in that form as they are maintained and put to use. More formal plans as used for business and military purposes, while initially created with and as an abstract thought, are likely to be written down, drawn up or otherwise stored in a form that is accessible to multiple people across time and space. This allows more reliable collaboration in the execution of the plan.

The term planning implies the working out of sub–components in some degree of detail. Broader–brush enunciations of objectives may qualify as metaphorical road–maps.

Planning literally just means the creation of a plan; it can be as simple as making a list. It has acquired a technical meaning, however, to cover the area of government legislation and regulations related to the use of resources.

Problem Solving

The *problem solving* forms part of thinking. Considered the most complex of all intellectual functions, problem solving has been defined as higher–order cognitive process that requires the modulation and control of more routine or fundamental skills. It occurs if an organism or an artificial intelligence system does not know how to proceed from a given state to a desired goal state. It is part of the larger problem process that includes problem finding and problem shaping.

The nature of human problem solving has been studied by psychologists over the past hundred years. There are several methods of studying problem solving, including: *introspection*, *behaviorism*, computer simulation and experimental methods.

Beginning with the early experimental work of the Gestaltists in Germany. Various reasons account for the choice of simple novel tasks: they had clearly defined optimal solutions, they were solvable within a relatively short time frame, researchers could trace participants' problem–solving steps, and so on. The researchers made the underlying assumption, of course, that simple tasks such as the Tower of Hanoi captured the main properties of 'real world' problems, and that the cognitive processes underlying participants' attempts to solve simple problems were representative of the processes engaged in when solving 'real world' problems. Thus researchers used simple problems for reasons of convenience, and thought generalizations to more complex problems would become possible. Perhaps the best–known and most impressive example of this line of research remains the work by Newell and Simon [NS72].

See more on problem solving below.

Learning

Learning is the process of acquiring knowledge, skills, attitudes, or values, through study, experience, or teaching, that causes a change of behavior that is persistent, measurable, and specified or allows an individual to formulate a new mental construct or revise a prior mental construct (conceptual knowledge such as attitudes or values). It is a process that depends on experience and leads to long–term changes in behavior potential. Behavior potential describes the possible behavior of an individual (not actual behavior) in a given situation in order to achieve a goal. But potential is not enough; if individual learning is not periodically reinforced, it becomes shallower and shallower, and eventually will be lost in that individual.

Short term changes in behavior potential, such as fatigue, do not constitute learning. Some long–term changes in behavior potential result from aging and development, rather than learning.

Education is the conscious attempt to promote learning in others. The primary function of 'teaching' is to create a safe, viable, productive learning environment. Management of the total learning environment to promote, enhance and motivate learning is a *paradigm shift* from a focus on teaching to a focus on learning.

The stronger the stimulation for the brain, the deeper the impression that is left in the neuronal network. Therefore a repeated, very intensive experience perceived through all of the senses (audition, sight, smell) of an individual will remain longer and prevail over other experiences. The complex interactions of neurons that have formed a network in the brain determine the direction of flow of the micro–voltage electricity that flows through the brain when a person thinks. The characteristics of the neuronal network shaped by previous impressions is what we call the person's 'character'.

The most basic learning process is *imitation*, one's personal repetition of an observed process, such as a smile. Thus an imitation will take one's time (attention to the details), space (a location for learning), skills (or practice), and other resources (for example, a protected area). Through copying, most infants learn how to hunt (i.e., direct one's attention), feed and perform most basic tasks necessary for survival.

The so–called *Bloom's Taxonomy* divides the learning process into a six–level hierarchy, where knowledge is the lowest order of cognition and evaluation the highest [Blo80]:

1. Knowledge is the memory of previously–learnt materials such as facts, terms, basic concepts and answers.
2. Comprehension is the understanding of facts and ideas by organization, comparison, translation, interpretation, and description.
3. Application is the use of new knowledge to solve problems.
4. Analysis is the examination and division of information into parts by identifying motives or causes. A person can analyze by making inferences and finding evidence to support generalizations.
5. Synthesis is the compilation of information in a new way by combining elements into patterns or proposing alternative solutions.
6. Evaluation is the presentation and defense of opinions by making judgments about information, validity of ideas or quality of work based on: *attention*,[1]

[1] The cognitive process of selectively concentrating on one thing while ignoring other things.

habituation,[2] *sensory adaptation*,[3] *sensitization*,[4] *Pavlovian conditioning*,[5] *operant conditioning*,[6] *social learning*[7] and *communication*.[8]

Language

Recall that a language is a system of signals, such as voice sounds, gestures or written symbols that encode or decode information. Human spoken and written languages can be described as a system of symbols (sometimes known as lexemes) and the grammars (rules) by which the symbols are manipulated. The word 'language' is also used to refer to common properties of languages. Human languages are usually referred to as natural languages, and the science of studying them is *linguistics*, with Ferdinand de Saussure and Noam Chomsky as the most influential figures. Besides symbols, languages also often conform to a

[2] An example of non–associative learning in which there is a progressive diminution of behavioral response probability with repetition of a stimulus.

[3] A change over time in the responsiveness of the sensory system to a constant stimulus.

[4] An example of non–associative learning in which the progressive amplification of a response follows repeated administrations of a stimulus.

[5] Learning of conditioned behavior as being formed by pairing two stimuli to condition an animal into giving a certain response.

[6] Operant conditioning is distinguished from Pavlovian conditioning in that operant conditioning deals with the modification of voluntary behavior through the use of consequences, while Pavlovian conditioning deals with the conditioning of involuntary reflexive behavior so that it occurs under new antecedent conditions. There are four contexts of operant conditioning:

- (i) *Positive reinforcement* occurs when a behavior (response) is followed by a favorable stimulus (commonly seen as pleasant) that increases the frequency of that behavior. In the Skinner box experiment, a stimulus such as food or sugar solution can be delivered when the rat engages in a target behavior, such as pressing a lever.
  (ii) *Negative reinforcement* occurs when a behavior (response) is followed by the removal of an aversive stimulus (commonly seen as unpleasant) thereby increasing that behavior's frequency. In the Skinner box experiment, negative reinforcement can be a loud noise continuously sounding inside the rat's cage until it engages in the target behavior, such as pressing a lever, upon which the loud noise is removed.
  (iii) *Positive punishment* (also called 'Punishment by contingent stimulation') occurs when a behavior (response) is followed by an aversive stimulus, such as introducing a shock or loud noise, resulting in a decrease in that behavior.
  (iv) *Negative punishment* (also called 'Punishment by contingent withdrawal') occurs when a behavior (response) is followed by the removal of a favorable stimulus, such as taking away a child's toy following an undesired behavior, resulting in a decrease in that behavior.

[7] Earning that occurs as a function of observing, retaining and replicating behavior observed in others.

[8] He process of symbolic activity, sometimes via a language.

rough grammar, or system of rules, used to manipulate the symbols. While a set of symbols may be used for expression or communication, it is primitive and relatively unexpressive, because there are no clear or regular relationships between the symbols. Some of the areas of the human brain involved in language processing are: Broca's area, Wernicke's area, Supramarginal gyrus, Angular gyrus, Primary Auditory Cortex.

Mathematics and computer science use artificial entities called *formal languages* (including programming languages and markup languages, but also some that are far more theoretical in nature). These often take the form of character strings, produced by some combination of formal grammar and semantics of arbitrary complexity.

Abstraction

Recall that *abstraction* is the process of reducing the information content of a concept, typically in order to retain only information which is relevant for a particular purpose. For example, abstracting a leather soccer ball to a ball retains only the information on general ball attributes and behavior. Similarly, abstracting an emotional state to happiness reduces the amount of information conveyed about the emotional state.

Abstraction typically results in complexity reduction leading to a simpler conceptualization of a domain in order to facilitate processing or understanding of many specific scenarios in a generic way.

In philosophical terminology, abstraction is the thought process wherein ideas are distanced from objects.

Abstraction uses a strategy of simplification, wherein formerly concrete details are left ambiguous, vague, or undefined; thus effective communication about things in the abstract requires an intuitive or common experience between the communicator and the communication recipient.

Abstractions sometimes have ambiguous referents; for example, 'happiness' (when used as an abstraction) can refer to as many things as there are people and events or states of being which make them happy. Likewise, 'architecture' refers not only to the design of safe, functional buildings, but also to elements of creation and innovation which aim at elegant solutions to construction problems, to the use of space, and at its best, to the attempt to evoke an emotional response in the builders, owners, viewers and users of the building.

Abstraction in philosophy is the process (or, to some, the alleged process) in concept–formation of recognizing some set of common features in individuals, and on that basis forming a concept of that feature. The notion of abstraction is important to understanding some philosophical controversies surrounding empiricism and the problem of universals. It has also recently become popular in formal logic under predicate abstraction.

Some research into the human brain suggests that the left and right hemispheres differ in their handling of abstraction. One side handles collections of examples (e.g., examples of a tree) whereas the other handles the concept itself.

Abstraction in mathematics is the process of extracting the underlying essence of a mathematical concept, removing any dependence on real world objects with which it might originally have been connected, and generalizing it so that it has wider applications.

Many areas of mathematics began with the study of real world problems, before the underlying rules and concepts were identified and defined as abstract structures. For example, geometry has its origins in the calculation of distances and areas in the real world; statistics has its origins in the calculation of probabilities in gambling; and algebra started with methods of solving problems in arithmetic.

Abstraction is an ongoing process in mathematics and the historical development of many mathematical topics exhibits a progression from the concrete to the abstract. Take the historical development of geometry as an example; the first steps in the abstraction of geometry were made by the ancient Greeks, with Euclid being the first person (as far as we know) to document the axioms of plane geometry. In the 17th century Descartes introduced Cartesian coordinates which allowed the development of analytic geometry. Further steps in abstraction were taken by Lobachevsky, Bolyai and Gauss who generalized the concepts of geometry to develop non–Euclidean geometries. Later in the 19th century mathematicians generalised geometry even further, developing such areas as geometry in n dimensions, projective geometry, affine geometry, finite geometry and differential geometry. Finally Felix Klein's 'Erlangen program' identified the underlying theme of all of these geometries, defining each of them as the study of properties invariant under a given group of symmetries. This level of abstraction revealed deep connections between geometry and abstract algebra.

The advantages of abstraction are:

(i) It reveals deep connections between different areas of mathematics;

(ii) Known results in one area can suggest conjectures in a related area; and

(iii) Techniques and methods from one area can be applied to prove results in a related area.

An abstract structure is a formal object that is defined by a set of laws, properties, and relationships in a way that is logically if not always historically independent of the structure of contingent experiences, for example, those involving physical objects. Abstract structures are studied not only in logic and mathematics but in the fields that apply them, as computer science, and in the studies that reflect on them, as philosophy and especially the philosophy of mathematics. Indeed, modern mathematics has been defined in a very general sense as the study of abstract structures by the *Bourbaki group*.

The main disadvantage of abstraction is that highly abstract concepts are more difficult to learn, and require a degree of mathematical maturity and experience before they can be assimilated.

In computer science, abstraction is a mechanism and practice to reduce and factor out details so that one can focus on a few concepts at a time.

The concept is by analogy with abstraction in mathematics. The mathematical technique of abstraction begins with mathematical definitions; this has the fortunate effect of finessing some of the vexing philosophical issues of abstraction. For example, in both computing and in mathematics, numbers are concepts in the programming languages, as founded in mathematics. Implementation details depend on the hardware and software, but this is not a restriction because the computing concept of number is still based on the mathematical concept.

Roughly speaking, abstraction can be either that of control or data. Control abstraction is the abstraction of actions while data abstraction is that of data structures. For example, control abstraction in structured programming is the use of subprograms and formatted control flows. Data abstraction is to allow for handling data bits in meaningful manners. For example, it is the basic motivation behind data–type. Object–oriented programming can be seen as an attempt to abstract both data and code.

Creativity

Now, recall that *creativity* is a mental process involving the generation of new ideas or concepts, or new associations between existing ideas or concepts. From a scientific point of view, the products of creative thought (sometimes referred to as divergent thought) are usually considered to have both originality and appropriateness. An alternative, more everyday conception of creativity is that it is simply the act of making something new. Although intuitively a simple phenomenon, it is in fact quite complex. It has been studied from the perspectives of behavioral psychology, social psychology, psychometrics, cognitive science, artificial intelligence, philosophy, history, economics, design research, business, and management, among others. The studies have covered everyday creativity, exceptional creativity and even artificial creativity. Unlike many phenomena in science, there is no single, authoritative perspective or definition of creativity. Unlike many phenomena in psychology, there is no standardized measurement technique.

Creativity has been attributed variously to divine intervention, cognitive processes, the social environment, personality traits, and chance ('accident', 'serendipity'). It has been associated with genius, mental illness and humor. Some say it is a trait we are born with; others say it can be taught with the application of simple techniques. Although popularly associated with art and literature, it is also an essential part of innovation and invention and is important in professions such as business, economics, architecture, industrial design, science and engineering.

Despite, or perhaps because of, the ambiguity and multi–dimensional nature of creativity, entire industries have been spawned from the pursuit of creative ideas and the development of creativity techniques. This mysterious phenomenon, though undeniably important and constantly visible, seems to lie tantalizingly beyond the grasp of scientific investigation.

J. Guilford's group developed the so–called 'Torrance Tests of Creative Thinking'. They involved simple tests of divergent thinking and other problem–solving skills, which were scored on [Gui67]:

1. Fluency: the total number of interpretable, meaningful, and relevant ideas generated in response to the stimulus;
2. Flexibility: the number of different categories of relevant responses;
3. Originality: the statistical rarity of the responses among the test subjects; and
4. Elaboration: the amount of detail in the responses.

Personality

On the other hand, *personality* is a *collection of emotional, thought and behavioral patterns unique to a person* that is consistent over time. Personality psychology is a branch of psychology which studies personality and individual different processes – that which makes us into a person. One emphasis is on trying to create a coherent picture of a person and all his or her major psychological processes. Another emphasis views it as the study of individual differences. These two views work together in practice. Personality psychologists are interested in broad view of the individual. This often leads to an interest in the most salient individual differences among people.

The word *personality* originates from the Latin *persona*, which means 'mask'. In the History of theater of the ancient Latin world, the mask was not used as a plot device to disguise the identity of a character, but rather was a convention employed to represent, or typify that character.

Gordon Allport delineated different kinds of traits, which he also called dispositions. Central traits are basic to an individual's personality, while secondary traits are more peripheral. Common traits are those recognized within a culture and thus may vary from culture to culture. Cardinal traits are those by which an individual may be strongly recognized.

Raymond Cattell's research propagated a two–tiered personality structure with sixteen 'primary factors' (16 Personality Factors) and five 'secondary factors' (see Table 23.1). Cattell referred to these 16 factors as *primary factors*, as opposed to the so–called 'Big Five' factors which he considered *global factors.* All of the primary factors correlate with global factors and could therefore be considered subfactors within them.

A different model was proposed by Hans Eysenck, who believed that just three traits: *extroversion*, *neuroticism* and *psychoticism* – were sufficient to describe human personality. Eysenck was one of the first psychologists to study personality with the method of *factor analysis*, a statistical technique introduced by Charles Spearman and expanded by Raymond Cattell. Eysenck's results suggested two main personality factors [Eys92a, Eys92b]. The first factor was the tendency to experience negative emotions, and Eysenck referred to it as 'neuroticism'. The second factor was the tendency enjoy positive events, especially social events, and Eysenck named it 'extraversion'. The two personality dimensions were described in his 1947 book 'Dimensions of Personality'. It is common practice in personality psychology to refer to the dimensions by the first letters, E and N. E and N provided a 2–dimensional space to describe individual differences in behavior. An analogy can be made to how latitude and longitude

Table 23.1. Cattell's 16 Personality Factors

Descriptors of Low Range	Primary Factor	Descriptors of High Range
Impersonal, distant, cool, reserved, detached, formal, aloof (Sizothymia)	Warmth	Warm, outgoing, attentive to others, kindly, easy going, participating, likes people (Affectothymia)
Concrete thinking, lower general mental capacity, less intelligent, unable to handle abstract problems (Lower Scholastic Mental Capacity)	Reasoning	Abstract–thinking, more intelligent, bright, higher general mental capacity, fast learner (Higher Scholastic Mental Capacity)
Reactive emotionally, changeable, affected by feelings, emotionally less stable, easily upset (Lower Ego Strength)	Emotional Stability	Emotionally stable, adaptive, mature, faces reality calm (Higher Ego Strength)
Deferential, cooperative, avoids conflict, submissive, humble, obedient, easily led, docile, accommodating (Submissiveness)	Dominance	Dominant, forceful, assertive, aggressive, competitive, stubborn, bossy (Dominance)
Serious, restrained, prudent, taciturn, introspective, silent (Desurgency)	Liveliness	Lively, animated, spontaneous, enthusiastic, happy go lucky, cheerful, expressive, impulsive (Surgency)
Expedient, nonconforming, disregards rules, self indulgent (Low Super Ego Strength)	Rule–Consciousness	Rule–conscious, dutiful, conscientious, conforming, moralistic, staid, rule bound (High Super Ego Strength)
Shy, threat–sensitive, timid, hesitant, intimidated (Threctia)	Social Boldness	Socially bold, venturesome, thick skinned, uninhibited (Parmia)
Utilitarian, objective, unsentimental, tough minded, self–reliant, no–nonsense, rough (Harria)	Sensitivity	Sensitive, aesthetic, sentimental, tender minded, intuitive, refined (Premsia)
Trusting, unsuspecting, accepting, unconditional, easy (Alaxia)	Vigilance	Vigilant, suspicious, skeptical, distrustful, oppositional (Protension)
Grounded, practical, prosaic, solution oriented, steady, conventional (Praxernia)	Abstractedness	Abstract, imaginative, absent minded, impractical, absorbed in ideas (Autia)
Forthright, genuine, artless, open, guileless, naive, unpretentious, involved (Artlessness)	Privateness	Private, discreet, nondisclosing, shrewd, polished, worldly, astute, diplomatic (Shrewdness)
Self–Assured, unworried, complacent, secure, free of guilt, confident, self satisfied (Untroubled)	Apprehension	Apprehensive, self doubting, worried, guilt prone, insecure, worrying, self blaming (Guilt Proneness)
Traditional, attached to familiar, conservative, respecting traditional ideas (Conservatism)	Openness to Change	Open to change, experimental, liberal, analytical, critical, free thinking, flexibility (Radicalism)
Group–oriented, affiliative, a joiner and follower dependent (Group Adherence)	Self–Reliance	Self–reliant, solitary, resourceful, individualistic, self sufficient (Self-Sufficiency)
Tolerated disorder, unexacting, flexible, undisciplined, lax, self-conflict, impulsive, careless of social rules, uncontrolled (Low Integration)	Perfectionism	Perfectionistic, organized, compulsive, self-disciplined, socially precise, exacting will power, control, self-sentimental (High Self–Concept Control)
Relaxed, placid, tranquil, torpid, patient, composed low drive (Low Ergic Tension)	Tension	Tense, high energy, impatient, driven, frustrated, over wrought, time driven. (High Ergic Tension)

describe a point on the face of the earth. Also, Eysenck noted how these two dimensions were similar to the four personality types first proposed by the ancient Greek physician Galen:

1. High N and High E = Choleric type;
2. High N and Low E = Melancholic type;
3. Low N and High E = Sanguine type; and
4. Low N and Low E = Phlegmatic type.

The third dimension, 'psychoticism', was added to the model in the late 1970s, based upon collaborations between Eysenck and his wife, Sybil B.G. Eysenck, the current editor of Personality and Individual Differences (see [Eys69, Eys76]).

The major strength of Eysenck's model was to provide detailed theory of the causes of personality (see his 1985 book 'Decline and Fall of the Freudian Empire'). For example, Eysenck proposed that extraversion was caused by variability in cortical arousal; "introverts are characterized by higher levels of activity than extraverts and so are chronically more cortically aroused than extraverts'. While it seems counterintuitive to suppose that introverts are more aroused than extraverts, the putative effect this has on behavior is such that the introvert seeks lower levels of stimulation. Conversely, the extravert seeks to heighten their arousal to a more optimal level (as predicted by the *Yerkes–Dodson Law*) by increased activity, social engagement and other stimulation–seeking behaviours.

Differences between Cattell and Eysenck emerged due to preferences for different forms of factor analysis, with Cattell using oblique, Eysenck orthogonal, rotation to analyze the factors that emerged when personality questionnaires were subject to statistical analysis. Today, the Big Five factors have the weight of a considerable amount of empirical research behind them. Building on the work of Cattell and others, Lewis Goldberg proposed a five–dimensional personality model, nicknamed the 'Big Five' personality traits:

Extroversion (i.e., 'extroversion vs. introversion' above; outgoing and physical–stimulation–oriented vs. quiet and physical–stimulation–averse);

1. Neuroticism (i.e., emotional stability; calm, unperturbable, optimistic vs. emotionally reactive, prone to negative emotions);
2. Agreeableness (i.e., affable, friendly, conciliatory vs. aggression aggressive, dominant, disagreeable);
3. Conscientiousness (i.e., dutiful, planful, and orderly vs. spontaneous, flexible, and unreliable); and
4. Openness to experience (i.e., open to new ideas and change vs. traditional and staid).

Character

A *character structure* is a system of relatively permanent motivational and other traits that are manifested in the characteristic ways that an individual relates to others and reacts to various kinds of challenges. The word 'structure' indicates that these several characteristics and/or learned patterns of behavior are linked in such a way as to produce a state that can be highly resistant to change. The idea has its roots in the work of Sigmund Freud and several of his followers, the most important of whom (in this respect) is Erich Fromm.

Among the earliest factors that determine an individual's eventual character structure are his or her genetic characteristics and early childhood nurture and education. A child who is well nurtured and taught in a relatively benign and consistent environment by loving adults who intend that the child should learn how to make objective appraisals regarding the environment will be likely to form a normal or productive character structure. On the other hand, a child whose nurture and/or education are not ideal, living in a treacherous environment and interacting with adults who do not take the long–term interests of the child to heart will be more likely to form a pattern of behavior that suits the child to avoid the challenges put forth by a malign social environment. The means that the child invents to make the best of a hostile environment. Although this may serve the child well while in that bad environment, it may also cause the child to react in inappropriate ways, ways damaging to his or her own interests, when interacting with people in a more ideal social context. Major trauma that occurs later in life, even in adulthood, can sometimes have a profound effect. However, character may also develop in a positive way according to how the individual meets the psychosocial challenges of the life cycle (Erikson).

Freud's first paper on character described the anal character consisting of stubbornness, stinginess and extreme neatness. He saw this as a reaction formation to the child's having to give up pleasure in anal eroticism.The positive version of this character is the conscientious, inner directed obsessive. Freud also described the erotic character as both loving and dependent. And the narcissistic character as the natural leader, aggressive and independent because of not internalizing a strong super ego.

For Erich Fromm, character develops as the way in which an individual structures modes of assimilation and relatedness. The character types are almost identical to Freud's but Fromm gives them different names, receptive, hoarding, exploitative. Fromm adds the marketing type as the person who continually adapts the self to succeed in the new service economy. For Fromm, character types can be productive or unproductive. Fromm notes that character structures develop in each individual to enable him or her to interact successfully within a given society, to adapt to its mode of production and social norms may be very counter–productive when used in a different society.

Wisdom

On the other hand, *wisdom* is the ability, developed through experience, insight and reflection, to discern truth and exercise good judgment. It is sometimes conceptualized as an especially well developed form of common sense. Most psychologists regard wisdom as distinct from the cognitive abilities measured by standardized intelligence tests. Wisdom is often considered to be a trait that can be developed by experience, but not taught. When applied to practical matters, the term wisdom is synonymous with prudence. Some see wisdom as a quality that even a child, otherwise immature, may possess independent of experience or complete knowledge. The status of wisdom or prudence as a virtue is recognized in cultural, philosophical and religious sources. Some define wisdom in a utilitarian sense, as foreseeing consequences and acting to maximize the long–term common good.

A standard philosophical definition says that wisdom consists of making the best use of available knowledge. As with all decisions, a wise decision may be made with incomplete information. The technical philosophical term for the opposite of wisdom is folly. For example, in his *Metaphysics*, Aristotle defines wisdom as knowledge of causes: why things exist in a particular fashion.

Beyond the simple expedient of experience (which may be considered the most difficult way to gain wisdom as through the 'school of hard knocks'), there are a variety of other avenues to gaining wisdom which vary according to different philosophies. For example, the so–called *freethinkers*believe that wisdom may come from pure reason and perhaps experience. Recall that *freethought* is a philosophical doctrine that holds that beliefs should be formed on the basis of science and logical principles and not be comprised by authority, tradition or any other dogmatic or otherwise fallacious belief system that restricts logical reasoning. The cognitive application of freethought is known as *freethinking*, and practitioners of freethought are known as freethinkers. Freethought holds that individuals should neither accept nor reject ideas proposed as truth without recourse to knowledge and reason. Thus, freethinkers strive to build their beliefs on the basis of facts, scientific inquiry, and logical principles, independent of the factual/logical fallacies and intellectually-limiting effects of authority, cognitive bias, conventional wisdom, popular culture, prejudice, sectarianism, tradition, urban legend and all other dogmatic or otherwise fallacious principles. When applied to religion, the philosophy of freethought holds that, given presently–known facts, established scientific theories, and logical principles, there is insufficient evidence to support the existence of supernatural phenomena. A line from 'Clifford's Credo' by the 19th Century British mathematician and philosopher William Clifford perhaps best describes the premise of freethought: "It is wrong always, everywhere, and for anyone, to believe anything upon insufficient evidence." Since many popular beliefs are based on dogmas, freethinkers' opinions are often at odds with commonly–established views.

On the other hand, there is also a common belief that wisdom comes from *intuition* or, 'superlogic', as it is called by Tony Buzan, inventor of *mind maps*. For example, *holists* believe that wise people sense, work with and align themselves and others to life. In this view, wise people help others appreciate the fundamental interconnectedness of life. Also, some religions hold that wisdom may be given as a gift from God. For example, *Buddha* taught that a wise person is endowed with good bodily conduct, good verbal conduct and good mental conduct and a wise person does actions that are unpleasant to do but give good results and doesn't do actions that are pleasant to do but give bad results; this is called *karma*. According to *Hindu scriptures*, spiritual wisdom – *jnana* alone can lead to liberation. *Confucius* stated that wisdom can be learned by three methods: (i) *reflection* (the noblest), (ii) *imitation* (the easiest) and (iii) *experience* (the bitterest).

Psychometric Definition of Intelligence and Its Criticisms

Despite the variety of concepts of intelligence, the most influential approach to understanding intelligence (i.e., with the most supporters and the most published research over the longest period of time) is based on *psychometric testing*, which regards intelligence as *cognitive ability*.

Recall that *psychometrics* is the field of study concerned with the theory and technique of psychological measurement, which includes the measurement of knowledge, abilities, attitudes, and personality traits. The field is primarily concerned with the study of differences between individuals. It involves two major research tasks, namely:

(i) the construction of instruments and procedures for measurement; and

(ii) the development and refinement of theoretical approaches to measurement.

Much of the early theoretical and applied work in psychometrics was undertaken in an attempt to measure intelligence.

The origin of psychometrics has connections to the related field of psychophysics. Charles Spearman, a pioneer in psychometrics who developed approaches to the measurement of intelligence, studied under Wilhelm Wundt and was trained in psychophysics. The psychometrician Louis Thurstone later developed and applied a theoretical approach to the measurement referred to as the law of comparative judgment, an approach which has close connections to the psychophysical theory developed by Ernst Weber and Gustav Fechner (see below). In addition, Spearman and Thurstone both made important contributions to the theory and application of factor analysis, a statistical method that has been used extensively in psychometrics. More recently, psychometric theory has been applied in the measurement of personality, attitudes and beliefs, academic achievement, and in health–related fields. Measurement of these unobservable phenomena is difficult, and much of the research and accumulated art in this discipline has been developed in an attempt to properly define and quantify such phenomena. Critics, including practitioners in the physical sciences and social activists, have argued that such definition and quantification is impossibly difficult, and that such measurements are often misused. Proponents of psychometric techniques can reply, though, that their critics often misuse data by not applying psychometric criteria, and also that various quantitative phenomena in the physical sciences, such as heat and forces, cannot be observed directly but must be inferred from their manifestations. Figures who made significant contributions to psychometrics include Karl Pearson, Louis Thurstone, Georg Rasch and Arthur Jensen.

Intelligence, narrowly defined by psychometrics, can be measured by intelligence tests, also called *intelligence quotient* (IQ)[9] tests. Such intelligence tests take many forms, but the common tests (*Stanford–Binet*,[10] *Raven's Progressive Matrices*,[11]

---

[9] An intelligence quotient or IQ is a score derived from a set of standardized tests of intelligence. Intelligence tests come in many forms, and some tests use a single type of item or question. Most tests yield both an overall score and individual sub–tests scores. Regardless of design, all IQ tests measure the same general intelligence. Component tests are generally designed and chosen because they are found to be predictable of later intellectual development, such as educational achievement. IQ also correlates with job performance, socioeconomic advancement, and 'social pathologies'. Recent work has demonstrated links between IQ and health, longevity, and functional literacy. However, IQ tests do not measure all meanings of 'intelligence', such as creativity. IQ scores are relative (like placement in a race), not absolute (like the measurement of a ruler). The average IQ scores for many populations were rising during the 20th century: a phenomenon called the *Flynn effect*. It is not known whether these changes in scores reflect real changes in intellectual abilities. On average, IQ scores are stable over a person's lifetime, but some individuals undergo large changes. For example, scores can be affected by the presence of learning disabilities.

[10] The modern field of intelligence testing began with the Stanford-Binet IQ test. The Stanford-Binet itself started with the French psychologist Alfred Binet who was charged by the French government with developing a method of identifying intellectually deficient children for placement in special education programs. As Binet indicated, case studies may be more detailed and at times more helpful, but the time required to test large numbers of people would be huge. Unfortunately, the tests he and his assistant Victor Henri developed in 1896 were largely disappointing.

[11] Raven's Progressive Matrices are widely used non–verbal intelligence tests. In each test item, one is asked to find the missing part required to complete a pattern. Each Set of items gets progressively harder, requiring greater cognitive capacity to encode and analyze. The test is considered by many intelligence experts to be one of the most **g**–loaded in existence. The matrices are offered in three different forms for different ability levels, and for age ranges from five through adult: (i) Colored Progressive Matrices (younger children and special groups); (ii) Standard Progressive Matrices (average 6 to 80 year olds); and (iii) Advanced Progressive Matrices (above average adolescents and adults). According to their author, Raven's Progressive Matrices and Vocabulary tests measure the two main components of general intelligence (originally identified by Spearman): the ability to think clearly and make sense of complexity, which is known as eductive ability (from the Latin root 'educere', meaning 'to draw out'; and the ability to store and reproduce information, known as reproductive ability. Adequate standardization, ease of use (without written or complex instructions), and minimal cost per person tested are the main reasons for its widespread international use in most countries of the world. It appears to measure a type of *reasoning ability* which is fundamental to making sense out of the 'booming buzzing confusion' in all walks of life. Thus, it has among the highest predictive validities of any test in most occupational groups and, even more importantly, in predicting social mobility... the level of job a person will attain and retain. Although it is sometimes criticized for being costly, this is based on a failure to calculate cost per person tested with re–usable test booklets that can be used up to 50 times each. The authors of the Manual recommend that, when used in selection, RPM scores are set in the context of information relating to Raven's framework for the assessment of Competence. Some of the most fundamental research

Wechsler Adult Intelligence Scale,[12] *Wechsler–Bellevue I*,[13] and others) all measure the same dominant form of intelligence, **g** or 'general intelligence factor'. The abstraction of **g** stems from the observation that scores on all forms of cognitive tests *positively correlate* with one another. **g** can be derived as the *principal intelligence factor* from *cognitive test scores* using the *multivariate correlation statistical method* of *factor analysis* (FA).

Correlation and Factor Analysis

Recall that *correlation*, also called *correlation coefficient*, indicates the strength and direction of a linear relationship between two random variables (see Figure 23.1). In other words, correlation is a measure of the relation between two or more statistical variables. In general statistical usage, correlation (or, co–rrelation) refers to the departure of two variables from independence, although correlation does not imply their *functional causal relation*. In this broad sense there are several coefficients, measuring the degree of correlation, adapted to the nature of data. A number of different coefficients are used for different situations. Correlation coefficients can range from −1.00 to +1.00. The value

in cognitive psychology has been carried out with the RPM. The tests have been shown to work–scale–measure the same thing – in a vast variety of cultural groups. There is no truth in the assertion that the low mean scores obtained in some groups arise from a general lack of familiarity with the way of thought measured by the test. Two remarkable, and relatively recent, findings are that, on the one hand, the actual scores obtained by people living in most countries with a tradition of literacy – from China, Russia, and India through Europe to Kuwait – are very similar at any point in time. On the other hand, in all countries, the scores have increased dramatically over time... such that 50% of our grandparents would be assigned to special education classes if they were judged against today's norms. Yet none of the common explanations (e.g., access to television, changes in education, changes in family size etc.) hold up. The explanation seems to have more in common with those put forward to explain the parallel increase in life expectancy... which has doubled over the same period of time.

[12] Wechsler Adult Intelligence Scale or WAIS is a general IQ test, published in February 1955 as a revision of the Wechsler–Bellevue test (1939), standardized for use with adults over the age of 16. In this test intelligence is quantified as the global capacity of the individual to act purposefully, to think rationally, and to deal effectively with the environment.

[13] David Wechsler (January 12, 1896, Lespedi, Romania – May 2, 1981, New York, New York) was a leading Romanian–American psychologist. He developed well–known intelligence scales, such as the Wechsler Adult Intelligence Scale (WAIS) and the Wechsler Intelligence Scale for Children (WISC). The Wechsler Adult Intelligence Scale (WAIS) was developed first in 1939 and then called the Wechsler–Bellevue Intelligence Test. From these he derived the Wechsler Intelligence Scale for Children (WISC) in 1949 and the Wechsler Preschool and Primary Scale of Intelligence (WPPSI) in 1967. Wechsler originally created these tests to find out more about his patients at the Bellevue clinic and he found the then–current *Binet IQ test* unsatisfactory. The tests are still based on his philosophy that intelligence is "the global capacity to act purposefully, to think rationally, and to deal effectively with (one's) environment."

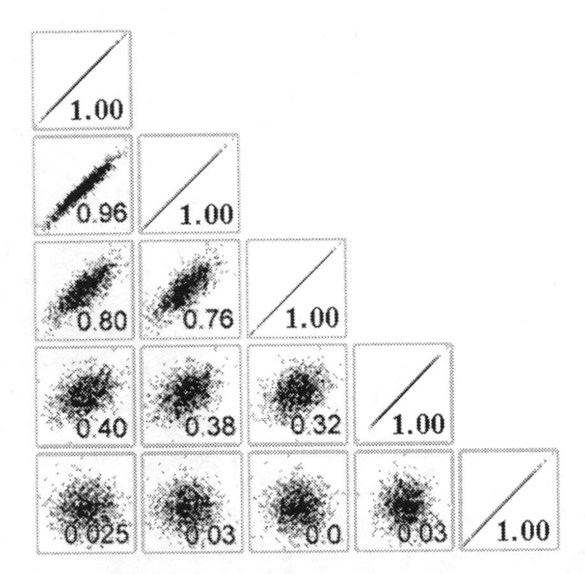

Fig. 23.1. Example of positive linear correlations between 1000 pairs of numbers. Note that each set of points correlates maximally with itself, as shown on the diagonal. Also, note that we have not plot the upper part of the correlation matrix as it is symmetrical.

of -1.00 represents a perfect negative correlation while a value of $+1.00$ represents a perfect positive correlation. The perfect correlation indicates an existence of functional relation between two statistical variables. A value of 0.00 represents a lack of correlation.Geometrically, the correlation coefficient can also be viewed as the cosine of the angle between the two vectors of samples drawn from the two random variables.

The most widely–used type of correlation simple linear coefficient is Pearson r, also called *linear* or *product–moment* correlation, which assumes that the two variables are measured on at least interval scales, and it determines the extent to which values of the two variables are 'proportional' to each other. The value of correlation coefficient does not depend on the specific measurement units used. Proportional means linearly related using regression line or least squares line. If the correlation coefficient is squared, then the resulting value (r^2, the *coefficient of determination*) will represent the proportion of common variation in the two variables (i.e., the 'strength' or 'magnitude' of the relationship). In order to evaluate the correlation between variables, it is important to know this 'magnitude' or 'strength' as well as the significance of the correlation.

The significance level calculated for each correlation is a primary source of information about the reliability of the correlation. The significance of correlation coefficient of particular magnitude will change depending on the size of the sample from which it was computed. The test of significance is based on the assumption that each of the two variables is normally distributed and that their bivariate ('combined') distribution is normal (which can be tested by examining the 3D bivariate distribution histogram). However, *Monte–Carlo* studies suggest that meeting those assumptions (especially the second one) is not absolutely crucial if our sample size is not very small and when the departure from normality is not very large. It is impossible to formulate precise recommendations based on those Monte–Carlo results, but many researchers follow a rule of thumb that if our sample size is 50 or more then serious biases are unlikely, and if our sample size is over 100 then you should not be concerned at all with the normality assumptions.

Recall that the *normal distribution*, also called *Gaussian distribution*, is an extremely important probability distribution in many fields. It is a family of distributions of the same general form, differing in their location and scale parameters: the *mean* ('average') μ and *standard deviation* ('variability') σ, respectively. The *standard normal distribution* is the normal distribution with a mean of zero and a standard deviation of one. It is often called the *bell curve* because the graph of its *probability density function pdf*, given by the *Gaussian function*

$$pdf = \frac{1}{\sigma\sqrt{2\pi}} \exp\left(-\frac{(x-\mu)^2}{2\sigma^2}\right),$$

resembles a bell shape (here, $\frac{1}{\sqrt{2\pi}}e^{-x^2/2}$ is the *pdf* for the standard normal distribution). The corresponding *cumulative distribution function cdf* is defined as the probability that a variable X has a value less than or equal to x, and it is expressed in terms of the *pdf* as

$$cdf = \frac{1}{\sigma\sqrt{2\pi}} \int_{-\infty}^{x} \exp\left(-\frac{(u-\mu)^2}{2\sigma^2}\right) du.$$

Now, the correlation $r_{X,Y}$ between two *normally distributed random variables* X and Y with expected values μX and μY and standard deviations σX and σY is defined as:

$$r_{XY} = \frac{\mathrm{cov}(X,Y)}{\sigma_X \sigma_Y} = \frac{E((X-\mu_X)(Y-\mu_Y))}{\sigma_X \sigma_Y},$$

where E denotes the expected value of the variable and cov means covariance. Since $\mu X = E(X)$, $\sigma_X^2 = E(X^2) - E^2(X)$ and similarly for Y, we can write (see, e.g., [CCW03])

$$r_{XY} = \frac{E(XY) - E(X)E(Y)}{\sqrt{E(X^2) - E^2(X)}\,\sqrt{E(Y^2) - E^2(Y)}}.$$

Assume that we have a data matrix $\mathbf{X} = \{x_{i\alpha}\}$ formed out of the *sample* $\{\mathbf{x_i}\}$ of n normally distributed simulator tests called observable–vectors or *manifest variables*, defined on the sample $\{\alpha = 1, \ldots, N\}$ of pilot (for the statistical significance the practical user's criterion is $N \geq 5n$). The *maximum likelihood estimator* of the Pearson correlation coefficient r_{ik} between any two manifest variables $\mathbf{x_i}$ and $\mathbf{x_k}$ is defined as[14]

[14] A time–dependent generalization $C_{\alpha\beta} = C_{\alpha\beta}(t)$ of the correlation coefficient r_{XY} is the *correlation function*, defined as follows. For the two time–series, $x_\alpha(t_i)$ and

$$r_{ik}=\frac{\sum_{\alpha=1}^{N}(x_{i\alpha}-\boldsymbol{\mu}_i)(x_{k\alpha}-\boldsymbol{\mu}_k)}{\sqrt{\sum_{\alpha=1}^{N}(x_{i\alpha}-\boldsymbol{\mu}_i)^2}\sqrt{\sum_{\alpha=1}^{N}(x_{k\alpha}-\boldsymbol{\mu}_k)^2}},$$

where

$$\mu_i=\frac{1}{N}\sum_{\alpha=1}^{N}x_{i\alpha}$$

is the arithmetic mean of the variable $\mathbf{x_i}$.[15] Correlation matrix $\mathbf{R}$ is the matrix $\mathbf{R}\equiv\mathbf{R}_{ik}=\{r_{ik}\}$ including $n\times n$ Pearson correlation coefficients r_{ik} calculated between n manifest variables $\{\mathbf{x_i}\}$. Therefore, $\mathbf{R}$ is symmetrical matrix with ones on the main diagonal. The correlation matrix $\mathbf{R}$ represents the total variability of all included manifest variables. In other words it stores all information about

---

$x_\beta(t_i)$ of the same length ($i=1,...,T$), one defines the correlation function by

$$C_{\alpha\beta}=\frac{\sum_i(x_\alpha(t_i)-\bar{x}_\alpha)(x_\beta(t_i)-\bar{x}_\beta)}{\sqrt{\sum_i(x_\alpha(t_i)-\bar{x}_\alpha)^2\sum_j(x_\beta(t_j)-\bar{x}_\beta)^2}},$$

where $\bar{x}$ denotes a time average over the period studied. For two sets of N time–series $x_\alpha(t_i)$ each ($\alpha,\beta=1,...,N$) all combinations of the elements $C_{\alpha\beta}$ can be used as entries of the $N\times N$ correlation matrix $\mathbf{C}$. By diagonalizing $\mathbf{C}$, i.e., solving the eigenvalue problem:

$$\mathbf{C}\mathbf{v}^k=\lambda_k\mathbf{v}^k,$$

one gets the eigenvalues λ_k ($k=1,...,N$) and the corresponding eigenvectors $\mathbf{v}^k=\{v_\alpha^k\}$.

[15] The following algorithm (in pseudocode) estimates bivariate correlation coefficient with good numerical stability:

```
Begin
  sum_sq_x = 0;
  sum_sq_y = 0;
  sum_coproduct = 0;
  mean_x = x[1];
  mean_y = y[1];
  for i in 2 to N:
    sweep = (i - 1.0) / i;
    delta_x = x[i] - mean_x;
    delta_y = y[i] - mean_y;
    sum_sq_x += delta_x * delta_x * sweep;
    sum_sq_y += delta_y * delta_y * sweep;
    sum_coproduct += delta_x * delta_y * sweep;
    mean_x += delta_x / i;
    mean_y += delta_y / i ;
  end_for;
  pop_sd_x = sqrt( sum_sq_x / N );
  pop_sd_y = sqrt( sum_sq_y / N );
  cov_x_y = sum_coproduct / N;
  correlation = cov_x_y / (pop_sd_x * pop_sd_y);
End.
```

all simulator tests and all pilot. Now, if the number of included simulator tests is small, this information is meaningful for the human mind. But if we perform one hundred tests (on five hundred pilot), then the correlation matrix contains ten thousand Pearson correlation coefficients. This is the reason for seeking the 'latent' factor structure, underlying the whole co–variability contained in the correlation matrix.

Therefore, the correlation is defined only if both of the standard deviations are finite and both of them are nonzero. It is a corollary of the *Cauchy–Schwarz inequality*[16] that the correlation cannot exceed 1 in absolute value. The correlation is 1 in the case of an increasing linear relationship, -1 in the case of a decreasing linear relationship, and some value in between in all other cases, indicating the degree of linear dependence between the variables. The closer the coefficient is to either -1 or 1, the stronger the correlation between the variables (see Figure 23.1). If the variables are independent then the correlation is 0, but the converse is not true because the correlation coefficient detects only linear dependencies between two variables. For example, suppose the random variable X is uniformly distributed on the interval from -1 to 1, and $Y = X^2$. Then Y is completely determined by X, so that X and Y are dependent, but their correlation is zero; this means that they are uncorrelated. The correlation matrix of n random variables $X_1, ..., X_n$ is the $n \times n$ matrix whose ij entry is $r_{X_i X_j}$. If the measures of correlation used are product–moment coefficients, the correlation matrix is the same as the covariance matrix of the standardized random variables X_i/σ_{X_i} (for $i = 1, ..., n$). Consequently it is necessarily a non–negative definite matrix. The correlation matrix is symmetrical (the correlation between X_i and X_j is the same as the correlation between X_j and X_i).

[16] The Cauchy–Schwarz inequality, named after Augustin Louis Cauchy (the father of complex analysis) and Hermann Amandus Schwarz, is a useful inequality encountered in many different settings, such as linear algebra applied to vectors, in analysis applied to infinite series and integration of products, and in probability theory, applied to variances and covariances. The Cauchy–Schwarz inequality states that if x and y are elements of real or complex inner product spaces then

$$|\langle x, y\rangle|^2 \leq \langle x, x\rangle \cdot \langle y, y\rangle.$$

The two sides are equal iff x and y are linearly dependent (or in geometrical sense they are parallel). This contrasts with a property that the inner product of two vectors is zero if they are orthogonal (or perpendicular) to each other. The inequality hence confers the notion of *the angle between the two vectors* to an inner product, where concepts of *Euclidean geometry* may not have meaningful sense, and justifies that the notion that inner product spaces are generalizations of *Euclidean space*.

An important consequence of the Cauchy–Schwarz inequality is that the inner product is a continuous function.

Another form of the Cauchy–Schwarz inequality is given using the notation of norm, as explained under norms on inner product spaces, as

$$|\langle x, y\rangle| \leq \|x\| \cdot \|y\|.$$

As a higher derivation of the correlation matrix analysis and its eigenvectors, the so–called principal components, the *factor analysis* (FA) is a multivariate statistical technique used to explain variability among a large set of observed random variables in terms of fewer unobserved random 'latent' variables, called *factors*. The observed, or 'manifested' variables are modelled as linear combinations of the factors, plus 'error terms'. According to FA, classical bivariate correlation analysis is an artificial extraction from a rial multivariate world, especially in human sciences. FA originated in psychometrics, and is used in social sciences, marketing, product management, operations research, and other applied sciences that deal with large multivariate quantities of data.

For example,[17] suppose a psychologist proposes a theory that there are two kinds of intelligence, 'verbal intelligence' and 'mathematical intelligence'. Note that these are inherently unobservable. Evidence for the theory is sought in the examination scores of 1000 students in each of 10 different academic fields. If a student is chosen randomly from a large population, then the student's 10 scores are random variables. The psychologist's theory may say that the average score in each of the 10 subjects for students with a particular level of verbal intelligence and a particular level of mathematical intelligence is a certain number times the level of verbal intelligence plus a certain number times the level of mathematical intelligence, i.e., it is a linear combination of those two 'factors'. The numbers by which the two 'intelligences' are multiplied are posited by the theory to be the same for all students, and are called 'factor loadings'. For example, the theory may hold that the average student's aptitude in the field of amphibology is

{ 10 × the student's verbal intelligence } + { 6 × the student's mathematical intelligence }.

The numbers 10 and 6 are the factor loadings associated with amphibology. Other academic subjects may have different factor loadings. Two students having identical degrees of verbal intelligence and identical degrees of mathematical intelligence may have different aptitudes in amphibology because individual aptitudes differ from average aptitudes. That difference is called the 'error' — an unfortunate misnomer in statistics that means the amount by which an individual differs from what is average. The observable data that go into factor analysis would be 10 scores of each of the 1000 students, a total of 10,000 numbers. The factor loadings and levels of the two kinds of intelligence of each student must be inferred from the data. Even the number of factors (two, in this example) must be inferred from the data.

In the example above, for $i = 1, ..., 1,000$ the ith student's scores are

$$\begin{array}{rcccccccc} x_{1,i} & = & \mu_1 & + & \ell_{1,1}v_i & + & \ell_{1,2}m_i & + & \varepsilon_{1,i} \\ \vdots & & \vdots & & \vdots & & \vdots & & \vdots \\ x_{10,i} & = & \mu_{10} & + & \ell_{10,1}v_i & + & \ell_{10,2}m_i & + & \varepsilon_{10,i} \end{array}$$

where $x_{k,i}$ is the ith student's score for the kth subject, μ_k is the mean of the students' scores for the kth subject, ν_i is the ith student's 'verbal intelligence',

[17] This oversimplified example should not be taken to be realistic. Usually we are dealing with many factors.

m_i is the ith student's 'mathematical intelligence', $\ell_{k,j}$ are the factor loadings for the kth subject, for $j = 1, 2$; $\epsilon_{k,i}$ is the difference between the ith student's score in the kth subject and the average score in the kth subject of all students whose levels of verbal and mathematical intelligence are the same as those of the ith student. In matrix notation, we have

$$X = \mu + LF + \epsilon,$$

where X is a $10 \times 1{,}000$ matrix of observable random variables, μ is a 10×1 column vector of unobservable constants (in this case constants are quantities not differing from one individual student to the next; and random variables are those assigned to individual students; the randomness arises from the random way in which the students are chosen), L is a 10×2 matrix of factor loadings (unobservable constants), F is a $2 \times 1{,}000$ matrix of unobservable random variables, ϵ is a $10 \times\ 1{,}000$ matrix of unobservable random variables.

Observe that by doubling the scale on which 'verbal intelligence', the first component in each column of F, is measured, and simultaneously halving the factor loadings for verbal intelligence makes no difference to the model. Thus, no generality is lost by assuming that the standard deviation of verbal intelligence is 1. Likewise for 'mathematical intelligence'. Moreover, for similar reasons, no generality is lost by assuming the two factors are uncorrelated with each other. The 'errors' ϵ are taken to be independent of each other. The variances of the 'errors' associated with the 10 different subjects are not assumed to be equal.

Mathematical basis of FA is *principal components analysis* (PCA), which is a technique for simplifying a dataset, by reducing multidimensional datasets to lower dimensions for analysis. Technically speaking, PCA is a *linear transformation*[18] that transforms the data to a new coordinate system such that the greatest variance by any projection of the data comes to lie on the first coordinate (called the first principal component), the second greatest variance on

[18] Recall that a *linear transformation* (also called *linear map* or *linear operator*) is a function between two vector spaces that preserves the operations of vector addition and scalar multiplication. In the language of abstract algebra, a linear transformation is a *homomorphism of vector spaces*, or a *morphism* in the category of vector spaces over a given field.

Let V and W be vector spaces over the same field K. A function (operator) $f : V \to W$ is said to be a *linear transformation* if for any two vectors $x, y \in V$ and any scalar $a \in K$, the following two conditions are satisfied:

$$\begin{aligned} \text{additivity} &: f(x+y) = f(x) + f(y), \qquad \text{and} \\ \text{homogeneity} &: f(ax) = af(x). \end{aligned}$$

This is equivalent to requiring that for any vectors $x_1, ..., x_m$ and scalars $a_1, ..., a_m$, the following equality holds:

$$f(a_1x_1 + \cdots + a_mx_m) = a_1f(x_1) + \cdots + a_mf(x_m).$$

the second coordinate, and so on. PCA can be used for *dimensionality reduction*[19] in a dataset while retaining those characteristics of the dataset that contribute most to its variance, by keeping lower–order principal components and ignoring higher–order ones. Such low–order components often contain the 'most important' aspects of the data. PCA is also called the (discrete) *Karhunen–Loève transform* (or KLT, named after Kari Karhunen and Michel Loève) or the *Hotelling transform* (in honor of Harold Hotelling[20]). PCA has the distinction of being the optimal linear transformation for keeping the subspace that has largest variance. This advantage, however, comes at the price of greater computational requirement if compared, for example, to the discrete cosine transform. Unlike other linear transforms, the PCA does not have a fixed set of basis vectors. Its basis vectors depend on the data set.

Assuming zero *empirical mean* (the empirical mean of the distribution has been subtracted from the data set), the principal component $\mathbf{w}_1$ of a dataset $\mathbf{x}$ can be defined as

$$\mathbf{w}_1 = \arg\max_{\|\mathbf{w}\|=1} E\left\{\left(\mathbf{w}^T\mathbf{x}\right)^2\right\}.$$

With the first $k-1$ components, the kth component can be found by subtracting the first $k-1$ principal components from $\mathbf{x}$,

$$\hat{\mathbf{x}}_{k-1} = \mathbf{x} - \sum_{i=1}^{k-1} \mathbf{w}_i \mathbf{w}_i^T \mathbf{x},$$

and by substituting this as the new dataset to find a principal component in

$$\mathbf{w}_k = \arg\max_{\|\mathbf{w}\|=1} E\left\{\left(\mathbf{w}^T\hat{\mathbf{x}}_{k-1}\right)^2\right\}.$$

[19] Dimensionality reduction in statistics can be divided into two categories: *feature selection* and *feature extraction*.

Feature selection approaches try to find a subset of the original features. Two strategies are *filter* (e.g., information gain) and *wrapper* (e.g., genetic algorithm) approaches. It is sometimes the case that data analysis such as regression or classification can be carried out in the reduced space more accurately than in the original space. On the other hand, feature extraction is applying a mapping of the multidimensional space into a space of fewer dimensions. This means that the original feature space is transformed by applying e.g., a linear transformation via a *principal components analysis*.

Dimensionality reduction is also a phenomenon discussed widely in physics, whereby a physical system exists in three dimensions, but its properties behave like those of a lower–dimensional system.

[20] Harold Hotelling (Fulda, Minnesota, September 29, 1895 - December 26, 1973) was a mathematical statistician. His name is known to all statisticians because of *Hotelling's T–square distribution* and its use in statistical hypothesis testing and confidence regions. He also introduced canonical correlation analysis, and is the eponym of *Hotelling's law*, *Hotelling's lemma*, and *Hotelling's rule* in economics.

Thefore, the Karhunen–Loève transform is equivalent to finding the *singular value decomposition*[21] of the data matrix $\mathbf{X}$,

$$\mathbf{X} = \mathbf{W}\boldsymbol{\Sigma}\mathbf{V}^T,$$

and then obtaining the reduced–space data matrix $\mathbf{Y}$ by projecting $\mathbf{X}$ down into the reduced space defined by only the first L singular vectors $\mathbf{W_L}$,

$$\mathbf{Y} = \mathbf{W_L}^T\mathbf{X} = \boldsymbol{\Sigma}_\mathbf{L}\mathbf{V_L}^T.$$

The matrix $\mathbf{W}$ of singular vectors of $\mathbf{X}$ is equivalently the matrix $\mathbf{W}$ of eigenvectors of the matrix of observed covariances $\mathbf{C} = \mathbf{X}\mathbf{X}^T$,

$$\mathbf{X}\mathbf{X}^T = \mathbf{W}\boldsymbol{\Sigma}^2\mathbf{W}^T.$$

The eigenvectors with the largest eigenvalues correspond to the dimensions that have the strongest correlation in the dataset.

Now, FA is performed as PCA[22] with subsequent orthogonal (non–correlated) or oblique (correlated) *factor rotation* for the simplest possible interpretation (see, e.g., [KM78a]).

---

[21] Recall that in linear algebra, the *singular value decomposition* (SVD) is an important factorization of a rectangular real or complex matrix, with several applications in signal processing and statistics. The SVD can be seen as a generalization of the *spectral theorem*, which says that normal matrices can be unitarily diagonalized using a basis of eigenvectors, to arbitrary, not necessarily square, matrices.

Suppose M is an $m \times n$ matrix whose entries come from the field K, which is either the field of real numbers, or the field of complex numbers. Then there exists a factorization of the form:

$$M = U\Sigma V^*,$$

where U is an $m \times m$ unitary matrix over K, the matrix Σ is $m \times n$ with nonnegative numbers on the diagonal and zeros off the diagonal, and V^* denotes the conjugate transpose of V, an $n \times n$ unitary matrix over K. Such a factorization is called a singular–value decomposition of M.

The matrix V thus contains a set of orthonormal 'input' or 'analyzing' basis vector directions for M. The matrix U contains a set of orthonormal 'output' basis vector directions for M. The matrix Σ contains the singular values, which can be thought of as scalar 'gain controls' by which each corresponding input is multiplied to give a corresponding output. A common convention is to order the values Σ_{ii} in non–increasing fashion. In this case, the diagonal matrix Σ is uniquely determined by M (although the matrices U and V are not).

[22] The alternative FA approach is the so–called *principal factor analysis* (PFA, also called *principal axis factoring*, PAF, and *common factor analysis*, PFA). PFA is a form of factor analysis which seeks the least number of factors which can account for the common variance (correlation) of a set of variables, whereas the more common principal components analysis (PCA) in its full form seeks the set of factors which can account for all the common and unique (specific plus error) variance in a set of variables. PFA uses a PCA strategy but applies it to a correlation matrix in which the diagonal elements are not 1's, as in PCA, but iteratively–derived estimates of the *communalities*.

In addition to PCA and PFA, there are other less–used extraction methods:

FA is used to uncover the latent structure (dimensions) of a set of variables. It reduces attribute space from a larger number of variables to a smaller number of factors and as such is a 'non–dependent' procedure (that is, it does not assume a dependent variable is specified). Factor analysis could be used for any of the following purposes:

1. To reduce a large number of variables to a smaller number of factors for modelling purposes, where the large number of variables precludes modelling all the measures individually. As such, factor analysis is integrated in *structural equation modelling* (SEM),[23] helping create the latent variables modeled by

---

1. Image factoring: based on the correlation matrix of predicted variables rather than actual variables, where each variable is predicted from the others using multiple regression.
2. Maximum likelihood factoring: based on a linear combination of variables to form factors, where the parameter estimates are those most likely to have resulted in the observed correlation matrix, using MLE methods and assuming multivariate normality. Correlations are weighted by each variable's uniqueness. (As discussed below, uniqueness is the variability of a variable minus its communality.) MLF generates a *chi–square goodness–of–fit test.* The researcher can increase the number of factors one at a time until a satisfactory goodness of fit is obtained. Warning: for large samples, even very small improvements in explaining variance can be significant by the goodness-of-fit test and thus lead the researcher to select too many factors.
3. Alpha factoring: based on maximizing the reliability of factors, assuming variables are randomly sampled from a universe of variables. All other methods assume cases to be sampled and variables fixed.
4. Unweighted least squares (ULS) factoring: based on minimizing the sum of squared differences between observed and estimated correlation matrices, not counting the diagonal.
5. Generalized least squares (GLS) factoring: based on adjusting ULS by weighting the correlations inversely according to their uniqueness (more unique variables are weighted less). Like MLF, GLS also generates a chi–square goodness–of–fit test. The researcher can increase the number of factors one at a time until a satisfactory goodness of fit is obtained.

[23] Structural equation modelling (SEM) grows out of and serves purposes similar to multiple regression, but in a more powerful way which takes into account the modelling of interactions, nonlinearities, correlated independents, measurement error, correlated error terms, multiple latent independents each measured by multiple indicators, and one or more latent dependents also each with multiple indicators. SEM may be used as a more powerful alternative to multiple regression, path analysis, factor analysis, time series analysis, and analysis of covariance. That is, these procedures may be seen as special cases of SEM, or, to put it another way, SEM is an extension of the general linear model (GLM) of which multiple regression is a part.

SEM is usually viewed as a confirmatory rather than exploratory procedure, using one of three approaches:

a) Strictly confirmatory approach: A model is tested using SEM goodness–of–fit tests to determine if the pattern of variances and covariances in the data is consistent

SEM. However, factor analysis can be and is often used on a standalone basis for similar purposes.

---

with a structural (path) model specified by the researcher. However as other unexamined models may fit the data as well or better, an accepted model is only a not–disconfirmed model.

b) Alternative models approach: One may test two or more causal models to determine which has the best fit. There are many goodness–of–fit measures, reflecting different considerations, and usually three or four are reported by the researcher. Although desirable in principle, this AM approach runs into the real-world problem that in most specific research topic areas, the researcher does not find in the literature two well-developed alternative models to test.

c) Model development approach: In practice, much SEM research combines confirmatory and exploratory purposes: a model is tested using SEM procedures, found to be deficient, and an alternative model is then tested based on changes suggested by SEM modification indexes. This is the most common approach found in the literature. The problem with the model development approach is that models confirmed in this manner are post–hoc ones which may not be stable (may not fit new data, having been created based on the uniqueness of an initial dataset). Researchers may attempt to overcome this problem by using a cross–validation strategy under which the model is developed using a calibration data sample and then confirmed using an independent validation sample.

Regardless of approach, SEM cannot itself draw causal arrows in models or resolve causal ambiguities. Theoretical insight and judgment by the researcher is still of utmost importance.

The SEM process centers around two steps: validating the measurement model and fitting the structural model. The former is accomplished primarily through confirmatory factor analysis, while the latter is accomplished primarily through path analysis with latent variables. One starts by specifying a model on the basis of theory. Each variable in the model is conceptualized as a latent one, measured by multiple indicators. Several indicators are developed for each model, with a view to winding up with at least three per latent variable after confirmatory factor analysis. Based on a large ($n > 100$) representative sample, factor analysis (common factor analysis or principal axis factoring, not principle components analysis) is used to establish that indicators seem to measure the corresponding latent variables, represented by the factors. The researcher proceeds only when the measurement model has been validated. Two or more alternative models (one of which may be the null model) are then compared in terms of *model fit*, which measures the extent to which the covariances predicted by the model correspond to the observed covariances in the data. The so–called modification indices and other coefficients may be used by the researcher to alter one or more models to improve fit.

Advantages of SEM compared to multiple regression include more flexible assumptions (particularly allowing interpretation even in the face of multicollinearity), use of confirmatory factor analysis to reduce measurement error by having multiple indicators per latent variable, the attraction of SEM's graphical modelling interface, the desirability of testing models overall rather than coefficients individually, the ability to test models with multiple dependents, the ability to model mediating variables, the ability to model error terms, the ability to test coefficients across multiple between–subjects groups, and ability to handle difficult data (time series with autocorrelated error, non–normal data, incomplete data).

2. To select a subset of variables from a larger set, based on which original variables have the highest correlations with the principal component factors.
3. To create a set of factors to be treated as uncorrelated variables as one approach to handling multi–collinearity in such procedures as multiple regression
4. To validate a scale or index by demonstrating that its constituent items load on the same factor, and to drop proposed scale items which cross–load on more than one factor.
5. To establish that multiple tests measure the same factor, thereby giving justification for administering fewer tests.
6. To identify clusters of cases and/or outliers.
7. To determine network groups by determining which sets of people cluster together.

The so–called *exploratory factor analysis* (EFA) seeks to uncover the underlying structure of a relatively large set of variables. The researcher's à priori assumption is that any indicator may be associated with any factor. This is the most common form of factor analysis. There is no prior theory and one uses factor loadings to intuit the factor structure of the data.

On the other hand, the so–called *confirmatory factor analysis* (CFA) seeks to determine if the number of factors and the loadings of measured (indicator) variables on them conform to what is expected on the basis of pre–established theory. Indicator variables are selected on the basis of prior theory and factor analysis is used to see if they load as predicted on the expected number of factors. The researcher's à priori assumption is that each factor (the number and labels of which may be specified à priori) is associated with a specified subset of indicator variables. A minimum requirement of confirmatory factor analysis is that one hypothesize beforehand the number of factors in the model, but usually also the researcher will posit expectations about which variables will load on which factors (see, e.g., [KM78b]). The researcher seeks to determine, for instance, if measures created to represent a latent variable really belong together.

The *factor loadings*, also called component loadings in PCA, are the correlation coefficients between the variables (rows) and factors (columns) in the *factor matrix*. Analogous to Pearson's r, the squared factor loading is the percent of variance in that variable explained by the factor. To get the percent of variance in all the variables accounted for by each factor, add the sum of the squared factor loadings for that factor (column) and divide by the number of variables (note that the number of variables equals the sum of their variances as the variance of a standardized variable is 1). This is the same as dividing the factor's eigenvalue by the number of variables.

The *factor scores*, also called component scores in PCA, factor scores are the scores of each case (row) on each factor (column). To compute the factor score for a given case for a given factor, one takes the case's standardized score on each variable, multiplies by the corresponding factor loading of the variable for the given factor, and sums these products. Computing factor scores allows one to look for factor outliers. Also, factor scores may be used as variables in subsequent modelling.

Rotation serves to make the output more understandable and is usually necessary to facilitate the interpretation of factors. The sum of eigenvalues is not affected by rotation, but rotation will alter the eigenvalues (and percent of variance explained) of particular factors and will change the factor loadings. Since alternative rotations may explain the same variance (have the same total eigenvalue) but have different factor loadings, and since factor loadings are used to intuit the meaning of factors, this means that different meanings may be ascribed to the factors depending on the rotation – a problem some cite as a drawback to factor analysis. If factor analysis is used, the researcher may wish to experiment with alternative rotation methods to see which leads to the most interpretable factor structure.

Varimax rotation is an orthogonal rotation of the factor axes to maximize the variance of the squared loadings of a factor (column) on all the variables (rows) in a factor matrix, which has the effect of differentiating the original variables by extracted factor. Each factor will tend to have either large or small loadings of any particular variable. A varimax solution yields results which make it as easy as possible to identify each variable with a single factor. This is the most common rotation option.

The oblique rotations allow the factors to be correlated, and so a factor correlation matrix is generated when oblique is requested. Two most common oblique rotation methods are:

Direct oblimin rotation – the standard method when one wishes a non–orthogonal solution , that is, one in which the factors are allowed to be correlated; this will result in higher eigenvalues but diminished interpretability of the factors; and

Promax rotation rotation – an alternative non–orthogonal rotation method which is computationally faster than the direct oblimin method and therefore is sometimes used for very large datasets.

FA advantages are:

1. Offers a much more objective method of testing intelligence in humans;
2. Allows for a satisfactory comparison between the results of intelligence tests; and
3. Provides support for theories that would be difficult to prove otherwise.

Charles Spearman pioneered the use of factor analysis in the field of psychology and is sometimes credited with the invention of factor analysis. He discovered that schoolchildren's scores on a wide variety of seemingly unrelated subjects were positively correlated, which led him to postulate that a general mental ability, or g, underlies and shapes human cognitive performance. His postulate now enjoys broad support in the field of intelligence research, where it is known as the g theory.

Raymond Cattell expanded on Spearman's idea of a two–factor theory of intelligence after performing his own tests and factor analysis. He used a multi–factor theory to explain intelligence. Cattell's theory addressed alternate factors in intellectual development, including motivation and psychology. Cattell also developed several mathematical methods for adjusting psychometric graphs, such as his 'scree' test and similarity coefficients. His research lead to the development of his theory of fluid and crystallized intelligence. Cattell was a strong advocate of factor analysis and psychometrics. He believed that all theory should be derived from research, which supports the continued use of empirical observation and objective testing to study human intelligence.

Factor Structure and Rotation

Starting with the correlation matrix **R** including the number of significant correlations, the goal of exploratory factor analysis (FA) is to detect latent underlying dimensions (i.e., the factor structure) among the set of all manifest variables. Instead of the correlation matrix, the factor analysis can start from the covariance matrix (see Figure 4), which is the symmetrical matrix with variances of all manifest variables on the main diagonal and their covariances in other matrix cells. For the purpose of the present project the correlation matrix is far more meaningful starting point. Three main applications of factor analytic techniques are (see, e.g. [CL71]):

1. to *reduce* the number of manifest variables,
2. to *classify* manifest variables, and
3. to *score* each individual soldier on the latent factor structure.

Factor analysis model expands each of the manifest variables $\mathbf{x}_i$ with the means $\boldsymbol{\mu}_i$ from the data matrix $\mathbf{X} = \{x_{i\alpha}\}$ as a linear vector–function

$$\mathbf{x}_i = \boldsymbol{\mu}_i + \mathbf{L}_{ij}\,\mathbf{f}_j + \mathbf{e}_i, \qquad (i = 1, \dots, n;\ j = 1, \dots, m) \tag{23.1}$$

where n and m denote the numbers of manifest and latent variables, respectively, $\mathbf{f}_j$ denotes the jth common–factor vector (with zero mean and unity–matrix covariance), $\mathbf{L} = \mathbf{L}_{ij}$ is the matrix of factor loadings l_{ij}, and $\mathbf{e}_i$ corresponds to the ith specific–factor vector (specific variance not explained by the common factors, with zero mean and diagonal–matrix covariance).

That portion of the variance of the ith manifest variable $\mathbf{x}_i$ contributed by the m common factors $\mathbf{f}_j$, the sum of squares of the loadings l_{ij}, is called the ith communality.

Now, in the correlation matrix **R** the variances of all variables are equal to 1.0. Therefore, the total variance in that matrix is equal to the number of variables. Extraction of factors is based on the solution of eigenvalue problem, i.e., characteristic equation for the correlation matrix **R**,

$$\mathbf{R}\mathbf{x}_i = \lambda_i \mathbf{x}_i,$$

where λ_i are eigenvalues of **R**, representing the variances extracted by the factors, and $\mathbf{x}_i$ now represent the corresponding eigenvectors, representing principal components or factors. The question then is, how many factors do we want to extract? Note that as we extract consecutive factors, they account for less and less variability. The decision of when to stop extracting factors basically depends on when there is only very little 'random' variability left. According to the widely used Kaiser criterion we can retain only factors with eigenvalues greater than 1. In essence this is like saying that, unless a factor extracts at least as much as the equivalent of one original variable, we drop it. The proportion of variance of a particular item that is due to common factors (shared with other items) is called communality. Therefore, an additional task facing us when applying this model is to estimate the communalities for each variable, that is, the proportion of variance that each item has in common with other items. The proportion of variance that is unique to each item is then the respective item's total variance minus the communality. A common starting point is to use the squared multiple correlation of an item with all other items as an estimate of the communality. The correlations between the manifest variables and the principal components are called factor loadings. The first factor is generally more highly correlated with the variables than the second, third and other factors, as these factors are extracted successively and will account for less and less variance overall.

Therefore, the principal component factor analysis of the sample correlation matrix $\mathbf{R}$ is specified in terms of its $m < n$ eigenvalue–eigenvector pairs $(\lambda_j, \mathbf{x}_j)$ where $\lambda_j \geq \lambda_{j+1}$. The matrix of estimated factor loadings l_{ij} is given by

$$\mathbf{L} = \left[\sqrt{\lambda_1}\mathbf{x}_1 | \sqrt{\lambda_2}\mathbf{x}_2 | \ldots | \sqrt{\lambda_m}\mathbf{x}_m\right].$$

Factor extraction can be performed also by other methods, collectively called *principal factors*, including: (i) Maximum likelihood factors, (ii) Principal axis method, (iii) Centroid method, (iv) Multiple R^2-communalities, and (v) Iterated Minres communalities. However, we shall stick on the principal components because of their obvious eigen–structure.

In any case, matrix of factor loadings $\mathbf{L}$ is determined only up to an orthogonal matrix $\mathbf{O}$. The communalities, given by the diagonal elements of $\mathbf{L}\mathbf{L}^T$ are also unaffected by the choice of $\mathbf{O}$. This ambiguity provides the rationale for 'factor rotation', since orthogonal matrices correspond to 'coordinate' rotations.

We could plot, theoretically, the factor loadings in a m–dimensional scatter–plot. In that plot, each variable is represented as a point. In this plot we could rotate the axes in any direction without changing the relative locations of the points to each other; however, the actual coordinates of the points, that is, the factor loadings would of course change. There are various rotational strategies that have been proposed. The goal of all of these strategies is to get a clear pattern of loadings, that is, factors that are somehow clearly marked by high loadings for some variables and low loadings for others. This general pattern is also sometimes referred to as simple structure (a more formalized definition can be found in most standard textbooks). Typical rotational strategies are Varimax, Quartimax, and Equimax (see Anderson, 1984). Basically, the extraction of principal components amounts to a variance maximizing Varimax–rotation of the original space of manifest–variables. We want to get a pattern of loadings on each factor that is as diverse as possible, lending itself to easier interpretation. After we have found the line on which the variance is maximal, there remains some variability around this line. In principal components analysis, after the first factor has been extracted, that is, after the first line has been drawn through the data, we continue and define another line that maximizes the remaining variability, and so on. In this manner, consecutive factors are extracted. Because each consecutive factor is defined to maximize the variability that is not captured by the preceding factor, consecutive factors are independent of each other. Put another way, consecutive factors are uncorrelated or orthogonal to each other.

Basically, the rotation of the matrix of the factor loadings $\mathbf{L}$ represents its post–multiplication, i.e. $\mathbf{L}^* = \mathbf{LO}$ by the rotation matrix $\mathbf{O}$, which itself resembles one of the matrices included in the classical rotational Lie groups $SO(m)$ (containing the specific $m-$fold combination of sinuses and cosinuses. The linear factor equation (23.1) represents the orthogonal factor model, provided that vectors $\mathbf{f}_j$ and $\mathbf{e}_i$ are independent (orthogonal to each other, i.e., having zero covariance).

The most frequently used Kaiser's Normal Varimax rotation procedure selects the orthogonal transformation $\mathbf{T}$ that 'spreads out' the squares of the loadings on each factor as much as possible, i.e., maximizes the total 'squared' variance

$$V = \frac{1}{n}\sum_{j=1}^{m}\left[\sum_{i=1}^{n}(l_{ij}^*)^4 - \frac{1}{n}\left(\sum_{i=1}^{n}(l_{ij}^*)^2\right)^2\right],$$

where l_{ij}^* denote the rotated factor loadings from the rotated factor matrix $\mathbf{L}^*$.

Besides orthogonal rotation, there is another concept of oblique (non-orthogonal, or correlated) factors, which could help to achieve more interpretable simple structure. Specifically, computational strategies have been developed to rotate factors so as to best represent clusters of manifest variables, without the constraint of orthogonality of factors. Oblique rotation produces the factor structure made from the smaller set of mutually correlated factors. An oblique rotation to the simple structure corresponds to *nonrigid* rotation of the factor-axes (i.e., principal components) in the factor space such that the rotated axes $\mathbf{l}_\mathbf{j}^* = \mathbf{L}_{\mathbf{obl}}^*$ (no longer perpendicular) pass (nearly) through the clusters of manifest variables. Although the purest mathematical background does not exist for the non–orthogonal factor rotation, the *parsimony principle*: "explain the maximum of the common variability of the data matrix $\mathbf{X} = \{x_{i\alpha}\}$ with the minimum number of factors", is fully developed only in this form of factor analysis, and the factor–correlation matrix $\mathbf{L}_{\mathbf{obl}}^*$ resembles the correlation matrix between manifest variables in the latent, factor space with double–reduced number of observables.

The linear factor equation (23.1) becomes now the *oblique factor model*

$$\mathbf{x}_i = \boldsymbol{\mu}_i + \mathbf{L}_{\mathbf{obl}}^*\,\mathbf{f}_j + \mathbf{e}_i, \quad (i = 1, \dots, n;\ j = 1, \dots, m),$$

where the vectors $\mathbf{f}_j$ and $\mathbf{e}_i$ are interdependent (correlated to each other). With oblique rotation, using common procedures, like Kaiser–Harris Orthoblique, Oblimin, Oblimax, Quartimin, Promax (see [And84]), we could

1. perform a hierarchical (iterated) factor analysis, obtaining second–order factors, third–order factors, etc., finishing with a single general factor (for example using principal component analysis of the factor–correlation matrix $\mathbf{L}^*_{\mathbf{obl}}$); and
2. develop the so–called 'cybernetic models': when two factors in the factor–correlation matrix $\mathbf{L}^*_{\mathbf{obl}}$ are highly correlated we can assume a linear functional link between them; connecting all correlated factors on the certain hierarchical level, we can make a block–diagram out of them depicting a linear system; this is the real point of the *exploratory* factor analysis.

The factor scores $S_{j\alpha}$ (where j labels factors and α labels individual pilot) are incidental parameters that characterize general performance of the individuals (see [CL71, And84]). Factor scores with zero mean and unity-matrix covariance are usually automatically evaluated in principal–component, orthogonal and oblique factor analysis, according to the formula:

$$S_{j\alpha} = (\mathbf{L}^T\mathbf{L})^{-1}\mathbf{L}^T(x_{j\alpha} - \bar{x}_{j\alpha}),$$

and replacing $\mathbf{L}$ by $\mathbf{L}^*$, and by $\mathbf{L}^*_{\mathbf{obl}}$, respectively. They represent an objective measure of the general performance of pilot on the battery of psycho–tests.[24]

---

[24] Here is the Mathematica algorithm for calculating the basic factor structure:
$Mean[x\_] := \frac{Plus@@x}{Length[x]};$
$Variance[x\_] := \frac{Plus@@(mean[x]-x)^2}{Length[x]};$
$StDev[x\_] := \sqrt{Variance[x]};$
$Covar[x1\_, x2\_] := \frac{Plus@@((mean[x1]-x1)((mean[x2]-x2))}{Length[x1]};$
$Corr[x1\_, x2\_] := \frac{Covar[x1,x2]}{StDev[x1]\ StDev[x2]};$
$CorrMat[X\_] := Table[Corr[X[[1,j]], X[[1,i]]]//N, \{i,m\}, \{j,m\}];$
Generate random data–matrix (m variables × n cases):
$NoVars = 10;\ NoCases = 50;\ m = NoVars;\ n = NoCases;$
$data = Array[x, \{NoCases, NoVars\}]//MatrixForm;$
$Table[x[i,j] = Random[Integer, \{1,5\}], \{i, NoCases\}, \{j, NoVars\}];$
Print["data = ",data//MatrixForm];
Calculate correlation matrix:
$R = CorrMat[data];\ Print[$"R=",$R//MatrixForm]$
Calculate eigenvalues of the correlation matrix:
$\lambda = Eigenvalues[R]//MatrixForm$
Corresponding eigenvectors:
$vec = Eigenvectors[R];\ Print[vec//Transpose//MatrixForm]$
Determine significant principal components
according to the criterion $\lambda \geq 2$:
$Print[$"PRINCIPAL COMPONENTS"
$\rightarrow \{vec[[1]], vec[[2]]\}//Transpose//MatrixForm]$
Define operator matrix:
$NoFact = 2;\ P = Array[p, NoVars, NoFact];$
$Table[p[i,j] = 1, \{i, NoVars\}, \{j, NoFact\}];$
$Table[p[i,j] = 0, \{i, 2, NoVars, 2\}, \{j, 2, NoFact, 2\}];$

The factor scores can be used further for multivariate regression in the latent space (instead in the original manifest space) for reducing the number of predictors in the general regression analysis (see [CL71]).

Quantum–Like Correlation and Factor Dynamics

To develop correlation and factor dynamics model, we are using geometrical analogy with *nonrelativistic quantum mechanics* (see [Dir82]). A time dependent state of a quantum system is determined by a normalized (complex), time–dependent, wave psi–function $\psi = \psi(t)$, i.e. a unit Dirac's 'ket' vector $|\psi(t)\rangle$, an element of the Hilbert space $L^2(\psi)$ with a coordinate basis (q^i), under the action of the Hermitian operators, obtained by the procedure of quantization of classical mechanical quantities, for which real eigenvalues are measured . The state–vector $|\psi(t)\rangle$, describing the motion of de Broglie's waves, has a statistical interpretation as the probability amplitude of the quantum system, for the square of its magnitude determines the density of the probability of the system detected at various points of space. The summation over the entire space must yield unity and this is the normalization condition for the psi–function, determining the unit length of the state vector $|\psi(t)\rangle$.

In the coordinate q–representation and the Schrödinger S–picture we consider an action of an evolution operator (in normal units Planck constant $\hbar = 1$)

$$\hat{S} \equiv \hat{S}(t, t_0) = \exp[-i\hat{H}(t - t_0)],$$

i.e., a one–parameter Lie–group of unitary transformations evolving a quantum system. The action represents an exponential map of the system's total energy operator – Hamiltonian $\hat{H} = \hat{H}(t)$. It moves the quantum system from one instant of time, t_0, to some future time t, on the state–vector $|\psi(t)\rangle$, rotating it: $|\psi(t)\rangle = \hat{S}(t, t_0)|\psi(t_0)\rangle$. In this case the Hilbert coordinate basis (q^i) is fixed,

---

$Table[p[i, j] = 0, \{i, 1, NoVars, 2\}, \{j, 1, NoFact, 2\}];$
$Print[\text{"}P = \text{"}, P//MatrixForm];$
Perform oblique rotation:
$Q = Transpose[P]; S = R.P; G = Q.S;$
$Do[k = \frac{1}{\sqrt{G\|i,i\|}}, \{i, NoFact\}];$
$F = Sk; Z = kG; C = Zk;$
$L = Inverse[C]; \Phi = F.L;$
Factor structure matrix:
$Print[\text{"}F = \text{"}, F//MatrixForm]$
Inter-factor correlation matrix:
$Print[\text{"}C = \text{"}, C//MatrixForm]$
Factor projection matrix:
$Print[\text{"}\Phi = \text{"}, \Phi//MatrixForm]$
Calculate factor scores for individual pilot:
$var[x\_] := x - mean[x];$
$Table[v[i] = var[X[[i]]//N], \{i, n\}];$
$T_F = Transpose[F]; F_F = Inverse[T_F.F].T_F;$
$Table[F_F.v[i], \{i, n\}]//MatrixForm.$

so the system operators do not evolve in time, and the system evolution is determined exclusively by the time–dependent Schrödinger equation

$$i\partial_t|\psi(t)\rangle = \hat{H}(t)|\psi(t)\rangle, \qquad (\partial_t = \partial/\partial t), \tag{23.2}$$

with initial condition given at one instant of time t_0 as $|\psi(t_0)\rangle = |\psi\rangle$.

If the Hamiltonian $\hat{H} = \hat{H}(t)$ does not explicitly depend on time (which is the case with the absence of variables of macroscopic fields), the state vector reduces to the exponential of the system energy:

$$|\psi(t)\rangle = \exp(-iE(t - t_0)|\psi\rangle,$$

satisfying the time–independent (i.e., stationary) Schrödinger equation

$$\hat{H}|\psi\rangle = E|\psi\rangle, \tag{23.3}$$

which represents the characteristic equation for the Hamiltonian operator $\hat{H}$ and gives its real eigenvalues (stationary energy states) E_n and corresponding orthonormal eigenfunctions (i.e., probability amplitudes) $|\psi_n\rangle$.

To model the correlation and factor dynamics we start with the characteristic equation for the correlation matrix

$$\mathbf{R}\mathbf{x} = \lambda\mathbf{x},$$

making heuristic analogy with the stationary Schrödinger equation (23.3). This analogy allows a 'physical' interpretation of the correlation matrix $\mathbf{R}$ as an operator of the 'total correlation or covariation energy' of the statistical system (the simulator–test data matrix $\mathbf{X} = \{x_{i\alpha}\}$), eigenvalues λ_n corresponding to the 'stationary energy states', and eigenvectors $\mathbf{x}_n$ resembling 'probability amplitudes' of the system.

So far we have considered one instant of time t_0. Including the time–flow into the stationary Schrödinger equation (23.3) we get the time–dependent Schrödinger equation (23.2) and returning back with our heuristic analogy, we get the basic equation of the n–dimensional correlation dynamics

$$\partial_t\mathbf{x}(t) = \mathbf{R}(t)\,\mathbf{x}_k(t), \tag{23.4}$$

with initial condition at time t_0 given as a stationary manifest–vectors $\mathbf{x}_k(t_0) = \mathbf{x}_k$ $(k = 1, \ldots, n)$.

In more realistic case of 'many' observables (i.e., very big n), instead of the correlation dynamics (23.4), we can use the reduced–dimension factor dynamics, represented by analogous equation in the factor space spanned by the extracted (oblique) factors $\mathbf{F} = \mathbf{f}_i$, with inter–factor–correlation matrix $\mathbf{C} = c_{ij}$ $(i, j = 1, \ldots,$ no. of factors)

$$\partial_t\mathbf{f}_i(t) = \mathbf{C}(t)\,\mathbf{f}_i(t), \tag{23.5}$$

subject to initial condition at time t_0 given as stationary vectors $\mathbf{f}_i(t_0) = \mathbf{f}_i$.

Now, according to the fundamental existence and uniqueness theorem for linear autonomous ordinary differential equations, if $A = A(t)$ is an $n \times n$ real matrix, then the initial value problem

$$\partial_t \mathbf{x}(t) = A\mathbf{x}(t), \qquad \mathbf{x}(0) = \mathbf{x}_0 \in \mathbb{R}^n,$$

has the unique solution

$$\mathbf{x}(t) = \mathbf{x}_0 \mathrm{e}^{tA}, \qquad \text{for all } t \in \mathbb{R}.$$

Therefore, analytical solutions of our correlation and factor–correlation dynamics equations (23.4) and (23.5) are given respectively by exponential maps

$$\mathbf{x}_k(t) = \mathbf{x}_k \exp[t\,\mathbf{R}],$$
$$\mathbf{f}_i(t) = \mathbf{f}_i \exp[t\,\mathbf{C}].$$

Thus, for each $t \in \mathbb{R}$, the matrix $\mathbf{x} \exp[t\,\mathbf{R}]$, respectively the matrix $\mathbf{f} \exp[t\,\mathbf{C}]$, maps

$$\mathbf{x}_k \mapsto \mathbf{x}_k \exp[t\,\mathbf{R}], \qquad \text{respectively} \qquad \mathbf{f}_i \mapsto \mathbf{f}_i \exp[t\,\mathbf{C}].$$

The sets $g^t_{corr} = \{\exp[t\,\mathbf{R}]\}_{t\in\mathbb{R}}$ and $g^t_{fact} = \{\exp[t\,\mathbf{C}]\}_{t\in\mathbb{R}}$ are 1–parameter families (groups) of linear maps of $\mathbb{R}^n$ into $\mathbb{R}^n$, representing the *correlation flow*, respectively the *factor–correlation flow* of simulator–tests. The linear flows g^t (representing both g^t_{corr} and g^t_{fact}) have two essential properties:

1. identity map: $g^0 = I$, and
2. composition: $g^{t_1+t_2} = g^{t_1} \circ g^{t_2}$.

They partition the state space $\mathbb{R}^n$ into subsets that we call 'correlation orbits', respectively 'factor–correlation orbits', through the initial states $\mathbf{x}_k$, and $\mathbf{f}_i$, of simulator tests, defined respectively by

$$\gamma(\mathbf{x}_k) = \{\mathbf{x}_k g^t | t \in \mathbb{R}\} \qquad \text{and} \qquad \gamma(\mathbf{f}_i) = \{\mathbf{f}_i g^t | t \in \mathbb{R}\}.$$

The correlation orbits can be classified as:

1. If $g^t\mathbf{x}_k = \mathbf{x}_k$ for all $t \in \mathbb{R}$, then $\gamma(\mathbf{x}_k) = \{\mathbf{x}_k\}$ and it is called a *point orbit*. Point orbits correspond to equilibrium points in the manifest and the factor space, respectively.
2. If there exists a $T > 0$ such that $g^T\mathbf{x}_k = \mathbf{x}_k$, then $\gamma(\mathbf{x}_k)$ is called a *periodic orbit*. Periodic orbits describe a system that evolves periodically in time in the manifest and the factor space, respectively.
3. If $g^t\mathbf{x}_k \neq \mathbf{x}_k$ for all $t \neq 0$, then $\gamma(\mathbf{x}_k)$ is called a *non–periodic orbit*.

Analogously, the factor–correlation orbits can be classified as:

1. If $g^t\mathbf{f}_i = \mathbf{f}_i$ for all $t \in \mathbb{R}$, then $\gamma(\mathbf{f}_i) = \{\mathbf{f}_i\}$ and it is called a point orbit. Point orbits correspond to equilibrium points in the manifest and the factor space, respectively.
2. If there exists a $T > 0$ such that $g^T\mathbf{f}_i = \mathbf{f}_i$, then $\gamma(\mathbf{f}_i)$ is called a periodic orbit. Periodic orbits describe a system that evolves periodically in time in the manifest and the factor space, respectively.
3. If $g^t\mathbf{f}_i \neq \mathbf{f}_i$ for all $t \neq 0$, then $\gamma(\mathbf{f}_i)$ is called a non–periodic orbit.

Now, to interpret properly the meaning of (really discrete) time in the correlation matrix $\mathbf{R} = \mathbf{R}(t)$ and factor–correlation matrix $\mathbf{C} = \mathbf{C}(t)$, we can perform a successive time–series $\{t, t + \Delta t, t + 2\Delta t, t + k\Delta t, \cdots\}$ of simulator tests (and subsequent correlation and factor analysis), and discretize our correlation (respectively, factor–correlation) dynamics, to get

$$\mathbf{x}_k(t + \Delta t) = \mathbf{x}_k(0) + \mathbf{R}(t)\,\mathbf{x}_k(t)\Delta t, \qquad \text{and}$$
$$\mathbf{f}_i(t + \Delta t) = \mathbf{f}_i(0) + \mathbf{C}(t)\,\mathbf{f}_i(t)\Delta t,$$

respectively. Finally we can represent the discrete correlation and factor–correlation dynamics in the form of the (computationally applicable) *three–point iterative dynamics equation*, respectively in the manifest space

$$\mathbf{x}_k^{s+1} = \mathbf{x}_k^{s-1} + \mathbf{R}_k^s\,\mathbf{x}_k^s,$$

and in the factor space

$$\mathbf{f}_i^{s+1} = \mathbf{f}_i^{s-1} + \mathbf{C}_i^s\,\mathbf{f}_i^s,$$

in which the time–iteration variable s labels the time occurrence of the simulator tests (and subsequent correlation and factor analysis), starting with the initial state, labelled $s = 0$.

FA–Based Intelligence

In the psychometric view, the concept of intelligence is most closely identified with Spearman's **g**, or Gf ('fluid **g**'). However, psychometricians can measure a wide range of abilities, which are distinct yet correlated. One common view is that these abilities are hierarchically arranged with **g** at the vertex (or top, overlaying all other cognitive abilities).[25]

On the other hand, critics of the psychometric approach, such as Robert Sternberg from Yale, point out that people in the general population have a somewhat different conception of intelligence than most experts. In turn, they argue that the psychometric approach measures only a part of what is commonly understood as intelligence. Other critics, such as Arthur Eddington,[26] argue that the equipment used in an experiment often determines the results and that proving

[25] Intelligence, IQ, and **g** are distinct terms. As already said above, intelligence is the term used in ordinary discourse to refer to cognitive ability. However, it is generally regarded as too imprecise to be useful for a scientific treatment of the subject. The intelligence quotient (IQ) is an index calculated from the scores on test items judged by experts to encompass the abilities covered by the term intelligence. IQ measures a multidimensional quantity: it is an amalgam of different kinds of abilities, the proportions of which may differ between IQ tests. The dimensionality of IQ scores can be studied by factor analysis, which reveals a single dominant factor underlying the scores on all IQ tests. This factor, which is a hypothetical construct, is called **g**. Variation in **g** corresponds closely to the intuitive notion of intelligence, and thus **g** is sometimes called *general cognitive ability* or *general intelligence*.

[26] Sir Arthur Stanley Eddington, OM (December 28, 1882 — November 22, 1944) was an astrophysicist of the early 20th century. The Eddington limit, the natural

that e.g., intelligence exists does not prove that current equipment measure it correctly. Sceptics often argue that so much scientific knowledge about the brain is still to be discovered that claiming the conventional IQ test methodology to be infallible is just a small step forward from claiming that *craniometry*[27] was the infallible method for measuring intelligence (which had scientific merits based on knowledge available in the nineteenth century).

A more fundamental criticism is that both the psychometric model used in these studies and the conceptualization of cognitive ability itself are fundamentally off beam. These views were expressed by none other than Charles Spearman, the 'discoverer' of **g** – himself. Thus he wrote: "Every normal man, woman, and child is a genius at something. It remains to discover at what. This must be a most difficult matter, owing to the very fact that it occurs in only a minute proportion of all possible abilities. It certainly cannot be detected by any of the testing procedures at present in current usage. But these procedures are capable, I believe, of vast improvement." In this context he noted that it is more important to ask 'What does this person think about?' than 'How well can he or she think?' Spearman went on to observe that the tests from which his **g** had emerged had no place in schools since they did not reflect the diverse talents of the children and thus deflected teachers from their fundamental educational role, which is to nurture and recognize these diverse talents.

He also noted, as paraphrased here, that the so–called 'cognitive ability' is not primarily cognitive but affective and conative. In constructing meaning out of confusion (Spearman's eductive ability) one first follows feelings that beckon or attract. One then has to engage in 'experimental interactions with the environment' to check out those, largely non–verbal, 'hunches'. This requires determination and persistence — *conation*. Now, all of these are difficult and demanding activities which will only be undertaken whilst one is undertaking activities one cares about. So the first question is: 'What kinds of activity is this person strongly motivated to undertake' (and the kinds of activity which people may be strongly motivated to undertake are legion and mostly unrelated to those assessed in conventional 'intelligence' tests). And the second question is: 'How many of the cumulative and substitutable components of competence required to carry out these activities effectively does this person display whilst carrying out that activity?' So one cannot, in reality, assess a person's intelligence, or even their eductive ability, except in relation to activities they care about. What one sees in e.g., the Raven Progressive Matrices is the cumulative effect of how well they do all these things in relation to a certain sort of task. The problem is that this is not — and cannot be — 'cognitive ability' in any general sense of the

limit to the luminosity that can be radiated by accretion onto a compact object, is named in his honor. He is famous for his work regarding the Theory of Relativity. Eddington wrote an article in 1919, Report on the relativity theory of gravitation, which announced Einstein's theory of general relativity to the English–speaking world.

[27] Craniometry is the technique of measuring the bones of the skull. Craniometry was once intensively practiced in anthropology/ethnology.

word but only in relation to this kind of task. As Roger Sperry[28] has observed, what is neurologically localized is not 'cognitive ability' in any general sense but the emotional predisposition to 'think' about a particular kind of thing (for more details, see e.g., papers of John Raven[29] [Rav02]).

Most experts accept the concept of a single dominant factor of intelligence, general mental ability or **g**, while others argue that intelligence consists of a set of relatively independent abilities [APS98]. The evidence for **g** comes from factor analysis of tests of cognitive abilities. The methods of factor analysis do not guarantee a single dominant factor will be discovered. Other *psychological tests*, which do not measure cognitive ability, such as *personality tests*, generate multiple factors.

Proponents of *multiple–intelligence theories* often claim that **g** is, at best, a measure of academic ability. Other types of intelligence, they claim, might be just as important outside of a school setting. Robert Sternberg has proposed a 'Triarchic Theory of Intelligence'. Howard Gardner's theory of multiple intelligences breaks intelligence down into at least eight different components: logical, linguistic, spatial, musical, kinesthetic, naturalist, intra–personal and inter–personal intelligences. Daniel Goleman and several other researchers have developed the concept of *emotional intelligence* and claim it is at least as important as more traditional sorts of intelligence. These theories grew from observations of human development and of brain injury victims who demonstrate an acute loss of a particular cognitive function (e.g., the ability to think numerically, or the ability to understand written language), without showing any loss in other cognitive areas.

---

[28] Roger Wolcott Sperry (August 20, 1913 – April 17, 1994) was a neuropsychologist who, together with David Hunter Hubel and Torsten Nils Wiesel, won the 1981 Nobel Prize in Medicine for his work with *split–brain* research. Before Sperry's experiments, some research evidence seemed to indicate that areas of the brain were largely undifferentiated and interchangeable. In his early experiments Sperry challenged this view by showing that after early development circuits of the brain are largely hardwired. In his Nobel–winning work, Sperry separated the *corpus callosum*, the area of the brain used to transfer signals between the right and left hemispheres, to treat epileptics. Sperry and his colleagues then tested these patients with tasks that were known to be dependent on specific hemispheres of the brain and demonstrated that the two halves of the brain may each contain consciousness. In his words, each hemisphere is "... indeed a conscious system in its own right, perceiving, thinking, remembering, reasoning, willing, and emoting, all at a characteristically human level, and . . . both the left and the right hemisphere may be conscious simultaneously in different, even in mutually conflicting, mental experiences that run along in parallel." This research contributed greatly to understanding the lateralization of brain functions. In 1989, Sperry also received National Medal of Science.

[29] John Carlyle Raven first published his Progressive Matrices in the United Kingdom in 1938. His three sons established Scotland–based test publisher JC Raven Ltd. in 1972. In 2004, Harcourt Assessment, Inc. a division of Harcourt Education acquired JC Raven Ltd.

In response, **g** theorists have pointed out that **g**'s *predictive validity*[30] has been repeatedly demonstrated, for example in predicting important non–academic outcomes such as job performance, while no multiple–intelligences theory has shown comparable validity. Meanwhile, they argue, the relevance, and even the existence, of multiple intelligences have not been borne out when actually tested [Hun01]. Furthermore, **g** theorists contend that proponents of multiple–intelligences (see, e.g., [Ste95]) have not disproved the existence of a general factor of intelligence [Kli00]. The fundamental argument for a general factor is that test scores on a wide range of seemingly unrelated cognitive ability tests (such as sentence completion, arithmetic, and memorization) are positively correlated: people who score highly on one test tend to score highly on all of them, and g thus emerges in a factor analysis. This suggests that the tests are not unrelated, but that they all tap a common factor.

Cognitive vs. Not–Cognitive Intelligence

Clearly, biologically realized 'cognitive intelligence' is the most complex property of human mind and can be perceived only by itself. Our problem is what we call or may call cognitive intelligence. From the formal, computational perspective, cognitive intelligence is one of ill defined concepts. Its definitions are immersed in numerous scientific contexts and mirrors their historical evolutions, as well as, different 'interests' of researchers. Its weakness is usually based on its abstract multifaces image and, on the other hand, a universal utility character.

The classical behavioral/biologists definition of intelligence reads: "Intelligence is the ability to adapt to new conditions and to successfully cope with life situations." This definition seems to be the best, but 'intelligence' here depends on available physical tools and specific life experience (individual hidden knowledge, preferences and access to information), therefore it is not enough selective to be measured, compared or designed. In general, cognitive intelligence is a human–like intelligence. Unfortunately there are many opinions what human–like intelligence means. For example, (i) cognitive intelligence uses a human mental introspective experience for the modelling of intelligent system thinking; and (ii) cognitive intelligence may use brain models to extract brain's intelligence property.

Therefore, cognitive intelligence can be seen as a product of human self–conscious recognition of efficient mental processes, defined a'priori as intelligent. In order to get a consensus on the notion of cognitive intelligence is useful to have an agreement on which intelligence is not cognitive. A not–cognitive intelligence could be considered as an intelligence being developed using not human

[30] In psychometrics, *predictive validity* is the extent to which a scale predicts scores on some criterion measure. For example, the validity of a cognitive test for job performance is the correlation between test scores and, say, supervisor performance ratings. Such a cognitive test would have predictive validity if the observed correlation were statistically significant. Predictive validity shares similarities with concurrent validity in that both are generally measured as correlations between a test and some criterion measure.

analogies; e.g., it is possible to construct very different models of flying objects starting from the observation of storks, balloons, beetles or clouds – maybe this observation can be useful.

The difference between human and artificial intelligence theories is similar to the difference between a birds theory of fly and the airplanes fly theory, the both can lead to a more general theory of fly but this last needs a goal–oriented and a higher abstraction level of the conceptualization/ontology.

According to the *TOGA meta–theory paradigms*,[31] for scientific and practical modelling purposes, it is reasonable to separate conceptually the following five concepts: *information, knowledge, preferences, intelligence and emotions.* If properly defined, all of them can be independently identified and designed.

---

[31] According to the *top–down object–based goal–oriented approach* (TOGA) standard, the Information–Preferences–Knowledge *cognitive architecture* consists of:

Data: everything what is/can be processed/transformed in computational and mental processes. Concept data is included in the ontology of 'elaborators', such as developers of methods, programmers and other computation service people. In this sense, data is a relative term and exists only in the couple (data, processing).

Information: data which represent a specific property of the domain of human or artificial agent's activity (such as: addresses, tel. numbers, encyclopedic data, various lists of names and results of measurements). Every information has always a source domain. It is a relative concept. Information is a concept from the ontology of modeler/problem–solver/decision–maker.

Knowledge: every abstract property of human/artificial agent which has ability to process/transform a quantitative/qualitative information into other information, or into another knowledge. It includes: instructions, emergency procedures, exploitation/user manuals, scientific materials, models and theories. Every knowledge has its reference domain where it is applicable. It has to include the source domain of the processed information. It is a relative concept.

Preference: an ordered relation among two properties of the domain of activity of a *cognitive agent*, it indicates a property with higher utility. Preference relations serve to establish an intervention goal of an agent. Cognitive preferences are relative. A preference agent which manages preferences of an intelligent agent can be external or its internal part.

Goal: a hypothetical state of the domain of activity which has maximal utility in a current situation. Goal serves to the choice and activate proper knowledge which process new information.

Document: a passive carrier of knowledge, information and/or preferences (with different structures), comprehensive for humans, and it has to be recognized as valid and useful by one or more human organizations, it can be physical or electronic.

Computer Program: (i) from the modelers and decision-makers perspective: an active carrier of different structures of knowledge expressed in computer languages and usually focused on the realization of predefined objectives (a design-goal). It may include build-in preferences and information and/or request specific IPK as data. (ii) from the software engineers perspective: a data-processing tool (more precise technical def. you may find on the Web).

Such conceptual modularity should enable to construct: *emotional intelligence, social intelligence, skill intelligence, organizational intelligence,* and many other X–intelligences, where X denotes a type of knowledge, preferences or a carrier system involved.

For example, business intelligence and emotional intelligence, rather are applications of intelligence either for business activities or for the second, under emotional/(not conscious) constrains and 'biological requests'.

In the above context, an *abstract intelligent agent* can be considered as a functional kernel of any natural or artificial intelligent system.

Intelligence and Cognitive Development

Although there is no general *theory of cognitive development*, the most historically influential theory was developed by Jean Piaget. *Piaget theory* provided many central concepts in the field of developmental psychology. His theory concerned the growth of intelligence, which for Piaget meant the ability to more accurately represent the world, and perform logical operations on representations of concepts grounded in the world. His theory concerns the emergence and acquisition of schemata, schemes of how one perceives the world, in 'developmental stages', times when children are acquiring new ways of mentally representing information. Piaget theory is considered 'constructivist, meaning that, unlike nativist theories (which describe cognitive development as the unfolding of innate knowledge and abilities) or empiricist theories (which describe cognitive development as the gradual acquisition of knowledge through experience), asserts that we construct our cognitive abilities through self–motivated action in the world.

Piaget divided schemes that children use to understand the world through four main stages, roughly correlated with and becoming increasingly sophisticated with age:

1. Sensorimotor stage (years 0–2),
2. Preoperational stage (years 2–7),
3. Concrete operational stage (years 7–11), and
4. Formal operational stage (years 11–adulthood).

Sensorimotor Stage

Infants are born with a set of congenital reflexes, according to Piaget, as well as a drive to explore their world. Their initial schemas are formed through differentiation of the congenital reflexes (see assimilation and accommodation, below).

The sensorimotor stage is the first of the four stages. According to Piaget, this stage marks the development of essential spatial abilities and understanding of the world in six sub–stages:

1. The first sub–stage occurs from birth to six weeks and is associated primarily with the development of reflexes. Three primary reflexes are described by Piaget: sucking of objects in the mouth, following moving or interesting

objects with the eyes, and closing of the hand when an object makes contact with the palm (palmar grasp). Over these first six weeks of life, these reflexes begin to become voluntary actions; for example, the palmar reflex becomes intentional grasping.
2. The second sub–stage occurs from six weeks to four months and is associated primarily with the development of habits. Primary circular reactions or repeating of an action involving only ones own body begin. An example of this type of reaction would involve something like an infant repeating the motion of passing their hand before their face. Also at this phase, passive reactions, caused by classical or operant conditioning, can begin.
3. The third sub–stage occurs from four to nine months and is associated primarily with the development of coordination between vision and prehension. Three new abilities occur at this stage: intentional grasping for a desired object, secondary circular reactions, and differentiations between ends and means. At this stage, infants will intentionally grasp the air in the direction of a desired object, often to the amusement of friends and family. Secondary circular reactions, or the repetition of an action involving an external object begin; for example, moving a switch to turn on a light repeatedly. The differentiation between means also occurs. This is perhaps one of the most important stages of a child's growth as it signifies the dawn of logic. Towards the late part of this sub–stage infants begin to have a sense of object permanence, passing the A–not–B error test.
4. The fourth sub-stage occurs from nine to twelve months and is associated primarily with the development of logic and the coordination between means and ends. This is an extremely important stage of development, holding what Piaget calls the 'first proper intelligence'. Also, this stage marks the beginning of goal orientation, the deliberate planning of steps to meet an objective.
5. The fifth sub–stage occurs from twelve to eighteen months and is associated primarily with the discovery of new means to meet goals. Piaget describes the child at this juncture as the 'young scientist', conducting pseudo–experiments to discover new methods of meeting challenges.
6. The sixth sub–stage is associated primarily with the beginnings of insight, or true creativity. This marks the passage into the preoperational stage.

Preoperational Stage

The Preoperational stage is the second of four stages of cognitive development. By observing sequences of play, Piaget was able to demonstrate that towards the end of the second year a qualitatively quite new kind of psychological functioning occurs. Operation in Piagetian theory is any procedure for mentally acting on objects. The hallmark of the preoperational stage is sparse and logically inadequate mental operations.

According to Piaget, the Sensorimotor stage of development is followed by this stage (2–7 years), which includes the following five processes:

1. Symbolic functioning, which is characterised by the use of mental symbols words or pictures which the child uses to represent something which is not physically present.
2. Centration, which is characterized by a child focusing or attending to only one aspect of a stimulus or situation. For example, in pouring a quantity of liquid from a narrow beaker into a shallow dish, a preschool child might judge the quantity of liquid to have decreased, because it is 'lower', that is, the child attends to the height of the water, but not to the compensating increase in the diameter of the container.
3. Intuitive thought, which occurs when the child is able to believe in something without knowing why she or he believes it.
4. Egocentrism, which is a version of centration, this denotes a tendency of child to only think from own point of view.
5. Inability to Conserve; Through Piaget's conservation experiments (conservation of mass, volume and number) Piaget concluded that children in the preoperational stage lack perception of conservation of mass, volume, and number after the original form has changed. For example, a child in this phase will believe that a string of beads set up in a 'O–O–O–O–O' pattern will have the same number of beads as a string which has a 'O–O–O–O–O' pattern, because they are the same length, or that a tall, thin 8-ounce cup has more liquid in it than a wide, fat 8–ounce cup.

Concrete Operational Stage

The concrete operational stage is the third of four stages of cognitive development in Piaget's theory. This stage, which follows the Preoperational stage and occurs from the ages of 7 to 11, is characterized by the appropriate use of logic. The six important processes during this stage are:

1. Decentering, where the child takes into account multiple aspects of a problem to solve it. For example, the child will no longer perceive an exceptionally wide but short cup to contain less than a normally-wide, taller cup.
2. Reversibility, where the child understands that numbers or objects can be changed, then returned to their original state. For this reason, a child will be able to rapidly determine that $4 + 4$ which they can answer to be 8, minus 4 will equal four, the original quantity.
3. Conservation: understanding that quantity, length or number of items is unrelated to the arrangement or appearance of the object or items. For instance, when a child is presented with two equally–sized, full cups they will be able to discern that if water is transferred to a pitcher it will conserve the quantity and be equal to the other filled cup.
4. Serialisation: the ability to arrange objects in an order according to size, shape, or any other characteristic. For example, if given different–shaded objects they may make a color gradient.
5. Classification: the ability to name and identify sets of objects according to appearance, size or other characteristic, including the idea that one set of

objects can include another. A child is no longer subject to the illogical limitations of animism (the belief that all objects are animals and therefore have feelings).

6. Elimination of Egocentrism: the ability to view things from another's perspective (even if they think incorrectly). For instance, show a child a comic in which Jane puts a doll under a box, leaves the room, and then Jill moves the doll to a drawer, and Jane comes back; a child in this stage will not say that Jane will think the doll is in the drawer.

Formal Operational Stage

The formal operational stage is the fourth and final of the stages of cognitive development of Piaget's theory. This stage, which follows the Concrete Operational stage, commences at around 11 years of age (puberty) and continues into adulthood. It is characterized by acquisition of the ability to think abstractly and draw conclusions from the information available. During this stage the young adult functions in a cognitively normal manner and therefore is able to understand such things as love, 'shades of gray', and values. Lucidly, biological factors may be traced to this stage as it occurs during puberty and marks the entering into adulthood in physiologically, cognitive, moral (Kohlberg), psychosexual (Freud), and social development (Erikson). Many people do not successfully complete this stage, but mostly remain in concrete operations.

Psychophysics

Recall that *psychophysics* is a subdiscipline of psychology, founded in 1860 by Gustav Fechner with the publication of 'Elemente der Psychophysik', dealing with the relationship between physical stimuli and their subjective correlates, or percepts. Fechner described research relating physical stimuli with how they are perceived and set out the philosophical foundations of the field. Fechner wanted to develop a theory that could relate matter to the mind, by describing the relationship between the world and the way it is perceived (Snodgrass, 1975). Fechner's work formed the basis of psychology as a science. Wilhelm Wundt, the founder of the first laboratory for psychological research, built upon Fechner's work.

The *Weber–Fechner law* attempts to describe the relationship between the physical magnitudes of stimuli and the perceived intensity of the stimuli. Ernst Weber was one of the first people to approach the study of the human response to a physical stimulus in a quantitative fashion. Gustav Fechner later offered an elaborate theoretical interpretation of Weber's findings, which he called simply Weber's law, though his admirers made the law's name a hyphenate. Fechner believed that Weber had discovered the fundamental principle of mind/body interaction, a mathematical analog of the function Rene Descartes once assigned to the pineal gland.

In one of his classic experiments, Weber gradually increased the weight that a blindfolded man was holding and asked him to respond when he first felt the increase. Weber found that the response was proportional to a relative increase in the weight. That is to say, if the weight is 1 kg, an increase of a few grams will not be noticed. Rather, when the mass is increased by a certain factor, an increase in weight is perceived. If the mass is doubled, the threshold is also doubled. This kind of relationship can be described by a linear ordinary differential equation as,

$$dp = k\frac{dS}{S},$$

where dp is the differential change in perception, dS is the differential increase in the stimulus and S is the stimulus at the instant. A constant factor k is to be determined experimentally. Integrating the above equation gives: $p = k \ln S + C$, where C is the constant of integration. To determine C, we can put $p = 0$, which means no perception; then we get, $C = -k \ln S_0$,where S_0 is that threshold of stimulus below which it is not perceived at all. In this way, we get the solution

$$p = k \ln \frac{S}{S_0}.$$

Therefore, the relationship between stimulus and perception is logarithmic. This logarithmic relationship means that if a stimulus varies as a geometric progression (i.e. multiplied by a fixed factor), the corresponding perception is altered in an arithmetic progression (i.e. in additive constant amounts). For example, if a stimulus is tripled in strength (i.e, 3×1), the corresponding perception may be two times as strong as its original value (i.e., $1 + 1$). If the stimulus is again tripled in strength (i.e., $3 \times 3 \times 1$), the corresponding perception will be three times as strong as its original value (i.e., $1 + 1 + 1$). Hence, for multiplications in stimulus strength, the strength of perception only adds. This logarithmic relationship is valid, not just for the sensation of weight, but for other stimuli and our sensory perceptions as well.

In case of vision, we have that the eye senses brightness logarithmically. Hence stellar magnitude is measured on a logarithmic scale. This magnitude scale was invented by the ancient Greek astronomer Hipparchus in about 150 B.C. He ranked the stars he could see in terms of their brightness, with 1 representing the brightest down to 6 representing the faintest, though now the scale has been extended beyond these limits. An increase in 5 magnitudes corresponds to a decrease in brightness by a factor 100.

In case of sound, we have still another logarithmic scale is the decibel scale of sound intensity. And yet another is pitch, which, however, differs from the other cases in that the physical quantity involved is not a 'strength'. In the case of perception of pitch, humans hear pitch in a logarithmic or geometric ratio–based fashion: For notes spaced equally apart to the human ear, the frequencies are related by a multiplicative factor. For instance, the frequency of corresponding notes of adjacent octaves differ by a factor of 2. Similarly, the perceived difference in pitch between 100 Hz and 150 Hz is the same as between 1000 Hz and 1500 Hz. Musical scales are always based on geometric relationships for this reason. Notation and theory about music often refers to pitch intervals in an additive way, which makes sense if one considers the logarithms of the frequencies, as $\log(a \times b) = \log a + \log b$.

Psychophysicists usually employ experimental stimuli that can be objectively measured, such as pure tones varying in intensity, or lights varying in luminance. All the senses have been studied:vision, hearing, touch (including skin and enteric perception), taste, smell, and the sense of time. Regardless of the sensory domain, there are three main topics in the psychophysical classification scheme: absolute thresholds, discrimination thresholds, and scaling.

The most common use of psychophysics is in producing scales of human experience of various aspects of physical stimuli. Take for an example the physical stimulus of frequency of sound. Frequency of a sound is measured in Hertz (Hz), cycles per second. But human experience of the frequencies of sound is not the same as the frequencies. For one thing, there is a frequency below which no sounds can be heard, no matter how intense they are (around 20 Hz depending on the individual) and there is a frequency above which no sounds can be heard, no matter how intense they are (around 20,000 Hz, again depending on the individual). For another, doubling the frequency of a sound (e.g., from 100 Hz to 200 Hz) does not lead to a doubling of experience. The perceptual experience of the frequency of sound is called pitch, and it is measured by psychophysicists in mels.

More analytical approaches allow the use of psychophysical methods to study neurophysiological properties and sensory processing mechanisms. This is of particular importance in human research, where other (more invasive) methods are not used due to ethical reasons. Areas of investigation include sensory thresholds, methods of measurement of sensitivity, and signal detection theory.

Perception is the process of acquiring, interpreting, selecting, and organizing sensory information. Methods of studying perception range from essentially biological or physiological approaches, through psychological approaches to the often abstract 'thought–experiments' of mental philosophy.

Experiments in psychophysics seek to determine whether the subject can detect a stimulus, identify it, differentiate between it and another stimulus, and describe the magnitude or nature of this difference [Sno75]. Often, the classic methods of experimentation are argued to be inefficient. This is because a lot of sampling and data has to be collected at points of the psychometric function that is known (the tails). Staircase procedures can be used to quickly estimate threshold. However, the cost of this efficiency, is that we do not get the same amount of information regarding the *psychometric function* as we can through classical methods; e.g., we cannot extract an estimate of the slope (derivative) of the function.

A psychometric function describes the relationship between a parameter of a physical stimulus and the responses of a person who has to decide about a certain aspect of that stimulus. The psychometric function usually resembles a sigmoid function with the percentage of correct responses (or a similar value) displayed on the ordinate and the physical parameter on the abscissa. If the stimulus parameter is very far towards one end of its possible range, the person will always be able to respond correctly. Towards the other end of the range, the person never perceives the stimulus properly and therefore the probability of correct responses is at chance level. In between, there is a transition range where the subject has an above–chance rate of correct responses, but does not always respond correctly. The inflection point of the sigmoid function or the point at which the function reaches the middle between the chance level and 100% is usually taken as sensory threshold. A common example is visual acuity testing with an eye chart. The person sees symbols of different sizes (the size is the relevant physical stimulus parameter) and has to decide which symbol it is. Usually, there is one line on the chart where a subject can identify some, but not all, symbols. This is equal to the transition range of the psychometric function and the sensory threshold corresponds to visual acuity.

On the other hand, a *sensory threshold* is a theoretical concept which states: "A stimulus that is less intense than the sensory threshold will not elicit any sensation." Whilst the concept can be applied to all senses, it is most commonly applied to the detection and perception of flavours and aromas. Several different sensory thresholds have been defined:

1. Absolute threshold: the lowest level at which a stimulus can be detected.
2. Recognition threshold: the level at which a stimulus can not only be detected but also recognised.
3. Differential threshold: the level at which an increase in a detected stimulus can be perceived.
4. Terminal threshold: the level beyond which a stimulus is no longer detected.

In other words, a threshold is the point of intensity at which the participant can just detect the presence of, or difference in, a stimulus. Stimuli with intensities below the threshold are considered not detectable, however stimuli at values close to threshold will often be detectable some proportion of the time. Due to this, a threshold is considered to be the point at which a stimulus, or change in a stimulus, is detected some proportion p of the time. An absolute threshold is the level of intensity of a stimulus at which the subject is able to detect the presence of the stimulus some proportion of the time (a p level of 50% is often used). An example of an absolute threshold is the number of hairs on the back of one's hand that must be touched before it can be felt, a participant may be unable to feel a single hair being touched, but may be able to feel two or three as this exceeds the threshold. A difference threshold is the magnitude of the difference between two stimuli of differing intensities that the participant is able to detect some proportion of the time (again, 50% is often used). To test this threshold, several difference methods are used. The subject may be asked to adjust one stimulus until it is perceived as the same as the other, may be asked to describe the magnitude of the difference between two stimuli, or may be asked to detect a stimulus against a background. Absolute and difference thresholds are sometimes considered similar because there is always background noise interfering with our ability to detect stimuli, however study of difference thresholds still occurs, for example in pitch discrimination tasks (see [Sno75]).

The *sensory analysis* applies principles of experimental design and statistical analysis to the use of human senses (sight, smell, taste, touch and hearing) for the purposes of evaluating consumer products. The discipline requires panels of human assessors, on whom the products are tested, and recording the responses made by them. By applying statistical techniques to the results it is possible to make inferences and insights about the products under test. Most large consumer goods companies have departments dedicated to sensory analysis. Sensory Analysis can generally be broken down into three sub–sections:

1. Effective Testing (dealing with objective facts about products);
2. Affective Testing (dealing with subjective facts such as preferences); and
3. Perception (the biochemical and psychological aspects of sensation).

The *signal detection theory* (SDT) is a means to quantify the ability to discern between signal and noise. It has applications in many fields such as quality control, telecommunications, and psychology (see [Abd06]). The concept is similar to the signal to noise ratio used in the sciences, and it is also usable in alarm management, where it is important to separate important events from background noise. According to the theory, there are a number of psychological determiners of how we will detect a signal, and where our threshold levels will be. Experience, expectations, physiological state (e.g, fatigue) and other factors affect thresholds. For instance, a sentry in wartime will likely detect fainter stimuli than the same sentry in peacetime. SDT is used when psychologists want to measure the way we make decisions under conditions of uncertainty, such as how we would perceive distances in foggy conditions. SDT assumes that 'the decision maker is not a passive receiver of information, but an active decision–maker who makes difficult perceptual judgements under conditions of uncertainty'. In foggy circumstances, we are forced to decide how far an object is away from us based solely upon visual stimulus which is impaired by the fog. Since the brightness of the object, such as a traffic light, is used by the brain to discriminate the distance of an object, and the fog reduces the brightness of objects, we perceive the object to be much further away than it actually is. To apply signal detection theory to a data set where stimuli were either present or absent, and the observer categorized each trial as having the stimulus present or absent, the trials are sorted into one of four categories, depending upon the stimulus and response:

	Respond 'Absent'	Respond 'Present'
Stimulus Present	Miss	Hit
Stimulus Absent	Correct Rejection	False Alarm

23.1.2 Problem Solving

Beginning in the 1970s, researchers became increasingly convinced that empirical findings and theoretical concepts derived from simple laboratory tasks did not necessarily generalize to more complex, real–life problems. Even worse, it appeared that the processes underlying creative problem solving in different domains differed from each other [Ste95]. These realizations have led to rather different responses in North America and in Europe.

In North America, initiated by the work of Herbert Simon on learning by doing in semantically rich domains, researchers began to investigate problem solving separately in different natural knowledge domains – such as physics, writing, or chess playing – thus relinquishing their attempts to extract a global theory of problem solving (see, e.g., [SF91]). Instead, these researchers have frequently focused on the development of problem solving within a certain domain, that is on the development of expertise.

Areas that have attracted rather intensive attention in North America include such diverse fields as: reading, writing, calculation, political decision making, managerial problem solving, lawyers' reasoning, personal problem solving, mathematical problem solving, mechanical problem solving, problem solving in electronics, computer skills, game playing, and social problem solving. In particular, George Pólya's 1945 book 'How to Solve It', is a small volume describing methods of problem–solving. It suggests the following steps when solving a mathematical problem:

1. First, you have to understand the problem.
2. After understanding, then make a plan.
3. Carry out the plan.
4. Look back on your work. How could it be better?

If this technique fails, Polya advises: "If you cannot solve a problem, then there is an easier problem you can solve: find it." Or, "If you cannot solve the proposed problem try to solve first some related problem. Could you imagine a more accessible related problem?"

His small book contains a dictionary–style set of heuristics, many of which have to do with generating a more accessible problem, like the ones given in the table below:

The technique 'have I used everything' is perhaps most applicable to formal educational examinations (e.g., n men digging m ditches, see footnote below) problems. The book has achieved 'classic' status because of its considerable influence. Marvin Minsky said in his influential paper 'Steps Toward Artificial Intelligence': "And everyone should know the work of George Polya on how to solve problems." Polya's book has had a large influence on mathematics textbooks. Most formulations of a problem solving framework in U.S. textbooks attribute some relationship to Polya's problem solving stages. Other books on problem solving are often related to less concrete and more creative techniques, like e.g., lateral thinking, mind mapping and brainstorming (see below).

On the other hand, in Europe, two main approaches have surfaced, one initiated by Donald Broadbent in the UK [Bro77, BB95] and the other one by Dietrich Dörner in Germany [Dor75, DV85, DW95]. The two approaches have in common an emphasis on relatively complex, semantically rich, computerized laboratory tasks, constructed to resemble 'real–life' problems. The approaches differ

Heuristic	Informal Description	Formal analogue
Analogy	can you find a problem analogous to your problem and solve that?	Map
Generalization	can you find a problem more general than your problem...?	Generalization
Induction	can you solve your problem by deriving a generalization from some examples?	Induction
Variation of the Problem	can you vary or change your problem to create a new problem (or set of problems) whose solution(s) will help you solve your original problem?	Search
Auxiliary Problem	can you find a subproblem or side problem whose solution will help you solve your problem?	Subgoal
Here is a problem related to yours and solved before	can you find a problem related to yours that has already been solved and use that to solve your problem?	Pattern recognition Pattern matching
Specialization	can you find a problem more specialized?	Specialization
Decomposing and Recombining	can you decompose the problem and "recombine its elements in some new manner"?	Divide and conquer
Working backward	can you start with the goal and work backwards to something you already know?	Backward chaining
Draw a Figure	can you draw a picture of the problem?	Diagrammatic Reasoning
Auxiliary Elements	can you add some new element to your problem to get closer to a solution?	Extension

somewhat in their theoretical goals and methodology, however. The tradition initiated by Broadbent emphasizes the distinction between cognitive problem–solving processes that operate under awareness versus outside of awareness, and typically employs mathematically well–defined computerized systems. The tradition initiated by Dörner, on the other hand, has an interest in the interplay of the cognitive, motivational, and social components of problem solving, and utilizes very complex computerized scenarios that contain up to 2,000 highly interconnected variables.

To sum up, researchers' realization that problem–solving processes differ across knowledge domains and across levels of expertise (see, e.g. [Ste95]) and that, consequently, findings obtained in the laboratory cannot necessarily generalize to problem–solving situations outside the laboratory, has during the past two decades led to an emphasis on real–world problem solving. This emphasis has been expressed quite differently in North America and Europe, however. Whereas North American research has typically concentrated on studying problem solving in separate, natural knowledge domains, much of the European research has focused on novel, complex problems, and has been performed with computerized scenarios (see [Fun95], for an overview).

Characteristics of Difficult Problems

As elucidated by D. Dorner and later expanded upon by J. Funke, difficult problems have some typical characteristics. Recategorized and somewhat reformulated from these original works, these characteristics can be summarized as follows:

Intransparency (lack of clarity of the situation), including commencement opacity and continuation opacity;

Polytely (multiple goals), including inexpressiveness, opposition and transience;

Complexity (large numbers of items, interrelations, and decisions), including enumerability, connectivity (hierarchy relation, communication relation, allocation relation), and heterogeneity;

Dynamism (time considerations), including temporal constraints, temporal sensitivity, phase effects, and dynamic unpredictability.

The resolution of difficult problems requires a direct attack on each of these characteristics that are encountered.

Some *standard problem–solving techniques*, also known as creativity techniques, include:

1. Trial–and–error;[32]

[32] Trial and error (also known in computer science literature as generate and test and as 'guess and check' when solving equations in elementary algebra) is a method of problem solving for obtaining knowledge, both propositional knowledge and know-how.

This approach can be seen as one of the two basic approaches to problem solving and is contrasted with an approach using insight and theory.

In trial and error, one selects (or, generates) a possible answer, applies it to the problem and, if it is not successful, selects (or generates) another possibility that is subsequently tried. The process ends when a possibility yields a solution.

In some versions of trial and error, the option that is a priori viewed as the most likely one should be tried first, followed by the next most likely, and so on until a solution is found, or all the options are exhausted. In other versions, options are simply tried at random.

This approach is most successful with simple problems and in games, and is often resorted to when no apparent rule applies. This does not mean that the approach need be careless, for an individual can be methodical in manipulating the variables in an effort to sort through possibilities that may result in success. Nevertheless, this method is often used by people who have little knowledge in the problem area.

Trial and error has a number of features:

solution-oriented: trial and error makes no attempt to discover why a solution works, merely that it is a solution.

problem-specific: trial and error makes no attempt to generalize a solution to other problems.

2. Brainstorming;[33]

---

non-optimal: trial and error is an attempt to find a solution, not all solutions, and not the best solution.

needs little knowledge: trial and error can proceed where there is little or no knowledge of the subject.

For example, trial and error has traditionally been the main method of finding new drugs, such as antibiotics. Chemists simply try chemicals at random until they find one with the desired effect.

The *scientific method* can be regarded as containing an element of trial and error in its formulation and testing of hypotheses. Also compare *genetic algorithms*, *simulated annealing* and *reinforcement learning* – all varieties of search which apply the basic idea of trial and error.

Biological Evolution is also a form of trial and error. Random mutations and sexual genetic variations can be viewed as trials and poor reproductive fitness as the error. Thus after a long time 'knowledge' of well–adapted genomes accumulates simply by virtue of them being able to reproduce.

Bogosort can be viewed as a trial and error approach to sorting a list.

In mathematics the method of trial and error can be used to solve formulae – it is a slower, less precise method than algebra, but is easier to understand.

[33] Brainstorming is a creativity technique of generating ideas to solve a problem. The main result of a brainstorm session may be a complete solution to the problem, a list of ideas for an approach to a subsequent solution, or a list of ideas resulting in a plan to find a solution. Brainstorming was originated in 1953 in the book 'Applied Imagination' by Alex Osborn, an advertising executive. Other methods of generating ideas are individual ideation and the morphological analysis approach.

Brainstorming has many applications but it is most often used in:

New product development – obtaining ideas for new products and improving existing products

Advertising – developing ideas for advertising campaigns

Problem solving – issues, root causes, alternative solutions, impact analysis, evaluation

Process management – finding ways of improving business and production processes

Project Management – identifying client objectives, risks, deliverables, work packages, resources, roles and responsibilities, tasks, issues

Team building – generates sharing and discussion of ideas while stimulating participants to think

Business planning – develop and improve the product idea.

Trial preparation by attorneys.

Brainstorming can be done either individually or in a group. In group brainstorming, the participants are encouraged, and often expected, to share their ideas with one another as soon as they are generated. Complex problems or brainstorm sessions with a diversity of people may be prepared by a chairman. The chairman is the leader and facilitator of the brainstorm session.

The key to brainstorming is to not interrupt the thought process. As ideas come to mind, they are captured and stimulate the development of better ideas. Thus a group brainstorm session is best conducted in a moderate–sized room, and participants sit so that they can all look at each–other. A flip chart, blackboard, or overhead

3. Morphological box;[34]
4. Method of focal objects;[35]
5. Lateral thinking;[36]

---

projector is placed in a prominent location. The room is free of telephones, clocks, or any other distractions.

[34] Morphological analysis was designed for multi-dimensional, non-quantifiable problems where causal modelling and simulation do not function well or at all. Fritz Zwicky developed this approach to seemingly non-reducible complexity [Zwi69]. Using the technique of cross consistency assessment (CCA), the system however does allow for reduction, not by reducing the number of variables involved, but by reducing the number of possible solutions through the elimination of the illogical solution combinations in a grid box.

[35] The technique of *focal objects* for problem solving involves synthesizing the seemingly non–matching characteristics of different objects into something new.

For example, to generate new solutions to gardening take some ideas at random, such swimming and a couch, and invent ways for them to merge. Swimming might be used with the idea of gardening to create a plant oxygen tank for underwater divers. A couch might be used with the idea of gardening to invent new genes that would grow plants into the shape of a couch. The larger the number of diverse objects included, the greater the opportunity for inventive solutions.

Another way to think of focal objects is as a memory cue: if you're trying to find all the different ways to use a brick, give yourself some random 'objects' (situations, concepts, etc.) and see if you can find a use. Given 'blender', for example, I would try to think of all the ways a brick could be used with a blender (as a lid?). Another concept for the brick game: find patterns in your solutions, and then break those patterns. If you keep finding ways to build things with bricks, think of ways to use bricks that don't involve construction. Pattern–breaking, combined with focal object cues, can lead to very divergent solutions.

[36] Lateral thinking is a term coined by Edward de Bono [Bon73], a Maltese psychologist, physician, and writer, although it may have been an idea whose time was ready; the notion of lateral truth is discussed by Robert M. Pirsig in Zen and the Art of Motorcycle Maintenance. de Bono defines Lateral Thinking as methods of thinking concerned with changing concepts and perception. For example:

It took two hours for two men to dig a hole five feet deep. How deep would it have been if ten men had dug the hole for two hours?

The answer appears to be 25 feet deep. This answer assumes that the thinker has followed a simple mathematical relationship suggested by the description given, but we can generate some lateral thinking ideas about what affects the size of the hole which may lead to different answers:

A hole may need to be of a certain size or shape so digging might stop early at a required depth.

The deeper a hole is, the more effort is required to dig it, since waste soil needs to be lifted higher to the ground level. There is a limit to how deep a hole can be dug by manpower without use of ladders or hoists for soil removal, and 25 feet is beyond this limit.

Deeper soil layers may be harder to dig out, or we may hit bedrock or the water table.

Each man digging needs space to use a shovel.

6. Mind mapping;[37]

It is possible that with more people working on a project, each person may become less efficient due to increased opportunity for distraction, the assumption he can slack off, more people to talk to, etc.

More men could work in shifts to dig faster for longer.

There are more men but are there more shovels?

The two hours dug by ten men may be under different weather conditions than the two hours dug by two men.

Rain could flood the hole to prevent digging.

Temperature conditions may freeze the men before they finish.

Would we rather have 5 holes each 5 feet deep?

The two men may be an engineering crew with digging machinery.

What if one man in each group is a manager who will not actually dig?

The extra eight men might not be strong enough to dig, or much stronger than the first two.

The most useful ideas listed above are outside the simple mathematics implied by the question. Lateral thinking is about reasoning that is not immediately obvious and about ideas that may not be obtainable by using only traditional step–by–step logic.

Techniques that apply lateral thinking to problems are characterized by the shifting of thinking patterns away from entrenched or predictable thinking to new or unexpected ideas. A new idea that is the result of lateral thinking is not always a helpful one, but when a good idea is discovered in this way it is usually obvious in hindsight, which is a feature lateral thinking shares with a joke.

Lateral thinking can be contrasted with critical thinking, which is primarily concerned with judging the truth value of statements and seeking error. Lateral Thinking is more concerned with the movement value of statements and ideas, how to move from them to other statements and ideas.

For example the statement 'cars should have square wheels' when considered with critical thinking would be evaluated as a poor suggestion, as there are many engineering problems with square wheels. The Lateral Thinking treatment of the same statement would be to see where it leads. Square wheels would produce predictable bumps. If bumps can be predicted then suspension can be designed to compensate. Another way to predict bumps would be a laser or sonar on the front of the car examining the road surface ahead. This leads to the idea of active suspension with a sensor on the car that has normal wheels. The initial statement has been left behind.

[37] Recall that a *mind map* is a diagram used to represent words, ideas, tasks or other items linked to and arranged radially around a central key word or idea. It is used to generate, visualize, structure and classify ideas, and as an aid in study, organization, problem solving, and decision making.

It is an image–centered diagram that represents semantic or other connections between portions of information. By presenting these connections in a radial, nonlinear graphical manner, it encourages a brainstorming approach to any given organizational task, eliminating the hurdle of initially establishing an intrinsically appropriate or relevant conceptual framework to work within.

A mind map is similar to a semantic network or cognitive map but there are no formal restrictions on the kinds of links used.

Most often the map involves images, words, and lines. The elements are arranged intuitively according to the importance of the concepts and they are organized into

7. Analogy with similar problems;[38] and

groupings, branches, or areas. The uniform graphic formulation of the semantic structure of information on the method of gathering knowledge, may aid recall of existing memories.

People have been using image centered radial graphic organization techniques referred to variably as mental or generic mind maps for centuries in areas such as engineering, psychology, and education, although the claim to the origin of the mind map has been made by a British popular psychology author, Tony Buzan.

The mind map continues to be used in various forms, and for various applications including learning and education (where it is often taught as 'Webs' or 'Webbing'), planning and in engineering diagramming.

When compared with the earlier original concept map (which was developed by learning experts in the 1960s) the structure of a mind map is a similar, but simplified, radial by having one central key word.

Mind maps have many applications in personal, family, educational, and business situations, including note-taking, brainstorming (wherein ideas are inserted into the map radially around the center node, without the implicit prioritization that comes from hierarchy or sequential arrangements, and wherein grouping and organizing is reserved for later stages), summarizing, revising and general clarifying of thoughts. For example, one could listen to a lecture and take down notes using mind maps for the most important points or keywords. One can also use mind maps as a mnemonic technique or to sort out a complicated idea. Mind maps are also promoted as a way to collaborate in color pen creativity sessions.

[38] Recall that analogy is either the cognitive process of transferring information from a particular subject (the analogue or source) to another particular subject (the target), or a linguistic expression corresponding to such a process. In a narrower sense, analogy is an inference or an argument from a particular to another particular, as opposed to deduction, induction, and abduction, where at least one of the premises or the conclusion is general. The word analogy can also refer to the relation between the source and the target themselves, which is often, though not necessarily, a similarity, as in the biological notion of analogy.

Niels Bohr's model of the atom made an analogy between the atom and the solar system.Analogy plays a significant role in problem solving, decision making, perception, memory, creativity, emotion, explanation and communication. It lies behind basic tasks such as the identification of places, objects and people, for example, in face perception and facial recognition systems. It has been argued that analogy is 'the core of cognition'. Specifically analogical language comprises exemplification, comparisons, metaphors, similes, allegories, and parables, but not metonymy. Phrases like and so on, and the like, as if, and the very word like also rely on an analogical understanding by the receiver of a message including them. Analogy is important not only in ordinary language and common sense, where proverbs and idioms give many examples of its application, but also in science, philosophy and the humanities. The concepts of association, comparison, correspondence, homomorphism, iconicity, isomorphism, mathematical homology, metaphor, morphological homology, resemblance, and similarity are closely related to analogy. In cognitive linguistics, the notion of conceptual metaphor may be equivalent to that of analogy.

Analogy has been studied and discussed since classical antiquity by philosophers, scientists and lawyers. The last few decades have shown a renewed interest in analogy, most notable in cognitive science.

8. Research.[39]

23.2 Situation Awareness and Decision Making

Decision making is possibly one of the most important aspects of human activities. And an adequate *situation awareness* (SA) is one of the necessary elements for optimal decision making [SI09]. Recall from The Law of Mental Training, that the most general SA–definition (due to M. Endsley) reads: "Situation awareness is the perception of the elements in the environment within a volume of time and space, the comprehension of their meaning, and projection of their status

---

With respect to the terms source and target, there are two distinct traditions of usage:

The logical and mathematical tradition speaks of an arrow, homomorphism, mapping, or morphism from what is typically the more complex domain or source to what is typically the less complex codomain or target, using all of these words in the sense of mathematical category theory.

The tradition that appears to be more common in cognitive psychology, literary theory, and specializations within philosophy outside of logic, speaks of a mapping from what is typically the more familiar area of experience, the source, to what is typically the more problematic area of experience, the target.

[39] Research is often described as an active, diligent, and systematic process of inquiry aimed at discovering, interpreting, and revising facts. This intellectual investigation produces a greater understanding of events, behaviors, or theories, and makes practical applications through laws and theories. The term research is also used to describe a collection of information about a particular subject, and is usually associated with science and the scientific method.

The word research derives from Middle French; its literal meaning is 'to investigate thoroughly'.

Thomas Kuhn, in his book 'The Structure of Scientific Revolutions', traces an interesting history and analysis of the enterprize of research.

Basic research (also called fundamental or pure research) has as its primary objective the advancement of knowledge and the theoretical understanding of the relations among variables. It is exploratory and often driven by the researcher's curiosity, interest, or hunch. It is conducted without any practical end in mind, although it may have unexpected results pointing to practical applications. The terms "basic" or "fundamental" indicate that, through theory generation, basic research provides the foundation for further, sometimes applied research. As there is no guarantee of short-term practical gain, researchers often find it difficult to get funding for basic research. Research is a subset of invention.

Applied research is done to solve specific, practical questions; its primary aim is not to gain knowledge for its own sake. It can be exploratory, but is usually descriptive. It is almost always done on the basis of basic research. Applied research can be carried out by academic or industrial institutions. Often, an academic institution such as a university will have a specific applied research program funded by an industrial partner interested in that program. Common areas of applied research include electronics, informatics, computer science, material science, process engineering, drug design...

in the near future." It is obvious that the environment and the characteristics of the space within it are dynamic and nonlinear. Therefore, SA is both dynamic and nonlinear. Further, there is a limit on maximum conceivable SA. Similar to any natural system, increase in SA cannot continue forever despite a continuous supply of new useful information. The factors that limit this continuous increase in SA include: (i) limitations in human perception and memory capacity, memory loss, mental fatigue and vigilance; and (ii) cognitive distractions. Based on these two assumptions, Sharma and Ivancevic (2009) have defined both a simple, intuitive SA-model and its $n-$dimensional generalization.

23.2.1 Simple, Intuitive SA–Model

The simple SA–model is defined by a discrete-time logistic equation:

$$S_{n+1} = BS_n(1 - S_n), \tag{23.6}$$

where S_{n+1} is the next SA, B is the SA growth-rate parameter, representing the amount of useful information flow per unit time, S_n is the present SA, and $(1 - S_n)$ keeps the maximum conceivable SA within limits by accommodating the influence of factors such as perceptual limitations, memory capacity, memory loss, mental fatigue, vigilance and cognitive distractions. The model (23.6) describes the next SA, S_{n+1}, as a function of the present SA, S_n, and the growth-rate parameter B. In this model, the term $(1 - S_n)$ keeps the SA within limits because as S_n increases, $(1 - S_n)$ decreases. In this simple model, SA is expressed as a fraction between zero and 1. Zero represents an absence of any SA, whereas 1 represents the maximum conceivable SA. It is worthwhile noting that the growth-rate parameter B is an important feature of the model.

The nonlinear dynamical characteristics of SA, embodied in the simple model (23.6) show that the supply rate of useful information is an important parameter that influences the quality of SA. A higher rate of information supply does not necessarily improve SA. On the contrary, if the supply rate is increased beyond a certain threshold, SA starts bifurcating. A further increase in this rate results in SA becoming chaotic. Such bifurcation and chaotic behavior result in SA degradation. Therefore, it is argued herein that the highest level of SA can be achieved by maintaining the supply rate of useful information just below the threshold where SA first starts bifurcating.

23.2.2 High–Dimensional and Time–Varying SA–Model

The simplistic SA–model (23.6) is further expanded to accommodate various SA dimensions, such as perceptual limitations, memory capacity, memory loss, mental fatigue, vigilance and cognitive distractions. The basic model (1) is generalized into an $n-$dimensional SA dynamics model, defined by the vector–matrix evolution equation in both discrete time (denoted by index t) and continuous time (denoted by a continuous–time growth–rate function $B_{ik}(t)$):

$$S_i^{t+1} = \sum_{k=1}^{n} B_{ik}(t) S_k^t (1 - S_i^t), \tag{23.7}$$

where upper index t denotes a discrete time–evolution step, while lower indices $i, k = 1, ..., n$ represent various SA dimensions. In (23.7), S_i^t represents the vector with normalized SA dimensions at a discrete time–step t:

$$\begin{pmatrix} S_1^t \\ S_2^t \\ ... \\ S_n^t \end{pmatrix} = \begin{pmatrix} \text{perceptual limitations}^t \\ \text{mental fatigue}^t \\ ... \\ \text{vigilance}^t \end{pmatrix},$$

while $S_i^{t+1}$is the next SA vector:

$$\begin{pmatrix} S_1^{t+1} \\ S_2^{t+1} \\ ... \\ S_n^{t+1} \end{pmatrix} = \begin{pmatrix} \text{perceptual limitations}^{t+1} \\ \text{mental fatigue}^{t+1} \\ ... \\ \text{vigilance}^{t+1} \end{pmatrix}.$$

$B_{ik}(t)$ is the $n \times n$ growth-parameter matrix with time-varying elements:

$$\begin{pmatrix} B_{11}(t) & B_{12}(t) & B_{1n}(t) \\ B_{21}(t) & B_{22}(t) & B_{2n}(t) \\ & & \\ B_{n1}(t) & B_{n2}(t) & B_{nn}(t) \end{pmatrix}.$$

23.2.3 Nonlinear Analysis of the SA–Model

The comprehensive study of the SA–models (23.6) and (23.7) by Sharma and Ivancevic [SI09] has suggested that the quality of SA depends on the growth-rate parameter, which represents the amount of useful information supplied per unit time. Further, the study showed the effect of variation of this parameter on SA using logistic maps and a bifurcation diagram. At the beginning, the increase in the parameter value resulted in a steady growth in SA. However, once the value attained a certain threshold, SA started bifurcating. An additional increase in this value resulted in more frequent bifurcations. When the value was raised even more, SA started behaving in a chaotic manner. A further increase in the value resulted in bifurcation and chaos again. The presence of such complex chaotic behavior was re-affirmed by computing the Lyapunov exponent for the logistic model of SA dynamics. One possible explanation for such bifurcation and chaos may be that the supply rate of useful information may have exceeded the human cognitive limitations. During bifurcation and chaos, SA is in an unstable state and keeps oscillating. The study suggested that such unstable SA represents uncertainty and, therefore, is undesirable. The results suggest that the supply of useful information at a higher rate (more information) does not always enhance SA. Instead, such supply may often disrupt or reduce SA. Therefore, it is argued that the growth-rate parameter just below the first bifurcation threshold should be selected to achieve the highest SA level. This is an important finding in SA, which may assist in improving the effectiveness of human decision making. The finding has a wide application ranging from commercial to military decision making. This new knowledge is invaluable to the individuals responsible for managing the community, enterprize or the nation to optimize their decisions. The knowledge is also useful in the design of command and control systems.

This knowledge is also crucial in modern tennis: it allows you to focus on the ball and yet to know what your opponent is doing or planning to do.

23.3 Quantum Brain and Cognition

In this section we present several aspects of the current research in quantum brain.

23.3.1 Biochemistry of Microtubules

Recent developments/efforts to understand aspects of the brain function at the *sub–neural* level are discussed in [II08c]. Microtubules (MTs), protein polymers constructing the cytoskeleton of a neuron, participate in a wide variety of dynamical processes in the cell. Of special interest for this subsection is the MTs participation in bio–information processes such as *learning* and *memory*, by possessing a well–known binary error–correcting code $[K_1(13, 2^6, 5)]$ with 64 words. In fact, MTs and DNA/RNA are *unique* cell structures that possess a code system. It seems that the MTs' code system is strongly related to a kind of *mental code* in the following sense. The MTs' periodic paracrystalline structure make them able to support a *superposition* of coherent quantum states, as it has been recently conjectured by Hameroff and Penrose [HP96], representing an *external* or *mental order*, for sufficient time needed for *efficient quantum computing* [II07b, II08c].

Living organisms are collective assemblies of cells which contain collective assemblies of organized material, including membranes, organelles, nuclei, and the *cytoplasm*, the bulk interior medium of living cells. Dynamic rearrangements of the cytoplasm within *eucaryotic cells*, the cells of all animals and almost all plants on Earth, account for their changing shape, movement, etc. This extremely important cytoplasmic structural and dynamical organization is due to the presence of networks of interconnected protein polymers, which are referred to as the *cytosceleton* due to their bone–like structure [HP96]. The cytoskeleton consists of MT's, actin micro-filaments, intermediate filaments and an *organizing complex*, the *centrosome* with its chief component the *centriole*, built from two bundles of microtubules in a separated **T** shape. Parallel–arrayed MTs are interconnected by cross–bridging proteins (*MT–Associated Proteins*: MAPs) to other MTs, organelle filaments and membranes to form *dynamic networks* [HP96]. MAPs may be contractile, structural, or enzymatic. A very important role is played by contractile MAPs, like dynein and kinesin, through their participation in cell movements as well as in intra-neural, or axoplasmic transport which moves material and thus is of fundamental importance for the *maintenance* and *regulation* of *synapses* (see, e.g., [Ecc64]). The structural bridges formed by MAPs stabilize MTs and prevent their disassembly. The MT–MAP 'complexes' or *cyto-sceletal networks* determine the cell architecture and dynamic functions, such a *mitosis*, or *cell division*, *growth*, *differentiation*, *movement*, and for us here the very crucial, *synapse formation and function*, all essential to the living state. It is usually said that *microtubules* are ubiquitous through the entire biology [HP96].

MTs are hollow cylinders comprised of an exterior surface of cross–section diameter 25 nm ($1\ nm = 10^{-9}$ meters) with 13 arrays (protofilaments) of protein dimers called tubulines. The interior of the cylinder, of cross–section diameter 14 nm, contains *ordered water* molecules, which implies the existence of an electric dipole moment and an electric field. The arrangement of the dimers is such that, if one ignores their size, they resemble triangular lattices on the MT surface. Each dimer consists of two hydrophobic protein pockets, and has an unpaired electron. There are two possible positions of the electron, called α and β *conformations*. When the electron is in the $\beta-$conformation there is a 29° distortion of the electric dipole moment as compared to the α conformation.

In standard models for the simulation of the MT dynamics (see [II07b, II08c] and references therein), the 'physical' DOF – relevant for the description of the energy transfer – is the projection of the electric dipole moment on the longitudinal symmetry axis ($x-$axis) of the MT cylinder. The 29° distortion of the $\beta-$conformation leads to a displacement u_n along the $x-$axis, which is thus the relevant physical DOF. There has been speculation for quite some time that MTs are involved in information processing: it has been shown that the particular geometrical arrangement (packing) of the tubulin proto-filaments obeys an error–correcting mathematical code known as the $K_2(13, 2^6, 5)-$code. Error correcting codes are also used in classical computers to protect against errors while in quantum computers special error correcting algorithms are used to protect against errors by preserving quantum coherence among qubits.

Information processing occurs via interactions among the MT proto–filament chains. The system may be considered as similar to a model of *interacting Ising chains* on a triangular lattice, the latter being defined on the plane stemming from filleting open and flattening the cylindrical surface of MT. Classically, the various dimers can occur in either α or β conformations. Each dimer is influenced by the neighboring dimers resulting in the possibility of a transition. This is the basis for classical information processing, which constitutes the picture of a (classical) cellular automaton.

23.3.2 Kink Soliton Model of MT–Dynamics

The *quantum nature* of an MT network results from the *assumption* that each dimer finds itself in a *superposition* of α and β conformations. Viewed as a *two–state quantum mechanical system*, the MT tubulin dimers couple to conformational changes with $10^{-9} - 10^{-11}$sec transitions, corresponding to an angular frequency $\omega \sim \mathcal{O}(10^{10}) - \mathcal{O}(10^{12})$ Hz (see [II07b, II08c]).

The *quantum computer* character of the MT network [Pen89] results from the assumption that each dimer finds itself in a superposition of α and β conformations [Ham87]. There is a macroscopic coherent state among the various chains, which lasts for $\mathcal{O}(1\,\mathrm{sec})$ and constitutes the 'preconscious' state. The interaction of the chains with (non–critical stringy) quantum gravity, then, induces self–collapse of the wave function of the coherent MT network, resulting in quantum computation [II07b, II08c].

Let u_n be the displacement field of the nth dimer in a MT chain. The continuous approximation proves sufficient for the study of phenomena associated with energy transfer in biological cells, and this implies that one can make the replacement

$$u_n \to u(x,t), \tag{23.8}$$

with x a spatial coordinate along the longitudinal symmetry axis of the MT. There is a time variable t due to fluctuations of the displacements $u(x)$ as a result of the dipole oscillations in the dimers.

The effects of the neighboring dimers (including neighboring chains) can be phenomenologically accounted for by an effective potential $V(u)$. In the kink–soliton model[40] a double–well potential was used, leading to a classical kink solution for the $u(x,t)$ field. More complicated interactions are allowed in this picture, where more generic polynomial potentials have been considered.

The effects of the surrounding water molecules can be summarized by a *viscous force* term that damps out the dimer oscillations,

$$F = -\gamma \partial_t u, \tag{23.9}$$

with γ determined phenomenologically at this stage. This friction should be viewed as an environmental effect, which however does not lead to energy dissipation, as a result of the non–trivial solitonic structure of the ground–state and the non–zero constant force due to the electric field. This is a well known result, directly relevant to energy transfer in biological systems.

In mathematical terms, the effective equation of motion for the relevant field DOF $u(x,t)$ reads:

$$u''(\xi) + \rho u'(\xi) = P(u), \tag{23.10}$$

where $\xi = x - vt$, $u'(\xi) = du/d\xi$, v is the velocity of the soliton, $\rho \propto \gamma$ [II07b, II08c], and $P(u)$ is a polynomial in u, of a certain degree, stemming from the variations of the potential $V(u)$ describing interactions among the MT chains. In the mathematical literature there has been a classification of solutions of equations of this form. For certain forms of the potential the solutions include *kink solitons* that may be responsible for dissipation–free energy transfer in biological cells:

$$u(x,t) \sim c_1 \left(\tanh[c_2(x - vt)] + c_3\right), \tag{23.11}$$

[40] Recall that kinks are solitary (non–dispersive) waves arising in various 1D (bio)physical systems.

where c_1, c_2, c_3 are constants depending on the parameters of the dimer lattice model. There are solitons of the form $u(x,t) = c_1' + \frac{c_2' - c_1'}{1+e^{c_3'(c_2'-c_1')(x-vt)}}$, where again c_i', $i = 1, \dots 3$ are appropriate constants.

A *semiclassical quantization* of such solitonic states has been considered by Ellis *et al.* (see [II07b, II08c]). The result of such a quantization yields a *modified soliton equation* for the (quantum corrected) field $u_q(x,t)$:

$$\partial_t^2 u_q(x,t) - \partial_x^2 u_q(x,t) + \mathcal{M}^{(1)}[u_q(x,t)] = 0, \tag{23.12}$$

with the notation

$$M^{(n)} = e^{\frac{1}{2}(G(x,y,t)-G_0(x,y))\frac{\partial^2}{\partial z^2}} U^{(n)}(z)|_{z=u_q(x,t)}, \qquad U^{(n)} \equiv d^n U/dz^n.$$

The quantity U denotes the potential of the original soliton Hamiltonian, and $G(x,y,t)$ is a bilocal field that describes quantum corrections due to the modified boson field around the soliton. The quantities $M^{(n)}$ carry information about the quantum corrections. For the kink soliton (23.11) the quantum corrections (23.12) have been calculated , thereby providing us with a concrete example of a large–scale quantum coherent state.

A typical propagation velocity of the kink solitons is $v \sim 2$ m/sec, although, models with $v \sim 20$ m/sec have also been considered. This implies that, for moderately long microtubules of length $L \sim 10^{-6}$ m, such kinks transport energy without dissipation in

$$t_F \sim 5 \times 10^{-7} \text{ s}. \tag{23.13}$$

Such time scales are comparable to, or smaller in magnitude than, the decoherence time scale of the above–described coherent (solitonic) states $u_q(x,t)$. This implies the possibility that fundamental quantum mechanical phenomena may then be responsible for frictionless energy (and signal) transfer across microtubular arrangements in the cell (see [II07b, II08c]).

23.3.3 Macro– and Microscopic Neurodynamical Self–similarity

Neuro– and psycho–dynamics have its physical behavior both on the macroscopic, classical, inter–neuronal level [II07d], and on the microscopic, quantum, intra–neuronal level [II07b, II08a]. On the macroscopic level, various models of neural networks (NNs, for short) have been proposed as goal–oriented models of the specific neural functions, like for instance, function–approximation, pattern–recognition, classification, or control. In the physically–based, Hopfield–type models of NNs [Hop82, Hop84] the information is stored as a content–addressable memory in which synaptic strengths are modified after the Hebbian rule (see [Heb49]). Its retrieval is made when the network with the symmetric couplings works as the point–attractor with the fixed points. Analysis of both activation and learning dynamics of Hopfield–Hebbian NNs using the techniques of statistical mechanics (see, e.g. [II08b]), provides us with the most important information of storage capacity, role of noise and recall performance [II06b, II07d].

Conversely, an indispensable role of quantum theory in the brain dynamics was emphasized recently. On the general microscopic intra–cellular level, energy transfer across the cells, without dissipation, had been first conjectured to occur in biological matter by [FK83]. The phenomenon conjectured by them was based on their 1D superconductivity model: in one dimensional electron systems with holes, the formation of solitonic structures due to electron–hole pairing results in the transfer of electric current without dissipation. In a similar manner, Frölich and Kremer conjectured that energy in biological matter could be transferred without dissipation, if appropriate solitonic structures are formed inside the cells. This idea has lead theorists to construct various models for the energy transfer across the cell, based on the formation of kink classical solutions.

The interior of living cells is structurally and dynamically organized by cytoskeletons, i.e., networks of protein polymers. Of these structures, microtubules (MTs, for short) appear to be the most fundamental. Their dynamics has been studied by a number of authors in connection with the mechanism responsible for dissipation-free energy transfer. Hameroff and his colleagues [HP96, HHT02] have conjectured another fundamental role for the MTs, namely being responsible for quantum computations in the human neurons. Penrose [Pen89, Pen94, Pen97, Pen98] further argued that the latter is associated with certain aspects of quantum theory that are believed to occur in the cytoskeleton MTs, in particular quantum superposition and subsequent collapse of the wave function of coherent MT networks. It has been suggested that the neural MTs are the microsites for the emergence of stable, macroscopic quantum coherent states, identifiable with the preconscious states; stringy–quantum space-time effects trigger an organized collapse of the coherent states down to a specific or conscious state.

In particular, MTs in the cytoskeletons of eukaryotic cells provide a wide range of micro–skeletal and micro–muscular functionalities. Some evidence has indicated that they can serve as a medium for intracellular signaling processing. For the inherent symmetry structures and the electric properties of tubulin dimers, the microtubule (MT) was treated as a 1D ferroelectric system (see [II07b, II08c]). The nonlinear dynamics of the dimer electric dipoles was described by virtue of the double–well potential and the physical problem was further mapped onto the pseudo–spin system, taking into account the effect of the external electric field on the MT.

More precisely, MTs are polymers of tubulin subunits (dimers) arranged on a hexagonal lattice. Each tubulin dimer comprises two monomers, the $\alpha-$tubulin and $\beta-$tubulin, and can be found in two states. In the first state a mobile negative charge is located into the $\alpha-$tubulin monomer and in the second into the $\beta-$tubulin monomer. Each tubulin dimer is modelled as an electrical dipole coupled to its neighbors by electrostatic forces. The location of the mobile charge in each dimer depends on the location of the charges in the dimer's neighborhood. Mechanical forces that act on the microtubule affect the distances between the dimers and alter the electrostatic potential. Changes in this potential affect the mobile negative charge location in each dimer and the charge distribution in the microtubule. The net effect is that mechanical forces affect the charge distribution in microtubules. Various models of the mind have been based on the idea that neuron MTs can perform computation. From this point of view, information processing is the fundamental issue for understanding the brain mechanisms that produce consciousness. The cytoskeleton polymers could store and process information through their dynamic coupling mediated by mechanical energy. The problem of information transfer and storage in brain microtubules was analyzed, considering them as a communication channel. Therefore, we have two space-time biophysical scales of neuro- and psycho-dynamics: classical and quantum. Naturally the question arises: are these two scales somehow inter-related, is there a space-time self-similarity between them?

The purpose of the present section is to prove the formal positive answer to the self-similarity question. We try to describe neurodynamics on both physical levels by the unique form of a single equation, namely open Liouville equation: NN–dynamics using its classical form, and MT–dynamics using its quantum form in the Heisenberg picture. If this formulation is consistent, that would prove the existence of the formal neuro-biological space-time self-similarity. Even more, this would prove the existence of a Neurodynamical Law, which acts on different scales of brain's functioning.

Open Liouville Equation

Hamiltonian Framework

Suppose that on the macroscopic NN–level we have a conservative Hamiltonian system acting in a $2ND$ symplectic phase space $T^*Q = \{q^i(t), p_i(t)\}$, $i = 1 \dots N$ (which is the cotangent bundle of the NN–configuration manifold $Q = \{q^i\}$), with a Hamiltonian function $H = H(q^i, p_i, t) : T^*Q \times \mathbb{R} \to \mathbb{R}$ (see [II06a, II06c]). The conservative dynamics is defined by classical Hamilton's canonical equations:

$$\begin{aligned} \dot{q}^i &= \partial_{p_i} H \quad \text{– contravariant velocity equation}, \\ \dot{p}_i &= -\partial_{q^i} H \quad \text{– covariant force equation}, \end{aligned} \tag{23.14}$$

(here and henceforth overdot denotes the total time derivative). Within the framework of the conservative Hamiltonian system (23.58) we can apply the formalism of classical Poisson brackets: for any two functions $A = A(q^i, p_i, t)$ and $B = B(q^i, p_i, t)$ their Poisson bracket is (using the summation convention) defined as [II06c, II07e]

$$[A, B] = (\partial_{q^i} A \, \partial_{p_i} B - \partial_{p_i} A \, \partial_{q^i} B).$$

Conservative Classical System

Any function $A(q^i, p_i, t)$ is called a constant (or integral) of motion of the conservative system (23.58) if [II06c, II07e]

$$\dot{A} \equiv \partial_t A + [A, H] = 0, \qquad \text{which implies} \qquad \partial_t A = -[A, H]. \tag{23.15}$$

For example, if $A = \rho(q^i, p_i, t)$ is a density function of ensemble phase–points (or, a probability density to see a state $\mathbf{x}(t) = (q^i(t), p_i(t))$ of ensemble at a moment t), then equation

$$\partial_t \rho = -[\rho, H] \qquad (23.16)$$

represents the *Liouville theorem*, which is usually derived from the continuity equation

$$\partial_t \rho + div(\rho\, \dot{\mathbf{x}}) = 0\,.$$

Conserved quantity here is the Hamiltonian function $H = H(q^i, p_i, t)$, which the sum of kinetic and potential energy. For example, in case of an ND harmonic oscillator, we have the phase space $M = T^*\mathbb{R}^N \simeq \mathbb{R}^{2N}$, with the symplectic form $\omega = dp_i \wedge dq^i$ and Hamiltonian (total energy) function:

$$H = \frac{1}{2}\sum_{i=1}^{N}\left[p_i^2 + (q^i)^2\right]. \qquad (23.17)$$

The corresponding Hamiltonian vector field X_H is given by

$$X_H = p_i \partial_{q^i} - q^i \partial_{p_i},$$

which gives canonical equations:

$$\dot{q}^i = p_i, \qquad \dot{p}_i = -\delta_{ij} q^j, \qquad \text{(where } \delta_{ij} \text{ is the Kronecker symbol).} \qquad (23.18)$$

In addition, for any two smooth ND functions $f, g : \mathbb{R}^{2N} \to \mathbb{R}$, the Poisson bracket is given by

$$[f, g]_\omega = \frac{\partial f}{\partial q^i}\frac{\partial g}{\partial p_i} - \frac{\partial f}{\partial p_i}\frac{\partial g}{\partial q^i},$$

which implieas that the particular functions $f = p_i p_j + q^i q^j$ and $g = p_i q^j + p_j q^i$ (for $i, j = 1, ..., N$) – are constants of motion. This system is integrable in an open set of $T^*\mathbb{R}^n$ with N integrability functions:

$$K_1 = H, \qquad K_2 = p_2^2 + (q^2)^2, \quad ..., \quad K_N = p_N^2 + (q^N)^2.$$

Conservative Quantum System

We perform the formal quantization of the conservative equation (23.16) in the Heisenberg picture: all variables become Hermitian operators (denoted by '$\wedge$'), the symplectic phase space $T^*Q = \{q^i, p_i\}$ becomes the Hilbert state space $\mathcal{H} = \mathcal{H}_{\hat{q}^i} \otimes \mathcal{H}_{\hat{p}_i}$ (where $\mathcal{H}_{\hat{q}^i} = \mathcal{H}_{\hat{q}^1} \otimes ... \otimes \mathcal{H}_{\hat{q}^N}$ and $\mathcal{H}_{\hat{p}_i} = \mathcal{H}_{\hat{p}_1} \otimes ... \otimes \mathcal{H}_{\hat{p}_N}$), the classical Poisson bracket $[\,,\,]$ becomes the quantum commutator $\{\,,\,\}$ multiplied by $-\mathrm{i}$ (in normal units) [II08a]

$$[\,,\,] \longrightarrow -\mathrm{i}\{\,,\,\}. \qquad (23.19)$$

In this way, the classical Liouville equation (23.16) becomes the quantum Liouville equation [II07e, II08a]

$$\partial_t \hat{\rho} = \mathrm{i}\{\hat{\rho}, \hat{H}\}\,, \tag{23.20}$$

where $\hat{H} = \hat{H}(\hat{q}^i, \hat{p}_i, t)$ is the Hamiltonian evolution operator, while

$$\hat{\rho} = \sum_a P(a)|\Psi_a><\Psi_a|, \qquad \text{where} \quad \mathrm{Tr}(\hat{\rho}) = 1$$

denotes the von Neumann density matrix operator, where each quantum state $|\Psi_a>$ occurs with probability $P(a)$; $\hat{\rho} = \hat{\rho}(\hat{q}^i, \hat{p}_i, t)$ is closely related to another von Neumann concept: entropy

$$S = -\mathrm{Tr}(\hat{\rho}[\ln\hat{\rho}]).$$

Open Classical System

We now move to the open (nonconservative) system: on the macroscopic NN–level the opening operation equals to the adding of a covariant vector of external (dissipative and/or motor) forces $F_i = F_i(q^i, p_i, t)$ to (the right-hand-side of) the covariant Hamilton's force equation, so that Hamilton's equations obtain the open (dissipative and/or forced) form [II06a, II06c]:

$$\dot{q}^i = \partial_{p_i} H, \qquad \dot{p}_i = -\partial_{q^i} H + F_i\,. \tag{23.21}$$

In the framework of the open Hamiltonian system (23.21) dynamics of any function $A(q^i, p_i, t)$ is defined by the open (dissipative and/or forced) evolution equation:

$$\partial_t A = -[A, H] + F_i[A, q^i]\,, \qquad ([A, q^i] = -\partial_{p_i} A)\,. \tag{23.22}$$

In particular, if $A = \rho(q^i, p_i, t)$ represents the density function of ensemble phase–points then its dynamics is given by the open (dissipative and/or forced) Liouville equation [II06c, II07e]:

$$\partial_t \rho = -[\rho, H] + F_i[\rho, q^i]\,. \tag{23.23}$$

Equation (23.23) represents the open classical model of our microscopic NN-dynamics.

For example, in case of our ND oscillator, Hamiltonian function (23.17) is not conserved any more, the canonical equations (23.18) become

$$\dot{q}^i = p_i, \qquad \dot{p}_i = F_i - \delta_{ij} q^j,$$

and the system is not integrable any more.

Classical NN-Dynamics

The generalized NN–dynamics, including two special cases of graded response neurons (GRN) and coupled neural oscillators (CNO), can be presented in the form of a Langevin stochastic equation [II07b, II08b]

$$\dot{\sigma}_i = f_i + \eta_i(t), \tag{23.24}$$

where $\sigma_i = \sigma_i(t)$ are the continual neuronal variables of ith neurons (representing either membrane action potentials in case of GRN, or oscillator phases in case of CNO); J_{ij} are individual synaptic weights; $f_i = f_i(\sigma_i, J_{ij})$ are the deterministic forces (given, in GRN-case, by

$$f_i = \sum_j J_{ij} \tanh[\gamma \sigma_j] - \sigma_i + \theta_i, \qquad \text{with} \quad \gamma > 0$$

and with the θ_i representing injected currents, and in CNO–case, by

$$f_i = \sum_j J_{ij} \sin(\sigma_j - \sigma_i) + \omega_i,$$

with ω_i representing the natural frequencies of the individual oscillators); the noise variables are given as

$$\eta_i(t) = \lim_{\Delta \to 0} \zeta_i(t) \sqrt{2T/\Delta},$$

where $\zeta_i(t)$ denote uncorrelated Gaussian distributed random forces and the parameter T controls the amount of noise in the system, ranging from $T = 0$ (deterministic dynamics) to $T = \infty$ (completely random dynamics).

More convenient description of the neural random process (23.24) is provided by the Fokker-Planck equation describing the time evolution of the probability density $P(\sigma_i)$ [II07a, II08b]

$$\partial_t P(\sigma_i) = -\sum_i \partial_{\sigma_i}[f_i P(\sigma_i)] + T \sum_i \partial_{\sigma_i^2} P(\sigma_i). \tag{23.25}$$

Now, in the case of deterministic dynamics $T = 0$, equation (23.25) can be easily put into the form of the conservative Liouville equation (23.16), by making the substitutions:

$$P(\sigma_i) \to \rho, \qquad f_i = \dot{\sigma}_i, \qquad \text{and} \qquad [\rho, H] = \operatorname{div}(\rho \,\dot{\sigma}_i) \equiv \sum_i \partial_{\sigma_i} (\rho \,\dot{\sigma}_i),$$

where $H = H(\sigma_i, J_{ij})$. Further, we can formally identify the stochastic forces, i.e., the second-order noise-term $T \sum_i \partial_{\sigma_i^2} \rho$ with $F^i[\rho, \sigma_i]$, to get the open Liouville equation (23.23).

Therefore, on the NN–level deterministic dynamics corresponds to the conservative system (23.16). Inclusion of stochastic forces corresponds to the system opening (23.23), implying the macroscopic arrow of time.

Open Quantum System

By formal quantization of equation (23.23), we obtain the quantum open Liouville equation [II08a, II08c]

$$\partial_t \hat{\rho} = \mathrm{i}\{\hat{\rho}, \hat{H}\} - \mathrm{i}\hat{F}_i\{\hat{\rho}, \hat{q}^i\}, \tag{23.26}$$

where $\hat{F}_i = \hat{F}_i(\hat{q}^i, \hat{p}_i, t)$ represents the covariant quantum operator of external friction forces in the Hilbert state space $\mathcal{H} = \mathcal{H}_{\hat{q}^i} \otimes \mathcal{H}_{\hat{p}_i}$.

Equation (23.26) represents the open quantum–decoherence model of our microscopic MT–dynamics.

Non–critical Stringy MT–Dynamics

In EMN–language of non-critical (SUSY) bosonic strings, our MT–dynamics equation (23.26) reads [II07e, II08a]

$$\partial_t \hat{\rho} = \mathrm{i}\{\hat{\rho}, \hat{H}\} - \mathrm{i}\hat{g}_{ij}\{\hat{\rho}, \hat{q}^i\}\dot{\hat{q}}^j, \tag{23.27}$$

where the target–space density matrix $\hat{\rho}(\hat{q}^i, \hat{p}_i)$ is viewed as a function of coordinates $\hat{q}^i$ that parameterize the couplings of the generalized $\sigma-$models on the bosonic string world-sheet, and their conjugate momenta $\hat{p}_i$, while $\hat{g}_{ij} = \hat{g}_{ij}(\hat{q}^i)$ is the quantum operator of the positive definite metric in the space of couplings. Therefore, the covariant quantum operator of external friction forces is in EMN–formulation given as $\hat{F}_i(\hat{q}^i, \dot{\hat{q}}^i) = \hat{g}_{ij}\,\dot{\hat{q}}^j$.

Equation (23.27) establishes the conditions under which a large–scale coherent state appearing in the MT-network, which can be considered responsible for loss–free energy transfer along the tubulins.

The system-independent properties of equation (23.27), are:
(i) Conservation of probability P

$$\partial_t P = \partial_t[\mathrm{Tr}(\hat{\rho})] = 0. \tag{23.28}$$

(ii) Conservation of energy E, on the average

$$\partial_t \langle\langle E\rangle\rangle \equiv \partial_t[\mathrm{Tr}(\hat{\rho}E)] = 0. \tag{23.29}$$

(iii) Monotonic increase in entropy

$$\partial_t S = \partial_t[-\mathrm{Tr}(\hat{\rho}\ln\hat{\rho})] = (\dot{\hat{q}}^i \hat{g}_{ij}\dot{\hat{q}}^j)S \geq 0, \tag{23.30}$$

due to the positive definiteness of the metric $\hat{g}_{ij}$, and thus automatically and naturally implying a microscopic arrow of time [II08a, II08c].

Equivalence of Neurodynamic Forms

Both the macroscopic NN–equation (23.23) and the microscopic MT–equation (23.26) have the same open Liouville form, which implies the arrow of time [II07e, II08a]. These demonstrates the existence of a formal neuro-biological space-time self–similarity.

Therefore, we have described neuro– and psycho–dynamics of both NN and MT ensembles, belonging to completely different biophysical space-time scales, brain's neural networks and brain's microtubules, by the unique form of the *open Liouville equation*, which implies the arrow of time. In this way the existence of the formal neuro-biological space-time self-similarity has been proved.

This proof implies the *existence of a unique Neurodynamical Law*, which acts on different scales of brain's functioning. In both cases of macroscopic continuous neural networks and microscopic discrete microtubules we have a process which is expressible in the open Liouville equation form.

23.3.4 Dissipative Quantum Brain Model

The *conservative brain* model was formulated within the framework of the quantum field theory (QFT) by [RU67]. The motivations at the basis of the formulation of the quantum brain model by Umezawa and Ricciardi trace back to the laboratory observations leading Lashley to remark (in 1940) that "masses of excitations... within general fields of activity, without regard to particular nerve cells are involved in the determination of behavior". In 1960's, K. Pribram, also motivated by experimental observations, started to formulate his *holographic hypothesis* (see [Pri71]). According to W. Freeman [Fre90, Fre91, Fre92, Fre96], "information appears indeed in such observations to be spatially uniform in much the way that the information density is uniform in a hologram". While the activity of the single neuron is experimentally observed in form of discrete and stochastic pulse trains and point processes, the 'macroscopic' activity of large assembly of neurons appears to be spatially coherent and highly structured in phase and amplitude.

Motivated by such an experimental situation, Umezawa and Ricciardi formulated in [RU67] the quantum brain model as a many–body physics problem, using the formalism of QFT with spontaneous breakdown of symmetry (which had been successfully tested in condensed matter experiments). Such a formalism provides the only available theoretical tool capable to describe long–range correlations such as the ones observed in the brain – presenting almost simultaneous responses in several regions to some external stimuli. The understanding of these long–range correlations in terms of modern biochemical and electrochemical processes is still lacking, which suggests that these responses could not be explained in terms of single neuron activity [Pri71].

Lagrangian dynamics in QFT is, in general, invariant under some group G of continuous transformations, as proposed by the famous Noether theorem. Now, spontaneous symmetry breakdown, one of the corner–stones of *Haken's synergetics* (see [Hak93]), occurs when the minimum energy state (the ground, or vacuum, state) of the system is not invariant under the full group G, but under one of its subgroups. Then it can be shown that collective modes, the so–called Nambu–Goldstone (NG) boson modes, are dynamically generated. Propagating over the whole system, these modes are the carriers of the *long–range correlation*, in which the order manifests itself as a global property dynamically generated. The long–range correlation modes are responsible for keeping the ordered pattern: they are coherently *condensed* in the ground state (similar to e.g., in the crystal case, where they keep the atoms trapped in their lattice sites). The long–range correlation thus forms a sort of net, extending over all the system volume, which traps the system components in the ordered pattern. This explains the "holistic" macroscopic collective behavior of the system components.

More precisely, according to the *Goldstone theorem* in QFT (see [II08a, II08c]), the spontaneous breakdown of the symmetry implies the existence of long–range correlation NG–modes in the ground state of the system. These modes are massless modes in the infinite volume limit, but they may acquire a finite, non-zero mass due to boundary or impurity effects. In the quantum brain model these modes are called dipole–wave–quanta (DWQ). The density of their condensation in the ground states acts as a *code* classifying the state and the memory there recorded. States with different code values are unitarily inequivalent states, i.e., there is no unitary transformation relating states of different codes.[41]

The *canonical quantization* procedure of a dissipative system requires to include in the formalism also the system representing the environment (usually the heat bath) in which the system is embedded. One possible way to do that is to depict the environment as the time–reversal image of the system [II08a, II08c]: the environment is thus described as the *double* of the system in the time–reversed dynamics (the system image in the mirror of time).

Within the framework of dissipative QFT, the brain system is described in terms of an *infinite collection of damped harmonic oscillators* A_κ (the simplest prototype of a dissipative system) representing the DWQ. Now, the collection of damped harmonic oscillators is ruled by the Hamiltonian [II08a]:

$$H = H_0 + H_I, \qquad \text{with}$$
$$H_0 = \Omega_\kappa(A_\kappa^\dagger A_\kappa - \tilde{A}_\kappa^\dagger \tilde{A}_\kappa), \qquad H_I = \mathrm{i}\Gamma_\kappa(A_\kappa^\dagger \tilde{A}_\kappa^\dagger - A_\kappa \tilde{A}_\kappa),$$

where Ω_κ is the frequency and Γ_κ is the damping constant. The $\tilde{A}_\kappa$ modes are the 'time–reversed mirror image' (i.e., the 'mirror modes') of the A_κ modes. They are the doubled modes, representing the environment modes, in such a way that κ generically labels their degrees–of–freedom. In particular, we consider the damped harmonic oscillator (DHO)

$$m\ddot{x} + \gamma\dot{x} + \kappa x = 0, \tag{23.31}$$

as a simple prototype for dissipative systems (with intention that thus get results also apply to more general systems). The damped oscillator (23.31) is a non–Hamiltonian system and therefore the customary canonical quantization procedure cannot be followed. However, one can face the problem by resorting to well known tools such as the *density matrix* ρ and the *Wigner function* $W = W(x, p, t)$.

Let us start with the special case of a *conservative particle* in the absence of friction γ, with the standard Hamiltonian (in natural units),

$$H = -(\partial_x)^2/2 + V(x).$$

Recall (from the previous subsection) that the *density matrix equation of motion*, i.e., *quantum Liouville equation*, is given by

$$\mathrm{i}\dot{\rho} = [H, \rho]. \tag{23.32}$$

[41] We remark that the spontaneous breakdown of symmetry is possible since in QFT there exist infinitely many ground states or vacua which are physically distinct (technically speaking, they are "unitarily inequivalent"). In quantum mechanics (QM), on the contrary, all the vacua are physically equivalent and thus there cannot be symmetry breakdown.

The density matrix function ρ is defined by

$$\langle x + \frac{1}{2}y|\rho(t)|x - \frac{1}{2}y\rangle = \psi^*(x + \frac{1}{2}y, t)\psi(x - \frac{1}{2}y, t) \equiv W(x, y, t),$$

with the associated standard expression for the *Wigner function* (see, e.g., [II07e]),

$$W(p, x, t) = \frac{1}{2\pi}\int W(x, y, t)\,\mathrm{e}^{(-\mathrm{i}py)}dy.$$

Now, in the coordinate $x-$representation, by introducing the notation

$$x_\pm = x \pm \frac{1}{2}y, \tag{23.33}$$

the Liouville equation (23.32) can be expanded as

$$\begin{aligned} &\mathrm{i}\,\partial_t\langle x_+|\rho(t)|x_-\rangle = \\ &\left\{-\frac{1^2}{2}\left[\partial^2_{x_+} - \partial^2_{x_-}\right] + [V(x_+) - V(x_-)]\right\}\langle x_+|\rho(t)|x_-\rangle, \end{aligned} \tag{23.34}$$

while the Wigner function $W(p, x, t)$ is now given by

$$\begin{aligned} \mathrm{i}\,\partial_t W(x, y, t) &= H_o W(x, y, t), \qquad \text{with} \\ H_o &= p_x p_y + V(x + \frac{1}{2}y) - V(x - \frac{1}{2}y), \\ \text{and} \qquad p_x &= -\mathrm{i}\partial_x, \qquad p_y = -\mathrm{i}\partial_y. \end{aligned} \tag{23.35}$$

The new Hamiltonian H_o (23.35) may be obtained from the corresponding Lagrangian

$$L_o = m\dot{x}\dot{y} - V(x + \frac{1}{2}y) + V(x - \frac{1}{2}y). \tag{23.36}$$

In this way, Vitiello concluded that the density matrix and the Wigner function formalism *required*, even in the conservative case (with zero mechanical resistance γ), the introduction of a 'doubled' set of coordinates, $x_\pm$, or, alternatively, x and y. One may understand this as related to the introduction of the 'couple' of indices *necessary* to label the density matrix elements (23.34).

Let us now consider the case of the *particle interacting* with a *thermal bath* at temperature T. Let f denote the *random force* on the particle at the position x due to the bath. The interaction Hamiltonian between the bath and the particle is written as

$$H_{int} = -fx. \tag{23.37}$$

Now, in the *Feynman–Vernon formalism* (see [II08a, II08c]), the *effective action* $A[x, y]$ for the particle is given by

$$A[x, y] = \int_{t_i}^{t_f} L_o(\dot{x}, \dot{y}, x, y)\,dt + I[x, y],$$

with L_o defined by (23.36) and

$$e^{iI[x,y]} = \langle (e^{-i\int_{t_i}^{t_f} f(t)x_-(t)dt})_- (e^{i\int_{t_i}^{t_f} f(t)x_+(t)dt})_+ \rangle, \tag{23.38}$$

where the symbol $\langle . \rangle$ denotes the average with respect to the thermal bath; '$(.)_+$' and '$(.)_-$' denote time ordering and anti–time ordering, respectively; the coordinates $x_\pm$ are defined as in (23.33). If the interaction between the bath and the coordinate x (23.37) were turned off, then the operator f of the bath would develop in time according to

$$f(t) = e^{iH_\gamma t} f e^{-iH_\gamma t},$$

where H_γ is the Hamiltonian of the isolated bath (decoupled from the coordinate x). $f(t)$ is then the force operator of the bath to be used in (23.38).

The interaction $I[x,y]$ between the bath and the particle has been evaluated in [SVW95] for a linear passive damping due to thermal bath by following Feynman–Vernon and Schwinger. The final result is:

$$\begin{aligned} I[x,y] &= \frac{1}{2}\int_{t_i}^{t_f} dt \, [x(t)F_y^{ret}(t) + y(t)F_x^{adv}(t)] \\ &+ \frac{i}{2}\int_{t_i}^{t_f}\int_{t_i}^{t_f} dtds \, N(t-s)y(t)y(s), \end{aligned}$$

where the retarded force on y, F_y^{ret}, and the advanced force on x, F_x^{adv}, are given in terms of the retarded and advanced Green functions $G_{ret}(t-s)$ and $G_{adv}(t-s)$ by

$$F_y^{ret}(t) = \int_{t_i}^{t_f} ds \, G_{ret}(t-s)y(s), \qquad F_x^{adv}(t) = \int_{t_i}^{t_f} ds \, G_{adv}(t-s)x(s),$$

respectively. In (23.39), $N(t-s)$ is the *quantum noise* in the fluctuating random force given by

$$N(t-s) = \frac{1}{2}\langle f(t)f(s) + f(s)f(t) \rangle.$$

The real and the imaginary part of the action are given respectively by

$$\mathrm{Re}\,(A[x,y]) = \int_{t_i}^{t_f} L \, dt, \tag{23.39}$$

$$L = m\dot{x}\dot{y} - \left[V(x+\frac{1}{2}y) - V(x-\frac{1}{2}y)\right] + \frac{1}{2}\left[xF_y^{ret} + yF_x^{adv}\right], \tag{23.40}$$

$$\text{and} \qquad \mathrm{Im}\,(A[x,y]) = \frac{1}{2}\int_{t_i}^{t_f}\int_{t_i}^{t_f} N(t-s)y(t)y(s)\, dtds. \tag{23.41}$$

Equations (23.39–23.41), are *exact* results for linear passive damping due to the bath. At quantum level nonzero y accounts for quantum noise effects in the fluctuating random force in the system–environment coupling arising from the imaginary part of the action (see [SVW95]). When in (23.40) we use

$$F_y^{ret} = \gamma\dot{y} \qquad \text{and} \qquad F_x^{adv} = -\gamma\dot{x} \qquad \text{we get,}$$

$$L(\dot{x}, \dot{y}, x, y) = m\dot{x}\dot{y} - V\left(x + \frac{1}{2}y\right) + V\left(x - \frac{1}{2}y\right) + \frac{\gamma}{2}(x\dot{y} - y\dot{x}). \qquad (23.42)$$

By using

$$V\left(x \pm \frac{1}{2}y\right) = \frac{1}{2}\kappa(x \pm \frac{1}{2}y)^2$$

in (23.42), the DHO equation (23.31) and its complementary equation for the y coordinate

$$m\ddot{y} - \gamma\dot{y} + \kappa y = 0. \qquad (23.43)$$

are derived. The y–oscillator is the time–reversed image of the x–oscillator (23.31). From the manifolds of solutions to equations (23.31) and (23.43), we could choose those for which the y coordinate is constrained to be zero, they simplify to

$$m\ddot{x} + \gamma\dot{x} + \kappa x = 0, \qquad y = 0.$$

Thus we get the classical damped oscillator equation from a Lagrangian theory at the expense of introducing an 'extra' coordinate y, later constrained to vanish. Note that the constraint $y(t) = 0$ is *not* in violation of the equations of motion since it is a true solution to (23.31) and (23.43).

Therefore, the general scheme of the dissipative quantum brain model can be summarized as follows. The starting point is that the brain is permanently coupled to the environment. Clearly, the specific details of such a coupling may be very intricate and changeable so that they are difficult to be measured and known. One possible strategy is to average the effects of the coupling and represent them, at some degree of accuracy, by means of some 'effective' interaction. Another possibility is to take into account the environmental influence on the brain by a suitable *choice* of the brain vacuum state. Such a choice is triggered by the external input (breakdown of the symmetry), and it actually is the end point of the internal (spontaneous) dynamical process of the brain (self–organization). The chosen vacuum thus carries the *signature* (memory) of the reciprocal brain–environment influence at a given time under given boundary conditions. A change in the brain–environment reciprocal influence then would correspond to a change in the choice of the brain vacuum: the brain state evolution or 'story' is thus the story of the trade of the brain with the surrounding world. The theory should then provide the equations describing the brain evolution 'through the vacua', each vacuum for each instant of time of its history.

The brain evolution is thus similar to a time–ordered sequence of photograms: each photogram represents the 'picture' of the brain at a given instant of time. Putting together these photograms in 'temporal order' one gets a movie, i.e. the story (the evolution) of open brain, which includes the brain–environment interaction effects.

The evolution of a memory specified by a given code value, say $\mathcal{N}$, can be then represented as a trajectory of given initial condition running over time–dependent vacuum states, denoted by $|0(t)\rangle_{\mathcal{N}}$, each one minimizing the free energy functional. These trajectories are known to be *classical* trajectories in the infinite volume limit: transition from one representation to another inequivalent one would be strictly forbidden in a quantum dynamics.

Since we have now two–modes (i.e., non–tilde and tilde modes), the memory state $|0(t)\rangle_{\mathcal{N}}$ turns out to be a two–mode coherent state. This is known to be an *entangled state*, i.e., it cannot be factorized into two single–mode states, the non–tilde and the tilde one. The physical meaning of such an entanglement between non-tilde and tilde modes is in the fact that the brain dynamics is permanently a dissipative dynamics. The entanglement, which is an unavoidable mathematical result of dissipation, represents the impossibility of cutting the links between the brain and the external world.[42]

In the dissipative brain model, noise and chaos turn out to be natural ingredients of the model. In particular, in the infinite volume limit the chaotic behavior of the trajectories in memory space may account for the high perceptive resolution in the recognition of the perceptual inputs. Indeed, small differences in the codes associated to external inputs may lead to diverging differences in the corresponding memory paths. On the other side, it also happens that codes differing only in a finite number of their components (in the momentum space) may easily be recognized as being the 'same' code, which makes possible that 'almost similar' inputs are recognized by the brain as 'equal' inputs (as in pattern recognition).

Therefore, the brain may be viewed as a complex system with (infinitely) many macroscopic configurations (the memory states). Dissipation is recognized to be the root of such a complexity.

23.4 Nonlinear Quantum Psychodynamics

Classical physics has provided a strong foundation for understanding brain function through measuring brain activity, modelling the functional connectivity of networks of neurons with algebraic matrices, and modelling the dynamics of neurons and neural populations with sets of coupled differential equations (see [II08a, II08c] and references therein). Various tools from classical physics enabled recognition and documentation of aspects of the physical states of the brain; the

[42] We remark that the entanglement is permanent in the large volume limit. Due to boundary effects, however, a unitary transformation could disentangle the tilde and non–tilde sectors: this may result in a pathological state for the brain. It is known that forced isolation of a subject produces pathological states of various kinds. We also observe that the tilde mode is not just a mathematical fiction. It corresponds to a real excitation mode (quasi–particle) of the brain arising as an effect of its interaction with the environment: the couples of non–tilde/tilde dwq quanta represent the correlation modes dynamically created in the brain as a response to the brain–environment reciprocal influence. It is the interaction between tilde and non–tilde modes that controls the irreversible time evolution of the brain: these collective modes are confined to live *in* the brain. They vanish as soon as the links between the brain and the environment are cut.

structures and dynamics of neurons, the operations of membranes and organelles that generate and channel electric currents; and the molecular and ionic carriers that implement the neural machineries of electrogenesis and learning. They support description of brain functions at several levels of complexity through measuring neural activity in the brains of animal and human subjects engaged in behavioral exchanges with their environments. One of the key properties of brain dynamics are the coordinated oscillations of populations of neurons that change rapidly in concert with changes in the environment [II06b, II07b]. Also, most experimental neurobiologists and neural theorists have focused on sensorimotor functions and their adaptations through various forms of learning and memory. Reliance has been placed on measurements of the rates and intervals of trains of action potentials of small numbers of neurons that are tuned to perceptual invariances and modelling neural interactions with discrete networks of simulated neurons. These and related studies have given a vivid picture of the cortex as a mosaic of modules, each of which performs a sensory or motor function; they have not given a picture of comparable clarity of the integration of modules.

According to [FV06], many–body quantum field theory appears to be the only existing theoretical tool capable to explain the dynamic origin of long–range correlations, their rapid and efficient formation and dissolution, their interim stability in ground states, the multiplicity of coexisting and possibly non–interfering ground states, their degree of ordering, and their rich textures relating to sensory and motor facets of behaviors. It is historical fact that many–body quantum field theory has been devised and constructed in past decades exactly to understand features like ordered pattern formation and phase transitions in condensed matter physics that could not be understood in classical physics, similar to those in the brain.

The domain of validity of the 'quantum' is not restricted to the microscopic world [Ume93]. There are macroscopic features of classically behaving systems, which cannot be explained without recourse to the quantum dynamics. This field theoretic model leads to the view of the phase transition as a condensation that is comparable to the formation of fog and rain drops from water vapor, and that might serve to model both the gamma and beta phase transitions. According to such a model, the production of activity with long–range correlation in the brain takes place through the mechanism of spontaneous breakdown of symmetry (SBS), which has for decades been shown to describe long-range correlation in condensed matter physics. The adoption of such a field theoretic approach enables modelling of the whole cerebral hemisphere and its hierarchy of components down to the atomic level as a fully integrated macroscopic quantum system, namely as a macroscopic system which is a quantum system not in the trivial sense that it is made, like all existing matter, by quantum components such as atoms and molecules, but in the sense that some of its macroscopic properties can best be described with recourse to quantum dynamics (see [FV06] and references therein).

It is well–known that non-equilibrium phase transitions [Hak93] are phenomena which bring about qualitative physical changes at the macroscopic level in presence of the same microscopic forces acting among the constituents of a system. Phase transitions can also be associated with autonomous robot competence levels, as informal specifications of desired classes of behaviors for robots over all environments they will encounter, as described by R. Brooks' subsumption architecture approach. The distributed network of augmented finite–state machines can exist in different phases or modalities of their state–space variables, which determine the systems intrinsic behavior. The phase transition represented by this approach is triggered by either internal (a set–point) or external (a command) control stimuli, such as a command to transition from a sleep mode to awake mode, or walking to running.

On the other hand, it is well–known that humans possess more degrees of freedom than are needed to perform any defined motor task, but are required to co-ordinate them in order to reliably accomplish high-level goals, while faced with intense motor variability. In an attempt to explain how this takes place, Todorov and Jordan [TJ02] have formulated an alternative theory of human motor co-ordination based on the concept of stochastic optimal feedback control. They were able to conciliate the requirement of goal achievement (e.g., grasping an object) with that of motor variability (biomechanical degrees of freedom). Moreover, their theory accommodates the idea that the human motor control mechanism uses internal 'functional synergies' to regulate task–irrelevant (redundant) movement.

Until recently, research concerning sensory processing and research concerning motor control have followed parallel but independent paths. The partitioning of the two lines of research in practice partly derived from and partly fostered a bipartite view of sensorimotor processing in the brain – that a sensory/perceptual system creates a general purpose representation of the world which serves as the input to the motor systems (and other cognitive systems) that generate action/behavior as an output. Recent results from research on vision in natural tasks have seriously challenged this view, suggesting that the visual system does not generate a general–purpose representation of the world, but rather extracts information relevant to the task at hand. At the same time, researchers in motor control have developed an increasing understanding of how sensory limitations and sensory uncertainty can shape the motor strategies that humans employ to perform tasks. Moreover, many aspects of the problem of sensorimotor control are specific to the mapping from sensory signals to motor outputs and do not exist in either domain in isolation. Sensory feedback control of hand movements, coordinate transformations of spatial representations and the influence of processing speed and attention on sensory contributions to motor control are just a few of these. In short, to understand how human (and animal) actors use sensory information to guide motor behavior, we must study sensory and motor systems as an integrated whole rather than as decomposable modules in a sequence of discrete processing steps.

Cognitive neuroscience investigations, including fMRI studies of human co–action, suggest that cognitive and neural processes supporting co–action include joint attention, action observation, task sharing, and action coordination (see [II07d]). For example, when two actors are given a joint control task (e.g., tracking a moving target on screen) and potentially conflicting controls (e.g., one person in charge of acceleration, the other – deceleration), their joint performance depends on how well they can anticipate each other's actions. In particular, better coordination is achieved when individuals receive real–time feedback about the timing of each other's actions.

A developing field in coordination dynamics involves the theory of social coordination, which attempts to relate the DC to normal human development of complex social cues following certain patterns of interaction. This work is aimed at understanding how human social interaction is mediated by meta-stability of neural networks. fMRI and EEG are particularly useful in mapping thalamocortical response to social cues in experimental studies. In particular, a new theory called the *Phi complex* has been developed by S. Kelso and collaborators, to provide experimental results for the theory of social coordination dynamics (see the recent nonlinear dynamics paper discussing social coordination and EEG dynamics. According to this theory, a pair of phi rhythms, likely generated in the mirror neuron system, is the hallmark of human social coordination. Using a dual-EEG recording system, the authors monitored the interactions of eight pairs of subjects as they moved their fingers with and without a view of the other individual in the pair.

23.4.1 Classical versus Quantum Probability

As *quantum probability* in human cognition and decision making has recently become popular, let us briefly describe this fundamental concept (see [II07e, II07d, II08a] for more details).

Classical Probability and Stochastic Dynamics

Recall that a *random variable* X is defined by its *distribution function* $f(x)$. Its *probabilistic description* is based on the following rules: (i) $P(X = x_i)$ is the probability that $X = x_i$; and (ii) $P(a \leq X \leq b)$ is the probability that X lies in a closed interval $[a, b]$. Its statistical description is based on: (i) μ_X or $E(X)$ is the mean or expectation of X; and (ii) σ_X is the standard deviation of X. There are two cases of random variables: discrete and continuous, each having its own probability (and statistics) theory.

A discrete random variable X has only a countable number of values $\{x_i\}$. Its distribution function $f(x_i)$ has the following properties:

$$P(X = x_i) = f(x_i), \qquad f(x_i) \geq 0, \qquad \sum_i f(x_i)\, dx = 1.$$

Statistical description of X is based on its discrete mean value μ_X and standard deviation σ_X, given respectively by

$$\mu_X = E(X) = \sum_i x_i f(x_i), \qquad \sigma_X = \sqrt{E(X^2) - \mu_X^2}.$$

Here $f(x)$ is a piecewise continuous function such that:

$$P(a \leq X \leq b) = \int_a^b f(x)\,dx, \qquad f(x) \geq 0,$$
$$\int_{-\infty}^{\infty} f(x)\,dx = \int_{\mathbb{R}} f(x)\,dx = 1.$$

Statistical description of X is based on its continuous mean μ_X and standard deviation σ_X, given respectively by

$$\mu_X = E(X) = \int_{-\infty}^{\infty} x f(x)\,dx, \qquad \sigma_X = \sqrt{E(X^2) - \mu_X^2}.$$

Now, let us observe the similarity between the two descriptions. The same kind of similarity between discrete and continuous quantum spectrum stroke P. Dirac when he suggested the combined integral approach, that he denoted by $\sum\!\!\!\!\!\!\int$ – meaning 'both integral and sum at once': summing over a discrete spectrum and integration over a continuous spectrum.

To emphasize this similarity even further, as well as to set–up the stage for the path integral, recall the notion of a *cumulative distribution function* of a random variable X, that is a function $F : \mathbb{R} \longrightarrow \mathbb{R}$, defined by

$$F(a) = P(X) \leq a.$$

In particular, suppose that $f(x)$ is the distribution function of X. Then

$$F(x) = \sum_{x_i \leq x} f(x_i), \qquad \text{or} \qquad F(x) = \int_{-\infty}^{\infty} f(t)\,dt,$$

according to as x is a discrete or continuous random variable. In either case, $F(a) \leq F(b)$ whenever $a \leq b$. Also,

$$\lim_{x \longrightarrow -\infty} F(x) = 0 \qquad \text{and} \qquad \lim_{x \longrightarrow \infty} F(x) = 1,$$

that is, $F(x)$ is monotonic and its limit to the left is 0 and the limit to the right is 1. Furthermore, its cumulative probability is given by

$$P(a \leq X \leq b) = F(b) - F(a),$$

and the Fundamental Theorem of Calculus tells us that, in the continuum case,

$$f(x) = \partial_x F(x).$$

Now, recall that *Markov stochastic process* is a random process characterized by a *lack of memory*, i.e., the statistical properties of the immediate future are uniquely determined by the present, regardless of the past [II06c, II08b].

For example, a *random walk* is an example of the *Markov chain*, i.e., a discrete–time Markov process, such that the motion of the system in consideration is viewed as a sequence of states, in which the transition from one state to another depends only on the preceding one, or the probability of the system being in state k depends only on the previous state $k-1$. The property of a Markov chain of prime importance in biomechanics is the existence of an *invariant distribution of states*: we start with an initial state x_0 whose absolute probability is 1. Ultimately the states should be distributed according to a specified distribution.

Between the pure deterministic dynamics, in which all DOF of the system in consideration are explicitly taken into account, leading to classical dynamical equations, for example in Hamiltonian form (using $\partial_x \equiv \partial/\partial x$),

$$\dot{q}^i = \partial_{p_i} H, \qquad \dot{p}_i = -\partial_{q^i} H, \tag{23.44}$$

(where q^i, p_i are coordinates and momenta, while $H = H(q,p)$ is the total system energy) – and pure stochastic dynamics (Markov process), there is so–called *hybrid dynamics*, particularly *Brownian dynamics*, in which some of DOF are represented only through their *stochastic influence* on others. As an example, suppose a system of particles interacts with a viscous medium. Instead of specifying a detailed interaction of each particle with the particles of the viscous medium, we represent the medium as a *stochastic force* acting on the particle. The stochastic force *reduces the dimensionally* of the dynamics.

Recall that the Brownian dynamics represents the phase–space trajectories of a collection of particles that individually obey *Langevin rate equations* in the field of force (i.e., the particles interact with each other via some deterministic force). For a free particle, the Langevin equation reads [II06c, II08b]:

$$m\dot{v} = R(t) - \beta v,$$

where m denotes the mass of the particle and v its velocity. The right–hand side represent the coupling to a *heat bath*; the effect of the random force $R(t)$ is to heat the particle. To balance overheating (on the average), the particle is subjected to *friction* β. In humanoid dynamics this is performed with the Rayleigh–Van der Pol's *dissipation*. Formally, the solution to the Langevin equation can be written as

$$v(t) = v(0)\exp\left(-\frac{\beta}{m}t\right) + \frac{1}{m}\int_0^t \exp[-(t-\tau)\beta/m]\, R(\tau)\, d\tau,$$

where the integral on the right–hand side is a *stochastic integral* and the solution $v(t)$ is a random variable. The stochastic properties of the solution depend significantly on the stochastic properties of the random force $R(t)$. In the Brownian dynamics the random force $R(t)$ is Gaussian distributed. Then the problem boils down to finding the solution to the Langevin stochastic differential equation with the supplementary condition (zero and mean variance)

$$< R(t) >= 0, \qquad < R(t)\, R(0) >= 2\beta k_B T \delta(t),$$

where $< . >$ denotes the mean value, T is temperature, k_B-*equipartition* (i.e., uniform distribution of energy) coefficient, Dirac $\delta(t)-$function.

Algorithm for computer simulation of the Brownian dynamics (for a single particle) can be written as [II06c, II08b]:

1. Assign an initial position and velocity.
2. Draw a random number from a Gaussian distribution with mean zero and variance.
3. Integrate the velocity to get v^{n+1}.
4. Add the random component to the velocity.

Another approach to taking account the coupling of the system to a heat bath is to subject the particles to collisions with *virtual particles.* Such collisions are imagined to affect only momenta of the particles, hence they affect the kinetic energy and introduce fluctuations in the total energy. Each stochastic collision is assumed to be an instantaneous event affecting only one particle.

The collision–coupling idea is incorporated into the Hamiltonian model of dynamics (23.44) by adding a stochastic force $R_i = R_i(t)$ to the $\dot{p}$ equation

$$\dot{q}^i = \partial_{p_i} H, \qquad \dot{p}_i = -\partial_{q^i} H + R_i(t).$$

On the other hand, the so–called *Ito stochastic integral* represents a kind of classical Riemann–Stieltjes integral from linear functional analysis, which is (in 1D case) for an arbitrary time–function $G(t)$ defined as the *mean square limit*

$$\int_{t_0}^{t} G(t)dW(t) = \text{ms} \lim_{n\to\infty} \{\sum_{i=1}^{n} G(t_{i-1}[W(t_i) - W(t_{i-1}]\}.$$

Now, the general ND Markov process can be defined by *Ito* stochastic differential equation (SDE),

$$\begin{aligned} dx_i(t) &= A_i[x^i(t), t]dt + B_{ij}[x^i(t), t]\, dW^j(t), \\ x^i(0) &= x_{i0}, \qquad (i, j = 1, \ldots, N) \end{aligned}$$

or corresponding *Ito stochastic integral equation*

$$x^i(t) = x^i(0) + \int_0^t ds\, A_i[x^i(s), s] + \int_0^t dW^j(s)\, B_{ij}[x^i(s), s],$$

in which $x^i(t)$ is the variable of interest, the vector $A_i[x(t), t]$ denotes deterministic *drift*, the matrix $B_{ij}[x(t), t]$ represents continuous stochastic *diffusion fluctuations*, and $W^j(t)$ is an $N-$ variable *Wiener process* (i.e., generalized Brownian motion), and $dW^j(t) = W^j(t + dt) - W^j(t)$.

Now, there are three well–known special cases of the *Chapman–Kolmogorov equation* (see [II06c, II08b]):

1. When both $B_{ij}[x(t), t]$ and $W(t)$ are zero, i.e., in the case of pure deterministic motion, it reduces to the *Liouville equation*

$$\partial_t P(x', t'|x'', t'') = -\sum_i \frac{\partial}{\partial x^i} \left\{A_i[x(t), t]\, P(x', t'|x'', t'')\right\}.$$

2. When only $W(t)$ is zero, it reduces to the *Fokker–Planck equation*

$$\partial_t P(x',t'|x'',t'') = -\sum_i \frac{\partial}{\partial x^i} \{A_i[x(t),t]\, P(x',t'|x'',t'')\} + \frac{1}{2}\sum_{ij} \frac{\partial^2}{\partial x^i \partial x^j} \{B_{ij}[x(t),t]\, P(x',t'|x'',t'')\}.$$

3. When both $A_i[x(t),t]$ and $B_{ij}[x(t),t]$ are zero, i.e., the state–space consists of integers only, it reduces to the *Master equation* of discontinuous jumps

$$\partial_t P(x',t'|x'',t'') = \int dx\, W(x'|x'',t)\, P(x',t'|x'',t'') - \int dx\, W(x''|x',t)\, P(x',t'|x'',t'').$$

The *Markov assumption* can now be formulated in terms of the conditional probabilities $P(x^i,t_i)$: if the times t_i increase from right to left, the conditional probability is determined entirely by the knowledge of the most recent condition. Markov process is generated by a set of conditional probabilities whose probability–density $P = P(x',t'|x'',t'')$ evolution obeys the general *Chapman–Kolmogorov integro–differential equation*

$$\partial_t P = -\sum_i \frac{\partial}{\partial x^i} \{A_i[x(t),t]\, P\} + \frac{1}{2}\sum_{ij} \frac{\partial^2}{\partial x^i \partial x^j} \{B_{ij}[x(t),t]\, P\} + \int dx\, \{W(x'|x'',t)\, P - W(x''|x',t)\, P\}$$

including *deterministic drift*, *diffusion fluctuations* and *discontinuous jumps* (given respectively in the first, second and third terms on the r.h.s.).

It is this general Chapman–Kolmogorov integro–differential equation, with its conditional probability density evolution, $P = P(x',t'|x'',t'')$, that we are going to model by the Feynman path integral *Feynman path integral* $\oint\!\!\!\!\!\Sigma$, providing us with the physical insight behind the abstract (conditional) probability densities.

23.4.2 The Life Space Foam

General nonlinear attractor dynamics, both deterministic and stochastic, as well as possibly chaotic, developed in the framework of Feynman path integrals, have recently [IA07] been applied to Lewinian field–theoretic psychodynamics (see [Lew97]), resulting in the development of a new concept of life–space foam (LSF) as a natural medium for motivational and cognitive psychodynamics. According to the LSF–formalism, the classic Lewinian life space can be macroscopically represented as a smooth manifold with steady force–fields and behavioral paths, while at the microscopic level it is more realistically represented as a collection

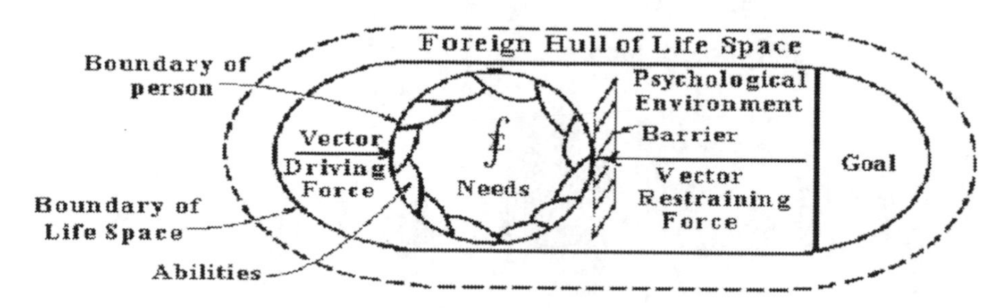

Fig. 23.2. Diagram of the *life space foam*: classical representation of Lewinian life space, with an adaptive path integral $\sum\!\!\!\!\!\!\int$ (denoting integration over continuous paths and summation over discrete Markov jumps) acting inside it and generating microscopic fluctuation dynamics.

of wildly fluctuating force–fields, (loco)motion paths and local geometries (and topologies with holes).

We have used the new LSF concept to develop modelling framework for motivational dynamics (MD) and induced cognitive dynamics (CD) (see [IA07]). Motivation processes both precede and coincide with every goal–directed action. Usually these motivation processes include the sequence of the following four feedforward *phases* [IA07]: (*)

1. *Intention Formation* $\mathcal{F}$, including: decision making, commitment building, etc.
2. *Action Initiation* $\mathcal{I}$, including: handling conflict of motives, resistance to alternatives, etc.
3. *Maintaining the Action* $\mathcal{M}$, including: resistance to fatigue, distractions, etc.
4. *Termination* $\mathcal{T}$, including parking and avoiding addiction, i.e., staying in control.

With each of the phases $\{\mathcal{F},\mathcal{I},\mathcal{M},\mathcal{T}\}$ in (*), we can associate a *transition propagator* – an ensemble of (possibly crossing) feedforward paths propagating through the 'wood of obstacles' (including topological holes in the LSF, see Figure 23.3), so that the complete transition is a product of propagators (as well as sum over paths). All the phases–propagators are controlled by a unique *Monitor* feedback process.

A set of least–action principles is used to model the smoothness of global, macro–level LSF paths, fields and geometry, according to the following prescription. The action $S[\Phi]$, psycho–physical dimensions of $Energy \times Time = Effort$ and depending on macroscopic paths, fields and geometries (commonly denoted by an abstract field symbol Φ^i) is defined as a temporal integral from the initial time instant t_{ini} to the final time instant t_{fin},

$$S[\Phi] = \int_{t_{ini}}^{t_{fin}} \mathfrak{L}[\Phi]\,dt, \tag{23.45}$$

with Lagrangian density given by

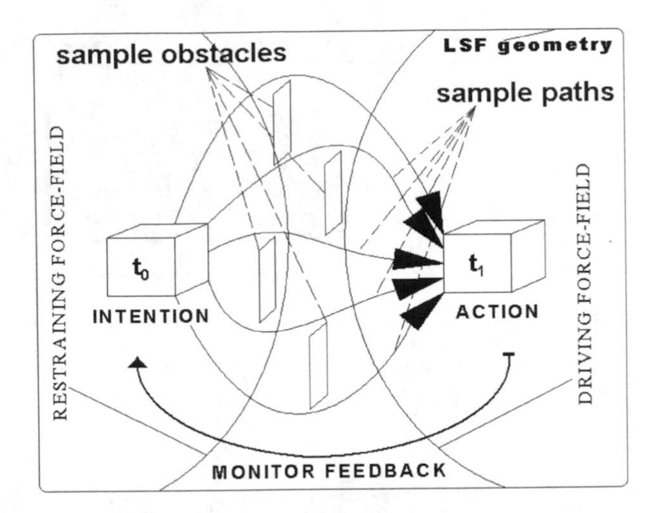

Fig. 23.3. *Transition–propagator* corresponding to each of the motivational phases $\{\mathcal{F}, \mathcal{I}, \mathcal{M}, \mathcal{T}\}$, consisting of an ensemble of feedforward paths propagating through the 'wood of obstacles'. The paths affected by driving and restraining force–fields, as well as by the local LSF–geometry. Transition goes from *Intention*, occurring at a sample time instant t_0, to *Action*, occurring at some later time t_1. Each propagator is controlled by its own *Monitor* feedback.

$$\mathfrak{L}[\Phi] = \int d^n x\, \mathcal{L}(\Phi_i, \partial_{x^j} \Phi^i),$$

where the integral is taken over all n coordinates $x^j = x^j(t)$ of the LSF, and $\partial_{x^j}\Phi^i$ are time and space partial derivatives of the Φ^i–variables over coordinates. The standard least action principle

$$\delta S[\Phi] = 0, \tag{23.46}$$

gives, in the form of the so–called Euler–Lagran-gian equations, a shortest (loco)motion path, an extreme force–field, and a life–space geometry of minimal curvature (and without holes). In this way, we effectively derive a unique globally smooth transition map

$$F\, :\, INTENTION_{t_{ini}} \longrightarrow ACTION_{t_{fin}}, \tag{23.47}$$

performed at a macroscopic (global) time–level from some initial time t_{ini} to the final time t_{fin}. In this way, we have obtained macro–objects in the global LSF: a single path described by Newtonian–like equation of motion, a single force–field described by Maxwellian–like field equations, and a single obstacle–free Riemannian geometry (with global topology without holes).

To model the corresponding local, micro–level LSF structures of rapidly fluctuating cognitive dynamics, an adaptive path integral is formulated, defining a multi–phase and multi–path (multi–field and multi–geometry) transition amplitude from the state of *Intention* to the state of *Action*,

$$\langle Action|Intention\rangle_{total} := \sum\!\!\!\!\!\!\int \mathcal{D}[w\Phi]\,\mathrm{e}^{\mathrm{i}S[\Phi]}, \tag{23.48}$$

where the Lebesgue integration is performed over all continuous $\Phi^i_{con} = paths + fields + geometries$, while summation is performed over all discrete processes and regional topologies Φ^j_{dis}. The symbolic differential $\mathcal{D}[w\Phi]$ in the general path integral (23.48), represents an adaptive path measure, defined as a weighted product (with $i = 1,...,n = con + dis$)

$$\mathcal{D}[w\Phi] = \lim_{N\longrightarrow\infty} \prod_{s=1}^{N} w_s d\Phi^i_s. \tag{23.49}$$

The adaptive path integral (23.48)–(23.49) represents an ∞–dimensional neural network, with weights w updating by the general rule [IA07]

$$new\ value(t+1) \;=\; old\ value(t) \;+\; innovation(t).$$

We remark here that the traditional neural networks approaches are known for their classes of functions they can represent. This limitation has been attributed to their low-dimensionality (the largest neural networks are limited to the order of 10^5 dimensions. The proposed path integral approach represents a new family of function-representation methods, which potentially offers a basis for a fundamentally more expansive solution.

On the macro–level in LSF we have the (loco)*motion action principle*

$$\delta S[x] = 0,$$

with the *Newtonian–like action* $S[x]$ given by

$$S[x] = \int_{t_{ini}}^{t_{fin}} dt\,[\frac{1}{2}g_{ij}\,\dot{x}^i\dot{x}^j + \varphi^i(x^i)], \tag{23.50}$$

where $\dot{x}^i$ represents motivational (loco)motion velocity vector with cognitive *processing speed*. The first bracket term in (23.50) represents the kinetic energy T,

$$T = \frac{1}{2}g_{ij}\,\dot{x}^i\dot{x}^j,$$

generated by the *Riemannian metric tensor* g_{ij}, while the second bracket term, $\varphi^i(x^i)$, denotes the family of potential force–fields, driving the (loco)motions $x^i = x^i(t)$ (the *strengths* of the fields $\varphi^i(x^i)$ depend on their positions x^i in LSF. The corresponding Euler–Lagrangian equation gives the Newtonian–like equation of motion

$$\frac{d}{dt}T_{\dot{x}^i} - T_{x^i} = -\varphi^i_{x^i}, \tag{23.51}$$

(subscripts denote the partial derivatives), which can be put into the standard Lagrangian form

$$\frac{d}{dt}L_{\dot{x}^i} = L_{x^i}, \qquad \text{with} \qquad L = T - \varphi^i(x^i).$$

Now, according to Lewin, the life space also has a sophisticated topological structure. As a Riemannian smooth n–manifold, the LSF–manifold $\mathbf{\Sigma}$ gives rise to its fundamental n– *groupoid*, or n–category $\Pi_n(\Sigma)$ (see [II06c, II07e]). In $\Pi_n(\Sigma)$, 0–cells are *points* in $\mathbf{\Sigma}$; 1–cells are *paths* in $\mathbf{\Sigma}$ (i.e., parameterized smooth maps $f : [0,1] \to \Sigma$); 2–cells are *smooth homotopies* (denoted by $\simeq$) *of paths* relative to endpoints (i.e., parameterized smooth maps $h : [0,1] \times [0,1] \to \Sigma$); 3–cells are *smooth homotopies of homotopies* of paths in $\mathbf{\Sigma}$ (i.e., parameterized smooth maps $j : [0,1] \times [0,1] \times [0,1] \to \Sigma$). Categorical *composition* is defined by *pasting* paths and homotopies. In this way, the following *recursive homotopy dynamics* emerges on the LSF–manifold $\mathbf{\Sigma}$ (**):

$$\texttt{0 - cell}: x_0 \bullet \qquad x_0 \in M; \qquad \text{in the higher cells below: } t, s \in [0,1];$$

$$\texttt{1 - cell}: x_0 \bullet \xrightarrow{f} \bullet x_1 \qquad f : x_0 \simeq x_1 \in M,$$

$f : [0,1] \to M,\ f : x_0 \mapsto x_1,\ x_1 = f(x_0),\ f(0) = x_0,\ f(1) = x_1;$

e.g., linear path: $f(t) = (1-t)\,x_0 + t\,x_1;$ or

Euler–Lagrangian f – dynamics with endpoint conditions (x_0, x_1) :

$$\frac{d}{dt}f_{\dot{x}^i} = f_{x^i}, \quad \text{with} \quad x(0) = x_0, \quad x(1) = x_1, \quad (i = 1, ..., n);$$

$$\texttt{2 - cell}: x_0 \bullet \underset{g}{\overset{f}{\Downarrow h}} \bullet x_1 \qquad h : f \simeq g \in M,$$

$h : [0,1] \times [0,1] \to M,\ h : f \mapsto g,\ g = h(f(x_0)),$

$h(x_0, 0) = f(x_0),\ h(x_0, 1) = g(x_0),\ h(0,t) = x_0,\ h(1,t) = x_1$

e.g., linear homotopy: $h(x_0, t) = (1-t)\,f(x_0) + t\,g(x_0);$ or

homotopy between two Euler–Lagrangian (f, g) – dynamics

with the same endpoint conditions (x_0, x_1) :

$$\frac{d}{dt}f_{\dot{x}^i} = f_{x^i}, \quad \text{and} \quad \frac{d}{dt}g_{\dot{x}^i} = g_{x^i} \quad \text{with} \quad x(0) = x_0, \quad x(1) = x_1;$$

$$\texttt{3 - cell}: x_0 \bullet \underset{g}{\overset{f}{h \overset{j}{\Rrightarrow} i}} \bullet x_1 \qquad j : h \simeq i \in M,$$

$j : [0,1] \times [0,1] \times [0,1] \to M,\ j : h \mapsto i,\ i = j(h(f(x_0)))$

$j(x_0, t, 0) = h(f(x_0)),\ j(x_0, t, 1) = i(f(x_0)),$

$j(x_0, 0, s) = f(x_0)$, $j(x_0, 1, s) = g(x_0)$,
$j(0, t, s) = x_0$, $j(1, t, s) = x_1$
e.g., linear composite homotopy: $j(x_0, t, s) = (1 - t)\, h(f(x_0)) + t\, i(f(x_0))$;
or, homotopy between two homotopies between above two Euler-Lagrangian (f, g) – dynamics with the same endpoint conditions (x_0, x_1).

On the micro–LSF level, instead of a single path defined by the Newtonian–like equation of motion (23.51), we have an ensemble of fluctuating and crossing paths with weighted probabilities (of the unit total sum). This ensemble of micro–paths is defined by the simplest instance of our adaptive path integral (23.48), similar to the Feynman's original *sum over histories*,

$$\langle Action|Intention\rangle_{paths} = \sum\!\!\!\!\!\!\int \mathcal{D}[wx]\, \mathrm{e}^{\mathrm{i}S[x]}, \tag{23.52}$$

where $\mathcal{D}[wx]$ is a functional measure on the *space of all weighted paths*, and the exponential depends on the action $S[x]$ given by (23.50). This procedure can be redefined in a mathematically cleaner way if we Wick–rotate the time variable t to imaginary values, $t \mapsto \tau = \mathrm{i}t$, thereby making all integrals real:

$$\sum\!\!\!\!\!\!\int \mathcal{D}[wx]\, \mathrm{e}^{\mathrm{i}S[x]} \xrightarrow{Wick} \sum\!\!\!\!\!\!\int \mathcal{D}[wx]\, \mathrm{e}^{-S[x]}. \tag{23.53}$$

Discretization of (23.53) gives the standard *thermo-dynamic–like partition function*

$$Z = \sum_j \mathrm{e}^{-w_j E^j/T}, \tag{23.54}$$

where E^j is the motion energy eigenvalue (reflecting each possible motivational energetic state), T is the temperature–like environmental control parameter, and the sum runs over all motion energy eigenstates (labelled by the index j). From (23.54), we can further calculate all thermodynamic–like and statistical properties of MD and CD, as for example, *transition entropy*, $S = k_B \ln Z$, etc.

Noisy Decision Making in the LSF

From CD–perspective, our adaptive path integral (23.52) calculates all (alternative) pathways of information flow during the transition $Intention \longrightarrow Action$. In the connectionist language, (23.52) represents *activation dynamics*, to which our $Monitor$ process gives a kind of *backpropagation* feedback, a common type of supervised learning

$$w_s(t+1) = w_s(t) - \eta \nabla J(t), \tag{23.55}$$

where η is a small constant, called the *step size*, or the *learning rate*, and $\nabla J(n)$ denotes the gradient of the 'performance hyper–surface' at the t–th iteration.

Now, the basic question about our local decision making process, occurring under uncertainty at the intention formation faze $\mathcal{F}$, is: Which alternative to choose? In our path–integral language this reads: Which path (alternative) should be given the highest probability weight w? This problem can be either iteratively solved by the learning process (23.55), controlled by the $MONITOR$ feedback, which we term *algorithmic approach*, or by the local decision making process under uncertainty, which we term *heuristic approach* [IA07]. This qualitative analysis is based on the micro–level interpretation of the Newtonian–like action $S[x]$, given by (23.50) and figuring both processing speed $\dot{x}$ and LTM (i.e., the force–field $\varphi(x)$, see next subsection). Here we consider three different cases:

1. If the potential $\varphi(x)$ is not very dependent upon position $x(t)$, then the more direct paths contribute the most, as longer paths, with higher mean square velocities $[\dot{x}(t)]^2$ make the exponent more negative (after Wick rotation (23.53)).
2. On the other hand, suppose that $\varphi(x)$ does indeed depend on position x. For simplicity, let the potential increase for the larger values of x. Then a direct path does not necessarily give the largest contribution to the overall transition probability, because the integrated value of the potential is higher than over another paths.
3. Finally, consider a path that deviates widely from the direct path. Then $\varphi(x)$ decreases over that path, but at the same time the velocity $\dot{x}$ increases. In this case, we expect that the increased velocity $\dot{x}$ would more than compensate for the decreased potential over the path.

Therefore, the most important path (i.e., the path with the highest weight w) would be the one for which any smaller integrated value of the surrounding field potential $\varphi(x)$ is more than compensated for by an increase in kinetic–like energy $\frac{m}{2}\dot{x}^2$. In principle, this is neither the most direct path, nor the longest path, but rather a middle way between the two. Formally, it is the path along which the average Lagrangian is minimal,

$$< \frac{m}{2}\dot{x}^2 + \varphi(x) > \quad \longrightarrow \quad \min, \tag{23.56}$$

i.e., the *path that requires minimal memory* (both LTM and WM) and *processing speed*. This mechanical result is consistent with the 'cognitive filter theory' of *selective attention* [Bro77], which postulates a low level filter that allows only a limited number of percepts to reach the brain at any time. In this theory, the importance of conscious, directed attention is minimized. The type of attention involving low level filtering corresponds to the concept of *early selection*.

Although we termed this 'heuristic approach' in the sense that we can instantly feel both the processing speed $\dot{x}$ and the LTM field $\varphi(x)$ involved, there is clearly a psycho–physical rule in the background, namely the averaging minimum relation (23.56).

From the decision making point of view, all possible paths (alternatives) represent the *consequences* of decision making. They are, by default, *short–term consequences*, as they are modelled in the micro–time–level. However, the path integral formalism allows calculation of the *long–term consequences*, just by extending the integration time, $t_{fin} \rightarrow \infty$. Besides, this *averaging decision mechanics* – choosing the optimal path – actually performs the 'averaging lift' in the LSF: from the micro–level to the macro–level.

For example, one of the simplest types of performance–degrading disturbances in the LSF is what we term motivational fatigue – a motivational drag factor that slows the actors' progress towards their goal. There are two fundamentally different sources of this motivational drag, both leading to apparently the same reduction in performance: (a) tiredness / exhaustion and (b) satiation (e.g., boredom). Both involve the same underlying mechanism (the raising valence of the alternatives to continuing the action) but the alternatives will differ considerably, depending on the properties of the task, from self–preservation / recuperation in the exhaustion case through to competing goals in the satiation case.

The spatial representation of this motivational drag is relatively simple: uni–dimensional LSF–coordinates may be sufficient for most purposes, which makes it attractive for the initial validation of our predictive model. Similarly uncomplicated spatial representations can be achieved for what we term motivational boost derived from the proximity to the goal (including the well–known phenomenon of 'the home stretch'): the closer the goal (e.g., a finishing line) is perceived to be, the stronger its 'pulling power' [Lew97]. Combinations of motivational drag and motivational boost effects may be of particular interest in a range of applications. These combinations can be modelled within relatively simple uni–dimensional LSF–coordinate systems.

23.4.3 Geometric Chaos and Topological Phase Transitions

In this section we extend the LSF–formalism to incorporate geometrical chaos [IJP08, II07a, II08b] and associated topological phase transitions.

It is well–known that on the basis of the ergodic hypothesis, statistical mechanics describes the physics of many-degrees of freedom systems by replacing time averages of the relevant observables with ensemble averages. Therefore, instead of using statistical ensembles, we can investigate the Hamiltonian (microscopic) dynamics of a system undergoing a phase transition. The reason for tackling dynamics is twofold. First, there are observables, like Lyapunov exponents, that are intrinsically dynamical. Second, the geometrization of Hamiltonian dynamics in terms of Riemannian geometry provides new observables and, in general, an interesting framework to investigate the phenomenon of phase transitions. The geometrical formulation of the dynamics of conservative systems [II06c, II08b] was first used by Krylov in his studies on the dynamical foundations of statistical mechanics and subsequently became a standard tool to study abstract systems in ergodic theory.

The simplest, mechanical–like LSF–action in the individual's LSF–manifold $\mathbf{\Sigma}$ has a Riemannian locomotion form [IA07] (summation convention is always assumed)

$$S[q] = \frac{1}{2} \int_{t_{ini}}^{t_{fin}} [a_{ij}\, \dot{q}^i \dot{q}^j - V(q)]\, dt, \tag{23.57}$$

where a_{ij} is the 'material' metric tensor that generates the total 'kinetic energy' of cognitive (loco)motions defined by their configuration coordinates q^i

and velocities $\dot{q}^i$, with the motivational potential energy $V(q)$ and the standard Hamiltonian

$$H(p,q) = \sum_{i=1}^{N} \frac{1}{2} p_i^2 + V(q), \qquad (23.58)$$

where p_i are the canonical (loco)motion momenta.

Dynamics of N DOF mechanical–like systems with action (23.57) and Hamiltonian (23.58) are commonly given by the set of geodesic equations [II06c, II07e]

$$\frac{d^2 q^i}{ds^2} + \Gamma^i_{jk} \frac{dq^j}{ds} \frac{dq^k}{ds} = 0, \qquad (23.59)$$

where Γ^i_{jk} are the Christoffel symbols of the affine Levi–Civita connection of the Riemannian LSF–manifold $\mathbf{\Sigma}$.

Alternatively, a description of the extrema of the Hamilton's action (23.57) can be obtained using the *Eisenhart metric* on an enlarged LSF space-time manifold (given by $\{q^0 \equiv t, q^1, \ldots, q^N\}$ plus one real coordinate q^{N+1}), whose arc–length is

$$ds^2 = -2V(q)(dq^0)^2 + a_{ij} dq^i dq^j + 2dq^0 dq^{N+1}. \qquad (23.60)$$

The manifold has a *Lorentzian structure* and the dynamical trajectories are those geode-sics satisfying the condition $ds^2 = Cdt^2$, where C is a positive constant. In this geometrical framework, the instability of the trajectories is the instability of the geodesics, and it is completely determined by the curvature properties of the LSF–manifold $\mathbf{\Sigma}$ according to the Jacobi equation of geodesic deviation [II06c, II07e]

$$\frac{D^2 J^i}{ds^2} + R^i{}_{jkm} \frac{dq^j}{ds} J^k \frac{dq^m}{ds} = 0, \qquad (23.61)$$

whose solution J, usually called Jacobi variation field, locally measures the distance between nearby geodesics; D/ds stands for the covariant derivative along a geodesic and $R^i{}_{jkm}$ are the components of the Riemann curvature tensor of the LSF–manifold $\mathbf{\Sigma}$.

Using the Eisenhart metric (23.60), the relevant part of the Jacobi equation (23.61) is given by the *tangent dynamics equation*:

$$\frac{d^2 J^i}{dt^2} + R^i{}_{0k0} J^k = 0, \qquad (i = 1, \ldots, N), \qquad (23.62)$$

where the only non-vanishing components of the curvature tensor of the LSF–manifold $\mathbf{\Sigma}$ are

$$R^i{}_{0k0} = \partial^2 V / \partial q^i \partial q^j.$$

The tangent dynamics equation (23.62) is commonly used to define *Lyapunov exponents* in dynamical systems given by the Riemannian action (23.57) and Hamiltonian (23.58), using the formula:

$$\lambda_1 = \lim_{t\to\infty} 1/2t \log(\Sigma_{i=1}^N [J_i^2(t) + \dot{J}_i^2(t)]/\Sigma_{i=1}^N [J_i^2(0) + \dot{J}_i^2(0)]). \quad (23.63)$$

Lyapunov exponents measure the strength of dynamical chaos.

Now, to relate these results to topological phase transitions within the LSF–manifold $\mathbf{\Sigma}$, recall that any two high–dimensional manifolds Σ_v and $\Sigma_{v'}$ have the same topology if they can be continuously and differentiably deformed into one another, that is if they are diffeomorphic. Thus by topology change the 'loss of diffeomorphicity is meant. In this respect, the so–called *topological theorem* [FP04] says that non–analyticity is the 'shadow' of a more fundamental phenomenon occurring in the system's configuration manifold (in our case the LSF–manifold): a topology change within the family of equipotential hypersurfaces

$$\Sigma_v = \{(q^1, \dots, q^N) \in \mathbb{R}^N |\, V(q^1, \dots, q^N) = v\},$$

where V and q^i are the microscopic interaction potential and coordinates respectively. This topological approach to PTs stems from the numerical study of the dynamical counterpart of phase transitions, and precisely from the observation of discontinuous or cuspy patterns displayed by the largest Lyapunov exponent λ_1 at the *transition energy*. Lyapunov exponents cannot be measured in laboratory experiments, at variance with thermodynamic observables, thus, being genuine dynamical observables they are only be estimated in numerical simulations of the microscopic dynamics. If there are critical points of V in configuration space, that is points $q_c = [\overline{q}_1, \dots, \overline{q}_N]$ such that $\nabla V(q)|_{q=q_c} = 0$, according to the *Morse lemma*, in the neighborhood of any critical point q_c there always exists a coordinate system $\tilde{q}(t) = [\tilde{q}^1(t), .., \tilde{q}^N(t)]$ for which

$$V(\tilde{q}) = V(q_c) - \tilde{q}_1^2 - \dots - \tilde{q}_k^2 + \tilde{q}_{k+1}^2 + \dots + \tilde{q}_N^2, \quad (23.64)$$

where k is the index of the critical point, i.e., the number of negative eigenvalues of the Hessian of the potential energy V. In the neighborhood of a critical point of the LSF–manifold $\mathbf{\Sigma}$, (23.64) yields

$$\partial^2 V/\partial q^i \partial q^j = \pm\delta_{ij},$$

which gives k unstable directions which contribute to the exponential growth of the norm of the tangent vector J.

This means that the strength of dynamical chaos within the individual's LSF–manifold $\mathbf{\Sigma}$, measured by the largest Lyapunov exponent λ_1 given by (23.63), is affected by the existence of critical points q_c of the potential energy $V(q)$. However, as $V(q)$ is bounded below, it is a good Morse function, with no vanishing eigenvalues of its Hessian matrix. According to *Morse theory*, the existence of critical points of V is associated with topology changes of the hypersurfaces $\{\Sigma_v\}_{v\in\mathbb{R}}$.

More precisely, let $V_N(q_1, \dots, q_N) : R^N \to R$, be a smooth, bounded from below, finite–range and confining potential[43]. Denote by $\Sigma_v = V^{-1}(v)$, $v \in R$,

[43] These requirements for V are fulfilled by standard interatomic and intermolecular interaction potentials, as well as by classical spin potentials.

its level sets, or equipotential hypersurfaces, in the LSF–manifold $\mathbf{\Sigma}$. Then let $\bar{v} = v/N$ be the potential energy per degree of freedom. If there exists N_0, and if for any pair of values $\bar{v}$ and $\bar{v}'$ belonging to a given interval $I_{\bar{v}} = [\bar{v}_0, \bar{v}_1]$ and for any $N > N_0$ then the sequence of the Helmoltz free energies $\{F_N(\beta)\}_{N\in\mathbb{N}}$ – where $\beta = 1/T$ (T is the temperature) and $\beta \in I_\beta = (\beta(\bar{v}_0), \beta(\bar{v}_1))$ – is uniformly convergent at least in $C^2(I_\beta)$ [the space of twice differentiable functions in the interval I_β], so that $\lim_{N\to\infty} F_N \in C^2(I_\beta)$ and neither first nor second order phase transitions can occur in the (inverse) temperature interval $(\beta(\bar{v}_0), \beta(\bar{v}_1))$, where the inverse temperature is defined as [FP04]

$$\beta(\bar{v}) = \partial S_N^{(-)}(\bar{v})/\partial\bar{v}, \qquad \text{while} \qquad S_N^{(-)}(\bar{v}) = N^{-1}\log\int_{V(q)\leq\bar{v}N} d^N q$$

is one of the possible definitions of the micro-canonical configurational entropy. The intensive variable $\bar{v}$ has been introduced to ease the comparison between quantities computed at different N–values.

This theorem means that a topology change of the $\{\Sigma_v\}_{v\in\mathbb{R}}$ at some v_c is a necessary condition for a phase transition to take place at the corresponding energy value. The topology changes implied here are those described within the framework of Morse theory through 'attachment of handles' to the LSF–manifold $\mathbf{\Sigma}$.

In the LSF path–integral language [IA07], we can say that suitable topology changes of equipotential submanifolds of the individual's LSF–manifold $\mathbf{\Sigma}$ can entail thermodynamic–like phase transitions [Hak93], according to the general formula:

$$\langle \text{phase out} \,|\, \text{phase in} \rangle := \sum\!\!\!\!\!\!\int_{\text{topology-change}} \mathcal{D}[w\Phi]\, \mathrm{e}^{\mathrm{i}S[\Phi]}.$$

The statistical behavior of the LSF–(loco)motion system (23.57) with the standard Hamiltonian (23.58) is encompassed, in the canonical ensemble, by its partition function, given by the phase–space path integral [II07e]

$$Z_N = \sum\!\!\!\!\!\!\int_{\text{top-ch}} \mathcal{D}[p]\mathcal{D}[q] \exp\{\mathrm{i}\int_t^{t'} [p\dot{q} - H(p,q)]\, d\tau\}, \tag{23.65}$$

where we have used the shorthand notation

$$\sum\!\!\!\!\!\!\int_{\text{top-ch}} \mathcal{D}[p]\mathcal{D}[q] \equiv \int \prod_\tau \frac{dq(\tau)dp(\tau)}{2\pi}.$$

The phase–space path integral (23.65) can be calculated as the *partition function*:

$$\begin{aligned} Z_N(\beta) &= \int \prod_{i=1}^N dp_i dq^i \mathrm{e}^{-\beta H(p,q)} = \left(\frac{\pi}{\beta}\right)^{\frac{N}{2}} \int \prod_{i=1}^N dq^i \mathrm{e}^{-\beta V(q)} \\ &= \left(\frac{\pi}{\beta}\right)^{\frac{N}{2}} \int_0^\infty dv\, \mathrm{e}^{-\beta v} \int_{\Sigma_v} \frac{d\sigma}{\|\nabla V\|}, \end{aligned} \tag{23.66}$$

where the last term is written using the so–called *co–area formula*, and v labels the equipotential hypersurfaces Σ_v of the LSF–manifold $\mathbf{\Sigma}$,

$$\Sigma_v = \{(q^1, \ldots, q^N) \in \mathbb{R}^N | V(q^1, \ldots, q^N) = v\}.$$

Equation (23.66) shows that the relevant statistical information is contained in the canonical configurational partition function

$$Z_N^C = \int \prod dq^i V(q) \mathrm{e}^{-\beta V(q)}.$$

Note that Z_N^C is decomposed, in the last term of (23.66), into an infinite summation of geometric integrals,

$$\int_{\Sigma_v} d\sigma \, / \|\nabla V\|,$$

defined on the $\{\Sigma_v\}_{v\in\mathbb{R}}$. Once the microscopic interaction potential $V(q)$ is given, the configuration space of the system is automatically foliated into the family $\{\Sigma_v\}_{v\in\mathbb{R}}$ of these equipotential hypersurfaces. Now, from standard statistical mechanical arguments we know that, at any given value of the inverse temperature β, the larger the number N, the closer to $\Sigma_v \equiv \Sigma_{u_\beta}$ are the microstates that significantly contribute to the averages, computed through $Z_N(\beta)$, of thermodynamic observables. The hypersurface Σ_{u_β} is the one associated with

$$u_\beta = (Z_N^C)^{-1} \int \prod dq^i V(q) \mathrm{e}^{-\beta V(q)},$$

the average potential energy computed at a given β. Thus, at any β, if N is very large the effective support of the canonical measure shrinks very close to a single $\Sigma_v = \Sigma_{u_\beta}$. Hence, the basic origin of a phase transition lies in a suitable topology change of the $\{\Sigma_v\}$, occurring at some v_c. This topology change induces the singular behavior of the thermodynamic observables at a phase transition. It is conjectured that the counterpart of a phase transition is a breaking of diffeomorphicity among the surfaces Σ_v, it is appropriate to choose a diffeomorphism invariant to probe if and how the topology of the Σ_v changes as a function of v. Fortunately, such a topological invariant exists, the Euler characteristic of the LSF–manifold $\mathbf{\Sigma}$, defined by [II06c, II07e]

$$\chi(\Sigma) = \sum_{k=0}^{N} (-1)^k b_k(\Sigma), \tag{23.67}$$

where the Betti numbers $b_k(\Sigma)$ are diffeomorphism invariants.[44] This homological formula can be simplified by the use of the Gauss–Bonnet–Hopf theorem, that relates $\chi(\Sigma)$ with the total Gauss–Kronecker curvature K_G of the LSF–manifold $\mathbf{\Sigma}$

---

[44] The Betti numbers b_k are the dimensions of the de Rham's cohomology vector spaces $H^k(\Sigma; \mathbb{R})$ (therefore the b_k are integers).

$$\chi(\Sigma) = \int_{\Sigma} K_G \, d\sigma, \qquad \text{where} \tag{23.68}$$
$$d\sigma = \sqrt{\det(a)} \, dx^1 dx^2 \cdots dx^n$$

is the invariant volume measure of the LSF–manifold $\boldsymbol{\Sigma}$ and a is the determinant of the LSF metric tensor a_{ij}. For technical details of this topological approach, see [II07e, II08b].

The domain of validity of the 'quantum' is not restricted to the microscopic world [Ume93]. There are macroscopic features of classically behaving systems, which cannot be explained without recourse to the quantum dynamics. This field theoretic model leads to the view of the phase transition as a condensation that is comparable to the formation of fog and rain drops from water vapor, and that might serve to model both the gamma and beta phase transitions. According to such a model, the production of activity with long-range correlation in the brain takes place through the mechanism of spontaneous breakdown of symmetry (SBS), which has for decades been shown to describe long-range correlation in condensed matter physics. The adoption of such a field theoretic approach enables modelling of the whole cerebral hemisphere and its hierarchy of components down to the atomic level as a fully integrated macroscopic quantum system, namely as a macroscopic system which is a quantum system not in the trivial sense that it is made, like all existing matter, by quantum components such as atoms and molecules, but in the sense that some of its macroscopic properties can best be described with recourse to quantum dynamics (see [FV06] and references therein).

23.4.4 Joint Action of Several Agents

In this section we propose an LSF–based model of the joint action of two or more actors, where actors can be both humans and robots. This joint action takes place in the joint LSF manifold Σ_J, composed of individual LSF manifolds $\Sigma_\alpha, \Sigma_\beta, \ldots$. It has a sophisticated geometrical and dynamical structure as follows.

To model the dynamics of the two–actor co–action, we propose to associate each of the actors with a set of their own time dependent trajectories, which constitutes an n–dimensional Riemannian LSF–manifold, $\Sigma_\alpha = \{x^i(t_i)\}$ and $\Sigma_\beta = \{y^j(t_j)\}$, respectively. Their associated tangent bundles contain their individual nD (loco)motion velocities, $T\Sigma_\alpha = \{\dot{x}^i(t_i) = dx^i/dt_i\}$ and $T\Sigma_\beta = \{\dot{y}^j(t_j) = dy^j/dt_j\}$. Further, following the general formalism of [IA07], outlined in the introduction, we use the modelling machinery consisting of: (i) Adaptive joint action at the top–master level, describing the externally–appearing deterministic, continuous and smooth dynamics, and (ii) Corresponding adaptive path integral (23.73) at the bottom–slave level, describing a wildly fluctuating dynamics including both continuous trajectories and Markov chains. This lower–level joint dynamics can be further discretized into a partition function of the corresponding statistical dynamics.

The smooth joint action with two terms, representing cognitive/motivational potential energy and physical kinetic energy, is formally given by:

$$A[x,y;t_i,t_j] = \frac{1}{2}\int_{t_i}\int_{t_j} \alpha_i\beta_j\,\delta(I_{ij}^2)\,\dot{x}^i(t_i)\,\dot{y}^j(t_j)\,dt_i dt_j + \frac{1}{2}\int_t g_{ij}\,\dot{x}^i(t)\dot{x}^j(t)\,dt,$$

$$\text{with} \qquad I_{ij}^2 = \left[x^i(t_i) - y^j(t_j)\right]^2, \qquad \text{where} \qquad IN \leq t_i, t_j, t \leq OUT. \tag{23.69}$$

The first term in (23.69) represents potential energy of the cognitive/ motivational interaction between the two agents α_i and β_j.[45] It is a double integral over a delta function of the square of interval I^2 between two points on the paths in their Life–Spaces; thus, interaction occurs only when this interval, representing the motivational cognitive distance between the two agents, vanishes. Note that the cognitive (loco) motions of the two agents $\alpha_i[x^i(t_i)]$ and $\beta_j[y^j(t_j)]$, generally occur at different times t_i and t_j unless $t_i = t_j$, when cognitive synchronization occurs.

The second term in (17.11) represents kinetic energy of the physical interaction. Namely, when the cognitive synchronization in the first term takes place, the second term of physical kinetic energy is activated in the common manifold, which is one of the agents' Life Spaces, say $\Sigma_\alpha = \{x^i(t_i)\}$.

The reason why we have chosen the action (17.11) as a macroscopic model for human joint action is that (17.11) naturally represents the transition map,

$$A[x,y;t_i,t_j] \ : \ \text{MENTAL INTENTION} \overset{Synch}{\Longrightarrow} \text{PHYSICAL ACTION},$$

from mutual cognitive intention to joint physical action, in which the joint action starts after the mutual cognitive intention is synchronized. In simple words, "we can efficiently act together only after we have tuned–up our intentions."

Conversely, if we have a need to represent coaction of three actors, say α_i, β_j and γ_k (e.g., α_i in charge of acceleration, β_j – deceleration and γ_k– steering), we can associate each of them with an nD Riemannian Life–Space manifold, $\Sigma_\alpha = \{x^i(t_i)\}$, $\Sigma_\beta = \{y^j(t_j)\}$, and $\Sigma_\gamma = \{z^k(t_k)\}$, respectively, with the corresponding tangent bundles containing their individual (loco) motion velocities, $T\Sigma_\alpha = \{\dot{x}^i(t_i) = dx^i/dt_i\}$, $T\Sigma_\beta = \{\dot{y}^j(t_j) = dy^j/dt_j\}$ and $T\Sigma_\gamma = \{\dot{z}^k(t_k) = dz^k/dt_k\}$. Then, instead of (17.11) we have

$$A[t_i,t_j,t_k;t] = \frac{1}{2}\int_{t_i}\int_{t_j}\int_{t_k} \alpha_i(t_i)\beta_j(t_j)\,\gamma_k(t_k) \times \delta(I_{ijk}^2)\,\dot{x}^i(t_i)\,\dot{y}^j(t_j)\,\dot{z}^k(t_k)\,dt_i dt_j dt_k + \frac{1}{2}\int_t W_{rs}^M(t,q,\dot{q})\,\dot{q}^r\dot{q}^s\,dt, \tag{23.70}$$

where $IN \leq t_i, t_j, t_k, t \leq OUT$, with
$I_{ijk}^2 = [x^i(t_i) - y^j(t_j)]^2 + [y^j(t_j) - z^k(t_k)]^2 + [z^k(t_k) - x^i(t_i)]^2$,

Due to an intrinsic chaotic coupling, the three–actor (or, n–actor, $n > 3$) joint action (23.70) has a considerably more complicated geometrical structure then

[45] Although, formally, this term contains cognitive velocities, it still represents 'potential energy' from the physical point of view.

the bilateral co–action (17.11).[46] It actually happens in the common $3n$D Finsler manifold $\Sigma_J = \Sigma_\alpha \cup \Sigma_\beta \cup \Sigma_\gamma$, parameterized by the local joint coordinates dependent on the common time t. That is, $\Sigma_J = \{q^r(t), r = 1, ..., 3n\}$. Geometry of the joint manifold Σ_J is defined by the Finsler metric function $ds = F(q^r, dq^r)$, defined by

$$F^2(q, \dot{q}) = g_{rs}(q, \dot{q})\dot{q}^r \dot{q}^s, \tag{23.71}$$

and the Finsler tensor $C_{rst}(q, \dot{q})$, defined by (see [II06c, II07e])

$$C_{rst}(q, \dot{q}) = \frac{1}{4}\frac{\partial^3 F^2(q, \dot{q})}{\partial \dot{q}^r \partial \dot{q}^s \partial \dot{q}^t} = \frac{1}{2}\frac{\partial g_{rs}}{\partial \dot{q}^r \partial \dot{q}^s}. \tag{23.72}$$

From the Finsler definitions (23.71)–(23.72), it follows that the partial interaction manifolds, $\Sigma_\alpha \cup \Sigma_\beta$, $\Sigma_\beta \cup \Sigma_y$ and $\Sigma_\alpha \cup \Sigma_y$, have Riemannian structures with the corresponding interaction kinetic energies,

$$T_{\alpha\beta} = \frac{1}{2}g_{ij}\dot{x}^i \dot{y}^j, \qquad T_{\alpha\gamma} = \frac{1}{2}g_{ik}\dot{x}^i \dot{z}^k, \qquad T_{\beta\gamma} = \frac{1}{2}g_{jk}\dot{y}^j \dot{z}^k.$$

At the slave level, the adaptive path integral (see [IA07]), representing an ∞–dimensional neural network, corresponding to the adaptive bilateral joint action (17.11), reads

$$\langle OUT|IN\rangle := \sum\!\!\!\!\!\!\int \mathcal{D}[w, x, y]\, \mathrm{e}^{\mathrm{i}A[x,y;t_i,t_j]}, \tag{23.73}$$

where the Lebesgue integration is performed over all continuous paths $x^i = x^i(t_i)$ and $y^j = y^j(t_j)$, while summation is performed over all associated discrete Markov fluctuations and jumps. The symbolic differential in the path integral (23.73) represents an adaptive path measure, defined as a weighted product

$$\mathcal{D}[w, x, y] = \lim_{N\to\infty} \prod_{s=1}^{N} w_{ij}^s dx^i dy^j, \qquad (i, j = 1, ..., n). \tag{23.74}$$

Similarly, in case of the triple joint action, the adaptive path integral reads,

$$\langle OUT|IN\rangle := \sum\!\!\!\!\!\!\int \mathcal{D}[w; x, y, z; q]\, \mathrm{e}^{\mathrm{i}A[t_i,t_j,t_k;t]}, \tag{23.75}$$

with the adaptive path measure defined by

$$\mathcal{D}[w; x, y, z; q] = \lim_{N\to\infty} \prod_{S=1}^{N} w_{ijkr}^S dx^i dy^j dz^k dq^r, \ (i, j, k = 1, ..., n;\ r = 1, ..., 3n). \tag{23.76}$$

The proposed path integral approach represents a new family of more expansive function-representation methods, which is now capable of representing input/output behavior of more than one actor. However, as we add the second and

[46] Recall that the necessary condition for chaos in continuous temporal or spatio-temporal systems is to have three variables with nonlinear couplings between them.

subsequent actors to the model, the requirements for the rigorous geometrical representations of their respective LSFs become nontrivial. For a single actor or a two–actor co–action the Riemannian geometry was sufficient, but it becomes insufficient for modelling the n–actor (with $n \geq 3$) joint action, due to an intrinsic chaotic coupling between the individual actors' LSFs. To model an n–actor joint LSF, we have to use the Finsler geometry, which is a generalization of the Riemannian one. This progression may seem trivial, both from standard psychological point of view, and from computational point of view, but it is not trivial from the geometrical perspective.

References

[Abd06] Abdi, H.: Signal detection theory. In: Salkind, N.J. (ed.) Encyclopedia of Measurement, Statistics. Sage, Thousand Oaks (2006)

[AT06] Anishchenko, A., Treves, A.: Autoassociative memory retrieval and spontaneous activity bumps in small-world networks of integrate-and-fire neurons. J. Physiol (Paris) 100(4), 225–236 (2006)

[APS98] American Psychological Association: Task force report. Gottfredson (1998)

[Ama77] Amari, S.: Dynamics of pattern formation in lateral-inhibition type neural fields. Biol. Cybern. 27, 77–87 (1977)

[And84] Anderson, T.W.: An Introduction to Multivariate Statistical Analysis, 2nd edn. Wiley, New York (1984)

[Arb98] Arbib, M. (ed.): Handbook of Brain Theory and Neural Networks, 2nd edn. MIT Press, Cambridge (1998)

[Bec08] Beck, F.: Synaptic quantum tunnelling. NeuroQuantology 6(2), 140–151 (2008)

[Bir15] Birkhoff, G.D.: The restricted problem of three–bodies, Rend. Circolo Mat. Palermo. 39, 255–334 (1915)

[Bir17] Birkhoff, G.D.: Dynamical Systems with Two Degrees of Freedom. Trans. Amer. Math. Soc. 18, 199–300 (1917)

[Bir27] Birkhoff, G.D.: Dynamical Systems. Amer. Math. Soc., Providence (1927)

[Blo80] Bloom, B.S.: All Our Children Learning. McGraw-Hill, New York (1980)

[Boh92] Bohm, D.: Thought as a System. Routledge, London (1992)

[Bon73] De Bono, E.: Lateral Thinking: Creativity Step by Step. Harper, Row (1973)

[Bro77] Broadbent, D.E.: Levels, hierarchies and the locus of control. Quarterly J. Exper. Psychology 29, 181–201 (1977)

[BB95] Berry, D.C., Broadbent, D.E.: Implicit learning in the control of complex systems: A reconsideration of some of the earlier claims. In: Frensch, P.A., Funke, J. (eds.) Complex problem solving: The European Perspective (131-150). Lawr. Erl. Assoc., Hillsdale (1995)

[Cha08] Chang, J.-J.: Studies and Discussion of Properties of Biophotons and Their Functions. 6(4), 420–430 (2008)

[CCW03] Cohen, J., Cohen, P., West, S.G., Aiken, L.S.: Applied multiple regression/correlation analysis for the behavioral sciences, 2nd edn. Lawr. Erl. Assoc., Hillsdale (2003)

[Con08] Conte, E.: Testing Quantum Consciousness. NeuroQuant. 6(2), 126–139 (2008)

[CL71] Cooley, W.W., Lohnes, P.R.: Multivariate Data Analysis. Wiley, New York (1971)

[CK90] Crick, F., Koch, C.: Towards a neurobiological theory of consciousness. Sem Neurosci. 2, 263–275 (1990)

[Cri94] Crick, F.: The Astonishing Hypothesis. Charles Scribner's Sons, New York (1994)

[CGP88] Cvitanovic, P., Gunaratne, G., Procaccia, I.: Topological and metric properties of Hénon-type strange attractors. Phys. Rev. A 38, 1503–1520 (1988)

[CAM05] Cvitanovic, P., Artuso, R., Mainieri, R., Tanner, G., Vattay, G.: Chaos: Classical and Quantum. ChaosBook.org, Niels Bohr Institute, Copenhagen (2005)

[DM04] Destexhe, A., Marder, E.: Plasticity in single neuron and circuit computations. Nature 431, 789–795 (2004)

[Dir82] Dirac, P.A.M.: Principles of Quantum Mechanics, 4th edn. Oxford Univ. Press, Oxford (1982)

[Dor75] Dorner, D.: Wie Menschen eine Welt verbessern wollten [How people wanted to improve the world]. Bild der Wissenschaft 12, 48–53 (1975)

[DV85] Dorner, D.: Verhalten, Denken und Emotionen [Behavior, thinking, emotions]. In: Eckensberger, L.H., Lantermann, E.D. (eds.) Emotion und Reflexivity (157-181). Urban, Schwarzenberg, Munchen, Germany (1985)

[DW95] Dorner, D., Wearing, A.: Complex problem solving: Toward a (computer-simulated) theory. In: Frensch, P.A., Funke, J. (eds.) Complex problem solving: The European Perspective (65-99). Lawr. Erl. Assoc., Hillsdale (1995)

[DKM95] Douglas, R.J., Koch, C., Mahowald, M., Martin, K., Suarez, H.: Recurrent excitation in neocortical circuits. Science 69, 981–985 (1995)

[Ecc64] Eccles, J.C.: The Physiology of Synapses. Springer, Berlin (1964)

[EIS67] Eccles, J.C., Ito, M., Szentagothai, J.: The Cerebellum as a Neuronal Machine. Springer, Berlin (1967)

[Ecc86] Eccles, J.C.: Do mental events cause neural events analogously to the probability fields of quantum mechanics? Proc. R. Soc. 227, 411–428 (1986)

[Ecc90] Eccles, J.C.: A unitary hypothesis of mind-brain interaction in the cerebral cortex. Proc. R. Soc. London B 240, 433–451 (1990)

[Ecc94] Eccles, J.C.: How the self controls its brain. Springer, Heidelberg (1994)

[Eys69] Eysenck, H.J., Eysenck, S.B.G.: Personality Structure and Measurement. Routledge, London (1969)

[Eys76] Eysenck, H.J., Eysenck, S.B.G.: Psychoticism as a Dimension of Personality. Hodder and Stoughton, London (1976)

[Eys92a] Eysenck, H.J.: A reply to Costa, McCrae. P or A, C - the role of theory. Personality and Individual Differences 13, 867–868 (1992)

[Eys92b] Eysenck, H.J.: Four ways five factors are not basic. Personality, Individual Differences 13, 667–673 (1992)

[Fei78] Feigenbaum, M.J.: Quantitative universality for a class of nonlinear transformations. J. Stat. Phys. 19, 25–52 (1978)

[Fei79] Feigenbaum, M.J.: The universal metric properties of nonlinear transformations. J. Stat. Phys. 21, 669–706 (1979)

[Fit61] FitzHugh, R.A.: Impulses and physiological states in theoretical models of nerve membrane. Biophys. J. 1, 445–466 (1961)

[Fre90] Freeman, W.J.: On the the problem of anomalous dispersion in chaotic phase transitions of neural masses, its significance for the management of perceptual information in brains. In: Haken, H., Stadler, M. (eds.) Synergetics of Cognition, vol. 45, pp. 126–143. Springer Verlag, Berlin (1990)

[Fre91] Freeman, W.J.: The physiology of perception. Sci. Am. 264(2), 78–85 (1991)

[Fre92] Freeman, W.J.: Tutorial on neurobiology: from single neurons to brain chaos. Int. J. Bif. Chaos 2(3), 451–482 (1992)

[Fre96] Freeman, W.J.: Random activity at the microscopic neural level in cortex sustains is regulated by low dimensional dynamics of macroscopic cortical activity. Int. J. of Neural Systems 7, 473 (1996)

[FP04] Franzosi, R., Pettini, M.: Theorem on the origin of Phase Transitions. Phys. Rev. Lett. 92(6), 60601 (2004)

[FV06] Freeman, W.J., Vitiello, G.: Nonlinear brain dynamics as macroscopic manifestation of underlying many–body field dynamics. Phys. Life Rev. 3(2), 93–118 (2006)

[FK83] Frölich, H., Kremer, F.: Coherent Excitations in Biological Systems. Springer, New York (1983)

[Fun95] Funke, J.: Solving complex problems: Human identification, control of complex systems. In: Sternberg, R.J., Frensch, P.A. (eds.) Complex problem solving: Principles, mechanisms (185-222), Lawr. Erl. Assoc., Hillsdale (1995)

[GMK94] Gabbiani, F., Midtgaard, J., Knoepfl, T.: Synaptic integration in a model of cerebellar granule cells. J. Neurophysiol. 72, 999–1009 (1994)

[Gle87] Gleick, J.: Chaos: Making a New Science. Penguin–Viking, New York (1987)

[Gra82] Gray, G.: Rehabilitating the dendritic spine. Trends Neurosci. 5, 5–6 (1982)

[Gui67] Guilford, J.P.: The Nature of Human Intelligence. McGraw-Hill, New York (1967)

[CG83] Cohen, M.A., Grossberg, S.: Absolute stability of global pattern formation, parallel memory storage by competitive neural networks. IEEE Trans. Syst., Man, Cybern. 13(5), 815–826 (1983)

[Gro03] Grossberg, S.: How does the cerebral cortex work? Development, learning, attention, and 3D vision by laminar circuits of visual cortex. Behav. Cogn. Neurosci. Rev. 2, 47–76 (2003)

[Gro69] Grossberg, S.: Embedding fields: A theory of learning with physiological implications. J. Math. Psych. 6, 209–239 (1969)

[Gro82] Grossberg, S.: Studies of Mind and Brain. Kluwer, Dordrecht (1982)

[Gro87] Grossberg, S.: Competitive learning: from interactive activation to adaptive resonance. Cog. Sci. 11, 23–63 (1987)

[Gro88] Grossberg, S.: Neural Networks and Natural Intelligence. MIT Press, Cambridge (1988)

[Gro99] Grossberg, S.: How does the cerebral cortex work? Learning, attention and grouping by the laminar circuits of visual cortex. Spatial Vision 12, 163–186 (1999)

[Hak93] Haken, H.: Advanced Synergetics: Instability Hierarchies of Self-Organizing Systems and Devices, 3rd edn. Springer, Berlin (1993)

[Hak02] Haken, H.: Brain Dynamics, Synchronization and Activity Patterns in Pulse-Codupled Neural Nets with Delays and Noise. Springer, New York (2002)

[HHT02] Hagan, S., Hameroff, S.R., Tuszynski, J.A.: Quantum computation in brain microtubules: Decoherence and biological feasibility. Phys. Rev. D 65, 61901 (2002)

[HKB85] Haken, H., Kelso, J.A.S., Bunz, H.: A theoretical model of phase transitions in human hand movements. Biol. Cybern. 51, 347–356 (1985)

[Ham87] Hameroff, S.R.: Ultimate Computing: Biomolecular Consciousness and Nanotechnology. North-Holland, Amsterdam (1987)

[HW83] Hameroff, S.R., Watt, R.C.: Do anesthetics act by altering electron mobility? Anesth. Anesth. Analg 62, 936–940 (1983)

[HP93] Hameroff, S.R., Penrose, R.: Conscious events as orchestrated spacetime selections. Journal of Consciousness Studies 3(1), 36–53 (1996)

[HP96] Hameroff, S.R., Penrose, R.: Orchestrated reduction of quantum coherence in brain microtubules: A model for consciousness. In: Hameroff, S.R., Kaszniak, A.W., Scott, A.C. (eds.) Toward a Science of Consciousness: the First Tucson Discussion and Debates, pp. 507–539. MIT Press, Cambridge (1996)

[Has00] Hasegawa, H.: Responses of a Hodgkin-Huxley neuron to various types of spike-train inputs. Phys. Rev. E 61, 718–726 (2000)

[Has02] Hasegawa, H.: Stochastic Resonance of Ensemble Neurons for Transient Spike Trains: A Wavelet Analysis. Phys. Rev. E 66, 21902 (2002)

[HHB01] Hasson, U., Hendler, T., Bashat, D.B., Malach, R.: Vase or face? A neural correlates of shape-selective grouping processes in the human brain. J. Cog. Neurosci. 13(6), 744–753 (2001)

[Hat77a] Hatze, H.: A myocybernetic control model of skeletal muscle. Biol. Cyber. 25, 103–119 (1977)

[Hat77b] Hatze, H.: A complete set of control equations for the human musculoskeletal system. J. Biomech. 10, 799–805 (1977b)

[Hat78] Hatze, H.: A general myocybernetic control model of skeletal muscle. Biol. Cyber. 28, 143–157 (1978)

[Hen66] Hénon, M.: Sur la topologie des lignes de courant dans un cas particulier. C. R. Acad. Sci. Paris A 262, 312–314 (1966)

[Hen69] Hénon, M.: Numerical study of quadratic area preserving mappings. Q. Appl. Math. 27 (1969)

[Hen76] Hénon, M.: A two-dimensional mapping with a strange attractor. Com. Math. Phys. 50, 69–77 (1976)

[Hay01] Haykin, S. (ed.): Kalman Filtering and Neural Networks. Wiley, New York (2001)

[Hay91] Haykin, S.: Adaptive Filter Theory. Prentice–Hall, Englewood Cliffs (1991)

[Hay98] Haykin, S.S.: Neural Networks: A Comprehensive Foundation, 2nd edn. Prentice-Hall, NJ (1998)

[Heb49] Hebb, D.O.: The Organization of Behavior. Wiley, New York (1949)

[Hec87] Hecht-Nielsen, R.: Counterpropagation networks. Applied Optics 26(23), 4979–4984 (1987)

[Hec90] Hecht-Nielsen, R.: NeuroComputing. Addison–Wesley, Reading (1990)

[Hil38] Hill, A.V.: The heat of shortening and the dynamic constants of muscle. Proc. Roy. Soc. B76, 136–195 (1938)

[HH52] Hodgkin, A.L., Huxley, A.F.: A quantitative description of membrane current and application to conduction and excitation in nerve. J. Physiol. 117, 500–544 (1952)

[HH52] Hodgkin, A.L., Huxley, A.F.: A qualitative description of membrane current and its application to conduction and excitation in nerve. J. Physiol. 117, 500 (1952)

[Hod64] Hodgkin, A.L.: The Conduction of the Nervous Impulse. Liverpool Univ. Press, Liverpool (1964)

[HN08a] Hong, S.L., Newell, K.M.: Entropy conservation in the control of human action. Nonl. Dyn. Psych. Life. Sci. 12(2), 163–190 (2008)

[HN08b] Hong, S.L., Newell, K.M.: Entropy compensation in human motor adaptation. Chaos 18(1), 13108 (2008)

[HCK02a] Hong, H., Choi, M.Y., Kim, B.J.: Synchronization on small-world networks. Phys. Rev. E 65, 26139 (2002)

[HCK02b] Hong, H., Choi, M.Y., Kim, B.J.: Phase ordering on small-world networks with nearest-neighbor edges. Phys. Rev. E 65, 047104 (2002)

[Hop82] Hopfield, J.J.: Neural networks, physical systems with emergent collective computational abilities. Proc. Natl. Acad. Sci. USA 79, 2554 (1982)

[Hop84] Hopfield, J.J.: Neurons with graded response have collective computational properties like those of two–state neurons. Proc. Natl. Acad. Sci. USA 81, 3088–3092 (1984)

[Hop95] Hopfield, J.J.: Pattern recognition computation using action potential timing for stimulus representation. Nature 376, 33–36 (1995)

[Hou67] Houk, J.C.: Feedback control of skeletal muscles. Brain Res. 5, 433–451 (1967)

[Hou78] Houk, J.C.: Participation of reflex mechanisms and reaction-time processes in the compensatory adjustments to mechanical disturbances. Progr. Clin. Neurophysiol. 4, 193–215 (1978)

[Hou79] Houk, J.C.: Regulation of stiffness by skeletomotor reflexes. Ann. Rev. Physiol. 41, 99–123 (1979)

[HBB96] Houk, J.C., Buckingham, J.T., Barto, A.G.: Models of the cerebellum, motor learning. Behavioral, Brain Sciences 19(3), 368–383 (1996)

[Hun01] Hunt, E.L.: Multiple views of multiple intelligence (Review of Intelligence Reframed: Multiple Intelligences for the 21st Century.) Contemp. Psych. 46, 5–7 (2001)

[Hux57] Huxley, A.F.: Muscle structure and theories of contraction. Progr. Biophys. Chem. 7, 255–328 (1957)

[HN54] Huxley, A.F., Niedergerke, R.: Changes in the cross-striations of muscle during contraction and stretch and their structural interpretation. Nature 173, 973–976 (1954)

[Ing82] Ingber, L.: Statistical mechanics of neocortical interactions. I. Basic formulation. Physica D 5, 83–107 (1982)

[Ing97] Ingber, L.: Statistical mechanics of neocortical interactions: Applications of canonical momenta indicators to electroencephalography. Phys. Rev. E 55(4), 4578–4593 (1997)

[Ing98] Ingber, L.: Statistical mechanics of neocortical interactions: Training, testing canonical momenta indicators of EEG. Mathl. Computer Modelling 27(3), 33–64 (1998)

[II06a] Ivancevic, V., Ivancevic, T.: Human–Like Biomechanics. Springer, Dordrecht (2005)

[II06b] Ivancevic, V., Ivancevic, T.: Natural Biodynamics. World Scientific, Singapore (2006)

[II06c] Ivancevic, V., Ivancevic, T.: Geometrical Dynamics of Complex Systems. Springer, Dordrecht (2006)

[II07a] Ivancevic, V., Ivancevic, T.: High-Dimensional Chaotic and Attractor Systems. Springer, Dordrecht (2007)

[IA07] Ivancevic, V., Aidman, E.: Life-space foam: A medium for motivational and cognitive dynamics. Physica A 382, 616–630 (2007)

[II07b] Ivancevic, V., Ivancevic, T.: Neuro-Fuzzy Associative Machinery for Comprehensive Brain and Cognition Modelling. Springer, Berlin (2007)

[II07d] Ivancevic, V., Ivancevic, T.: Computational Mind: A Complex Dynamics Perspective. Springer, Berlin (2007)

[II07e] Ivancevic, V., Ivancevic, T.: Applied Differential Geometry: A Modern Introduction. World Scientific, Singapore (2007)

[II08a] Ivancevic, V., Ivancevic, T.: Quantum Leap: From Dirac and Feynman, Across the Universe, to Human Body and Mind. World Scientific, Singapore (2008)

[II08b] Ivancevic, V., Ivancevic, T.: Complex Nonlinearity: Chaos, Phase Transitions, Topology Change and Path Integrals. Springer, Heidelberg (2008)

[II08c] Ivancevic, V., Ivancevic, T.: Human versus humanoid robot biodynamcis. Int. J. Hum. Rob. 5(4), 699–713 (2008)

[IJD09] Ivancevic, T., et al.: Complex Sports Biodynamics with Practical Applications in Tennis. Springer (Cognitive Systems Monographs, Vol. 2), Berlin (2009)

[Iva09] Ivancevic, T.: Jet Methods in Time–Dependent Lagrangian Biomechanics. Cent. Eur. J. Phys. (Online First, November 2009)

[Iva10] Ivancevic, T.: Jet Spaces in Modern Hamiltonian Biomechanics, Cent. Eur. J. Phys. (in press)

[IJP08] Ivancevic, T., Jain, L., Pattison, J., Hariz, A.: Nonlinear Dynamics and Chaos Methods in Neurodynamics and Complex Data Analysis. Nonl. Dyn 56(1-2) (2010)

[II08c] Ivancevic, V., Ivancevic, T.: Quantum Neural Computation. Springer, Dordrecht (2010)

[Izh99b] Izhikevich, E.M.: Weakly Connected Quasiperiodic Oscillators, FM Interactions and Multiplexing in the Brain. SIAM J. Appl. Math. 59(6), 2193–2223 (1999)

[Kan65] Kant, I.: Critique of Pure Reason (trans. Norman Kemp Smith) Discussions of Transcendental Unity of Apperception, 135–161; Transcendental Aesthetic, 74–81; Phenomena and Noumena, 257–275. St. Martin's Press, New York (1965)

[KM78a] Kim, J., Mueller, C.W.: Introduction to factor analysis: What it is and how to do it. Quantitative Applications in the Social Sciences Series, vol. 13. Sage Publications, Thousand Oaks (1978)

[KM78b] Kim, J., Mueller, C.W.: Factor Analysis: Statistical methods and practical issues. Quantitative Applications in the Social Sciences Series, vol. 14. Sage Publications, Thousand Oaks (1978)

[Kli00] Kline, P.: A Psychometrics Primer. Free Assoc. Books, London (2000)

[Kos92] Kosko, B.: Neural Networks, Fuzzy Systems, A Dynamical Systems Approach to Machine Intelligence. Prentice–Hall, New York (1992)

[Lew97] Lewin, K.: Resolving Social Conflicts, and, Field Theory in Social Science. Am. Psych. Assoc., Washington (1997)

[Lor63] Lorenz, E.N.: Deterministic Nonperiodic Flow. J. Atmos. Sci. 20, 130–141 (1963)

[LBM05] De Lucia, M., Bottaccio, M., Montuori, M., Pietronero, L.: A topological approach to neural complexity. Phys. Rev. E 71, 016114 (2005)

[MJS07] Maass, W., Joshi, P., Sontag, E.D.: Computational Aspects of Feedback in Neural Circuits. PLoS Comput. Biol. 3(1), e165 (2007)

[Man07] Manousakis, E.: Quantum theory, consciousness and temporal perception: binocular rivalry. arXiv:0709.4516 (2007)

[Mar98] Marieb, E.N.: Human Anatomy and Physiology, 4th edn., Benjamin/Cummings, Menlo Park, CA (1998)

[Mal81] Von der Mahlsburg, C.: The correlation theory of brain function. MPI Biophysical Chemistry, Internal Report, 81–82 (1981)

[Mal88] Von der Mahlsburg, C.: Pattern recognition by labelled graph matching. Neural Networks 7, 1019–1030 (1988)

[Mal798] Malthus, T.R.: An essay on the Principle of Population. Originally published in 1798, Penguin (1970)

[Mas04] Mashour, G.A.: The Cognitive Binding Problem: From Kant to Quantum Neurodynamics. NeuroQuant. 1, 29–38 (2004)

[May73] May, R.M. (ed.): Stability and Complexity in Model Ecosystems. Princeton Univ. Press, Princeton (1973)

[May76] May, R.: Simple Mathematical Models with Very Complicated Dynamics. Nature 261(5560), 459–467 (1976)

[May76] May, R.M. (ed.): Theoretical Ecology: Principles and Applications. Blackwell Sci. Publ., Malden (1976)

[MNL03] Motter, A.E., Nishikawa, T., Lai, Y.-C.: Large-scale structural organization of social networks. Phys. Rev. E 68, 036105 (2003)

[Mou80] Mountcastle, V.N.: Medical physiology. C.V. Mosby Comp., St. Louis (1980)

[Mur02] Murray, J.D.: Mathematical Biology, 3rd edn. An Introduction, vol. I. Springer, New York (2002)

[NAY60] Nagumo, J., Arimoto, S., Yoshizawa, S.: An active pulse transmission line simulating 1214-nerve axons. Proc. IRL 50, 2061–2070 (1960)

[NS72] Newell, A., Simon, H.A.: Human problem solving. Prentice-Hall, Englewood Cliffs (1972)

[NML02] Nishikawa, T., Motter, A.E., Lai, Y.-C., Hoppensteadt, F.C.: Smallest small-world network. Phys. Rev. E 66, 046139 (2002)

[NML03] Nishikawa, T., Motter, A.E., Lai, Y.C., Hoppensteadt, F.C.: Heterogeneity in Oscillator Networks: Are Smaller Worlds Easier to Synchronize? Phys. Rev. Lett. 91, 014101 (2003)

[Nob62] Noble, D.: A modification of the Hodgkin–Huxley equations applicable to Purkinie fibre action and peace–maker potentials. J. Physiol. 160, 317–330 (1962)

[PG97] Pakkenberg, B., Gundersen, H.J.: Neocortical neuron number in humans: effect of sex and age. J. Comp. Neurol. 384(2), 312–320 (1997)

[Pen89] Penrose, R.: The Emperor's New Mind. Oxford Univ. Press, Oxford (1989)

[Pen94] Penrose, R.: Shadows of the Mind. Oxford Univ. Press, Oxford (1994)

[Pen97] Penrose, R.: The Large, the Small and the Human Mind. Cambridge Univ. Press, Cambridge (1997)

[Pen98] Penrose, R.: Quantum computation, entanglement and state reduction. Phil. Trans. Roy. Soc. London 356, 1927–1939 (1998)

[Per03] Pereira, A.: The Quantum Mind/Classical Brain Problem. NeuroQuant. 1, 94–118 (2003)

[Pop08] Popp, F.-A.: Consciousness as Evolutionary Process based on Coherent States. NeuQuant. 6(4), 431–439 (2008)

[Pri71] Pribram, K.H.: Languages of the brain. Prentice-Hall, Englewood Cliffs (1971)

[Rav02] Raven, J.: Intelligence, Engineered Invisibility and the Destruction of Life on Earth. In: McKinzey, R.K. (ed.) WebPsychEmpiricist, WPE (2002)

[RU67] Ricciardi, L.M., Umezawa, H.: Brain physics and many-body problems. Kibernetik 4, 44 (1967)

[Roy08] Roy, A.: Connectionism, Controllers, and a Brain Theory. IEEE Tr. SMCA 38(6), 1434–1441 (2008)

[RT71] Ruelle, D., Takens, F.: On the nature of turbulence. Comm. Math. Phys. 20, 167–192 (1971)

[Sch07] Schöner, G.: Dynamical Systems Approaches to Cognition. In: Cambridge Handbook of Computational Cognitive Modeling. Cambridge Univ. Press, Cambridge (2007)

[SI09] Sharma, S., Ivancevic, V.: Nonlinear dynamical characteristics of situation awareness. The. Iss. Erg. Sci. 113, iFirst (2009)

[SG95] Singer, W., Gray, C.M.: Visual feature integration and temporal correlation hypothesis. Ann. Rev. Neurosci. 18, 555–586 (1995)

[Sma67] Smale, S.: Differentiable Dynamical Systems. Bull. AMS 73, 747–817 (1967)

[Sno75] Snodgrass, J.G.: Psychophysics. In: Scharf, B. (ed.) Experimental Sensory Psychology, pp. 17–67 (1975)

[Spa82] Sparrow, C.: The Lorenz Equations: Bifurcations, Chaos, and Strange Attractors. Springer, New York (1982)

[SF91] Sternberg, R.J., Frensch, P.A. (eds.): Complex problem solving: Principles, mechanisms. Lawr. Erl. Assoc., Hillsdale (1991)

[SVW95] Srivastava, Y.N., Vitiello, G., Widom, A.: Quantum dissipation and quantum noise. Annals Phys. 238, 200 (1995)

[Ste95] Sternberg, R.J.: Conceptions of expertise in complex problem solving: A comparison of alternative conceptions. In: Frensch, P.A., Funke, J. (eds.) Complex problem solving: The European Perspective (295-321), Lawr. Erl. Assoc., Hillsdale (1995)

[Str94] Strogatz, S.: Nonlinear Dynamics and Chaos. Addison-Wesley, Reading (1994)

[Str01] Strogatz, S.H.: Exploring complex networks. Nature 410, 268 (2001)

[Sze78] Szentagothai, J.: The neuron network of the cerebral cortex: a functional interpretation. Proc. R. Soc. London, B 201, 219–248 (1978)

[Teg00] Tegmark, M.: The importance of quantum decoherence in brain processes. Phys. Rev. E 61, 4194–4206 (2000)

[TS01] Thompson, J.M.T., Stewart, H.B.: Nonlinear Dynamics and Chaos: Geometrical Methods for Engineers and Scientists. Wiley, New York (2001)

[TJ02] Todorov, E., Jordan, M.I.: Optimal feedback control as a theory of motor coordination. Nat. Neurosci. 5(11), 1226–1235 (2002)

[LBM05] Tononi, G., Sporns, O., Edelman, G.M.: A measure for brain complexity: relating functional segregation and integration in the nervous system. Proc. Natl. Acad. Sci. USA 91(11), 5033–5037 (1994)

[Ume93] Umezawa, H.: Advanced Field Theory: Micro Macro and Thermal Concepts. Am. Inst. Phys., New York (1993)

[Neu58] Von Neumann, J.: The Computer and the Brain. Yale Univ. Press (1958)

[Ver838] Verhulst, P.F.: Notice sur la loi que la population pursuit dans son accroissement. Corresp. Math. Phys. 10, 113–121 (1838)

[Ver845] Verhulst, P.F.: Recherches Mathematiques sur La Loi D'Accroissement de la Population (Mathematical Researches into the Law of Population Growth Increase). Nouveaux Memoires de l'Academie Royale des Sciences et Belles-Lettres de Bruxelles 18(1), 1–45 (1845)

[Wat99] Watts, D.J.: Small Worlds. Princeton Univ. Press, Princeton (1999)

[WS98] Watts, D.J., Strogatz, S.H.: Collective dynamics of 'small-world' networks. Nature 393, 440–442 (1998)

[Wil56] Wilkie, D.R.: The mechanical properties of muscle. Brit. Med. Bull. 12, 177–182 (1956)

[Zee03] Zee, A.: Quantum Field Theory in a Nutshell. Princeton Univ. Press, Princeton (2003)

[Zwi69] Zwicky, F.: Discovery, Invention and Research – Through the Morphological Approach. Macmillian, Toronto (1969)

Index